A·N·N·U·A·L E·D·I·T·I·O·N·S

ENVIRONMENT *02/03*

Twenty-First Edition

EDITOR

John L. Allen

University of Wyoming

John L. Allen is professor of geography at the University of Wyoming. He received his bachelor's degree in 1963 and his M.A. in 1964 from the University of Wyoming, and in 1969 he received his Ph.D. from Clark University. His special area of interest is the impact of contemporary human societies on environmental systems.

McGraw-Hill/Dushkin

530 Old Whitfield Street, Guilford, Connecticut 06437

Visit us on the Internet
http://www.dushkin.com

Credits

1. **The Global Environment: An Emerging World View**
   Unit photo—Courtesy of NASA.
2. **The World's Population: People and Hunger**
   Unit photo—United Nations photo by J. P. Laffont/NJ.
3. **Energy: Present and Future Problems**
   Unit photo—© 2002 by Sweet By & By/Cindy Brown.
4. **Biosphere: Endangered Species**
   Unit photo—United Nations photo by M. Gonzalez.
5. **Resources: Land, Water, and Air**
   Unit photo—United Nations photo by Saw Lwin.
6. **Pollution: The Hazards of Growth**
   Unit photo—Courtesy of Daniel E. Hebding.

Copyright

Cataloging in Publication Data
Main entry under title: Annual Editions: Environment. 2002/2003.
1. Environment—Periodicals. 2. Ecology—Periodicals. I. Allen, John L., *comp.*
II. Title: Environment
ISBN 0–07–250682–2   301.31'05   79–644216   ISSN 0272–9008

Twenty-First Edition

Cover image © 2002 PhotoDisc, Inc.
Printed in the United States of America   1234567890BAHBAH5432   Printed on Recycled Paper

To the Reader

In publishing ANNUAL EDITIONS we recognize the enormous role played by the magazines, newspapers, and journals of the public press in providing current, first-rate educational information in a broad spectrum of interest areas. Many of these articles are appropriate for students, researchers, and professionals seeking accurate, current material to help bridge the gap between principles and theories and the real world. These articles, however, become more useful for study when those of lasting value are carefully collected, organized, indexed, and reproduced in a low-cost format, which provides easy and permanent access when the material is needed. That is the role played by ANNUAL EDITIONS.

At the beginning of our new millennium, environmental dilemmas long foreseen by natural and social scientists began to emerge in a number of guises: population/food imbalances, problems of energy scarcity, acid rain, toxic and hazardous wastes, ozone depletion, water shortages, massive soil erosion, global atmospheric pollution and possible climate change, forest dieback and tropical deforestation, and the highest rates of plant and animal extinction that the world has known in 65 million years.

These and other problems continue to worsen in spite of an increasing amount of national and international attention to environmental issues and increased environmental awareness and legislation. The problems have resulted from centuries of exploitation and unwise use of resources, accelerated recently by the short-sighted public policies that have favored the short-term, expedient approach to problem solving over longer-term economic and ecological good sense.

Part of the problem is unsustained efforts to deal with environmental issues. During the 1980s, economic problems generated by resource scarcity caused the relaxation of environmental quality standards and contributed to the refusal of many of the world's governments and international organizations to develop environmentally sound protective measures that were viewed as too costly. More recently, in the 1990s, as environmental protection policies were adopted, they were often cosmetic, designed for good press but little else. Even with these public relations policies, governments often lacked either the will or the means to implement them properly. The absence of effective environmental policy has been particularly apparent in those countries that are striving to become economically developed. But even in the more highly developed nations, economic concerns tend to favor a loosening of environmental controls. In the United States, the interests of maintaining jobs for the timber industry, for example, imperil many of the last areas of old-growth forests, and the desire to maintain agricultural productivity at all costs causes the use of destructive and toxic chemicals to continue on the nation's farmlands. In addition, concerns over energy availability have created the need for foreign policy and military action to protect the developed nations' access to cheap oil and have prompted increasing reliance on technological quick fixes, as well as the development of environmentally sensitive areas to new energy resource exploration and exploitation.

Despite the recent tendency of the U.S. government to turn its back on environmental issues, there is some reason to hope that, globally, a new environmental consciousness is awakening. International conferences have been held on global warming and other environmental issues, and, in spite of the recalcitrance of the United States on the global warming issue, there is some evidence of an increased international desire to do something about environmental quality before it is too late. Unfortunately, increasing globalization of internal conflict and the emergence of terrorism as an instrument of national policy—particularly where terrorism may employ environmental contamination as a weapon—may produce future environmental problems that are almost too frightening to think about.

The articles contained in *Annual Editions: Environment 02/03* have been selected for the light that they shed on these and other problems and issues. The selection process was aimed at including material that will be readily assimilated by the general reader. Additionally, every effort has been made to choose articles that encourage an understanding of the nature of the environmental problems that beset us and how, with wisdom and knowledge and the proper perspective, they can be solved or at least mitigated. Accordingly, the selections in this book have been chosen more for their intellectual content than for their emotional tone. They have been arranged into an order of topics—the global environment, population and food, energy, the biosphere, resources, and pollution—that lends itself to a progressive understanding of the causes and effects of human modifications of Earth's environmental systems.

Both the *World Wide Web* sites and *topic guide* in this edition can be used to further explore the issues raised. In addition, this edition contains both a newly refreshed *Environmental Information Retrieval* guide and *glossary*.

Readers can have input into the next edition of *Annual Editions: Environment* by completing and returning the postpaid *article rating form* at the back of the book.

John L. Allen
Editor

Contents

UNIT 1
The Global Environment: An Emerging World View

Four selections provide information on the current state of Earth and the changes we will face.

The concepts in bold italics are developed in the article. For further expansion, please refer to the Topic Guide and the Index.

UNIT 2
The World's Population: People and Hunger

Six unit selections examine the problems the world will have in feeding its ever-increasing population.

Unit Overview

The concepts in bold italics are developed in the article. For further expansion, please refer to the Topic Guide and the Index.

UNIT 3
Energy: Present and Future Problems

Five articles in this unit consider the problems of meeting present and future energy needs. Alternative energy sources are also examined.

UNIT 4
Biosphere: Endangered Species

Five articles examine the problems in the world's biosphere that include biological extinction, natural ecosystems, bioinvasion, endangered animals, and ecotourism.

The concepts in bold italics are developed in the article. For further expansion, please refer to the Topic Guide and the Index.

UNIT 5
Resources: Land, Water, and Air

Six selections discuss the environmental problems affecting our land, water, and air resources.

Unit Overview   134

The concepts in bold italics are developed in the article. For further expansion, please refer to the Topic Guide and the Index.

UNIT 6
Pollution: The Hazards of Growth

In this unit, four selections weigh the environmental impacts of the growth of human population.

The concepts in bold italics are developed in the article. For further expansion, please refer to the Topic Guide and the Index.

Topic Guide

This topic guide suggests how the selections in this book relate to the subjects covered in your course. You may want to use the topics listed on these pages to search the Web more easily.

On the following pages a number of Web sites have been gathered specifically for this book. They are arranged to reflect the units of this *Annual Edition*. You can link to these sites by going to the DUSHKIN ONLINE support site at *http://www.dushkin.com/online/*.

ALL THE ARTICLES THAT RELATE TO EACH TOPIC ARE LISTED BELOW THE BOLD-FACED TERM.

World Wide Web Sites

The following World Wide Web sites have been carefully researched and selected to support the articles found in this reader. The easiest way to access these selected sites is to go to our DUSHKIN ONLINE support site at *http://www.dushkin.com/online/*.

AE: Environment 02/03

The following sites were available at the time of publication. Visit our Web site—we update DUSHKIN ONLINE regularly to reflect any changes.

General Sources

Britannica's Internet Guide
http://www.britannica.com
This site presents extensive links to material on world geography and culture, encompassing material on wildlife, human lifestyles, and the environment.

EnviroLink
http://envirolink.netforchange.com
One of the world's largest environmental information clearinghouses, EnviroLink is a grassroots nonprofit organization that unites organizations and volunteers around the world and provides up-to-date information and resources.

Library of Congress
http://www.loc.gov
Examine this extensive Web site to learn about resource tools, library services/resources, exhibitions, and databases in many different subfields of environmental studies.

SocioSite: Sociological Subject Areas
http://www.pscw.uva.nl/sociosite/TOPICS/
This huge sociological site from the University of Amsterdam provides many discussions and references of interest to students of the environment, such as the links to information on ecology and consumerism.

U.S. Geological Survey
http://www.usgs.gov
This site and its many links are replete with information and resources in environmental studies, from explanations of El Niño to discussion of concerns about water resources.

UNIT 1: The Global Environment: An Emerging World View

Earth Science Enterprise
http://www.earth.nasa.gov
Information about NASA's Mission to Planet Earth program and its Science of the Earth System can be found here. Surf to learn about satellites, El Niño, and even "strategic visions" of interest to environmentalists.

National Geographic Society
http://www.nationalgeographic.com
Links to *National Geographic*'s huge archive are provided here. There is a great deal of material related to the atmosphere, the oceans, and other environmental topics.

Santa Fe Institute
http://acoma.santafe.edu
This home page of the Santa Fe Institute—a nonprofit, multidisciplinary research and education center—will lead to many interesting links related to its primary goal: to create a new kind of scientific research community, pursuing emerging science.

United Nations
http://www.unsystem.org
Visit this official Web site Locator for the United Nations System of Organizations to get a sense of the scope of international environmental inquiry today. Various UN organizations concern themselves with everything from maritime law to habitat protection to agriculture.

United Nations Environment Programme (UNEP)
http://www.unep.ch
Consult this home page of UNEP for links to critical topics of concern to environmentalists, including desertification, migratory species, and the impact of trade on the environment. The site will direct you to useful databases and global resource information.

UNIT 2: The World's Population: People and Hunger

The Hunger Project
http://www.thp.org
Browse through this nonprofit organization's site to explore the ways in which it attempts to achieve its goal: the sustainable end to global hunger through leadership at all levels of society. The Hunger Project contends that the persistence of hunger is at the heart of the major security issues that are threatening our planet.

Penn Library Resources
http://www.library.upenn.edu/resources/websitest.html
This vast site is rich in links to information about virtually every subject you can think of in environmental studies. Its extensive population and demography resources address such concerns as migration, family planning, and health and nutrition in various world regions.

World Health Organization
http://www.who.int
The home page of the World Health Organization provides links to a wealth of statistical and analytical information about health and the environment in the developing world.

WWW Virtual Library: Demography & Population Studies
http://demography.anu.edu.au/VirtualLibrary/
This is a definitive guide to demography and population studies. A multitude of important links to information about global poverty and hunger can be found here.

UNIT 3: Energy: Present and Future Problems

Alternative Energy Institute, Inc.
http://www.altenergy.org
On this site created by a nonprofit organization, learn about the impacts of the use of conventional fuels on the environment. Also learn about research work on new forms of energy.

Communications for a Sustainable Future
http://csf.colorado.edu
This site will lead to information on topics in international environmental sustainability. It pays particular attention to the political economics of protecting the environment.

www.dushkin.com/online/

Energy and the Environment: Resources for a Networked World

http://zebu.uoregon.edu/energy.html

An extensive array of materials having to do with energy sources—both renewable and nonrenewable—as well as other topics of interest to students of the environment is found on this site.

Institute for Global Communication/EcoNet

http://www.igc.org/igc/gateway/

This environmentally friendly site provides links to dozens of governmental, organizational, and commercial sites having to do with energy sources. Resources address energy efficiency, renewable generating sources, global warming, and more.

U.S. Department of Energy

http://www.energy.gov

Scrolling through the links provided by this Department of Energy home page will lead to information about fossil fuels and a variety of sustainable/renewable energy sources.

UNIT 4: Biosphere: Endangered Species

Friends of the Earth

http://www.foe.co.uk/index.html

Friends of the Earth, a nonprofit organization based in the United Kingdom, pursues a number of campaigns to protect the Earth and its living creatures. This site has links to many important environmental sites, covering such broad topics as ozone depletion, soil erosion, and biodiversity.

Smithsonian Institution Web Site

http://www.si.edu

Looking through this site, which will provide access to many of the enormous resources of the Smithsonian, offers a sense of the biological diversity that is threatened by humans' unsound environmental policies and practices.

World Wildlife Federation (WWF)

http://www.wwf.org

This home page of the WWF leads to an extensive array of information links about endangered species, wildlife management and preservation, and more. It provides many suggestions for how to take an active part in protecting the biosphere.

UNIT 5: Resources: Land, Water, and Air

Global Climate Change

http://www.puc.state.oh.us/consumer/gcc/index.html

The goal of this PUCO (Public Utilities Commission of Ohio) site is to serve as a clearinghouse of information related to global climate change. Its extensive links provide an explanation of the science and chronology of global climate change, acronyms, definitions, and more.

National Oceanic and Atmospheric Administration (NOAA)

http://www.noaa.gov

Through this home page of NOAA, find information about coastal issues, fisheries, climate, and more.

National Operational Hydrologic Remote Sensing Center (NOHRSC)

http://www.nohrsc.nws.gov

Flood images are available at this site of the NOHRSC, which works with the U.S. National Weather Service to track weather-related information.

Virtual Seminar in Global Political Economy/Global Cities & Social Movements

http://csf.colorado.edu/gpe/gpe95b/resources.html

Links to subjects of interest in regional environmental studies, covering topics such as sustainable cities, megacities, and urban planning are available here. Many international nongovernmental organizations are included.

Websurfers Biweekly Earth Science Review

http://home.rmi.net/~michaelg/index.html

This is a biweekly compilation of Internet sites devoted to the terrestrial and planetary sciences. It includes a list of hyperlinks to related earth science sites and news items.

UNIT 6: Pollution: The Hazards of Growth

IISDnet

http://iisd1.iisd.ca

The International Institute for Sustainable Development's site presents information through links on business and sustainable development, developing ideas, and Hot Topics.

Persistant Organic Pollutants (POP)

http://irptc.unep.ch/pops/

Visit this site to learn more about persistant organic pollutants (POPs) and the issues and concerns surrounding them.

School of Labor and Industrial Relations (SLIR): Hot Links

http://www.lir.msu.edu/hotlinks/

The Michigan State University's SLIR page connects to industrial relations sites throughout the world. It has links to U.S. government statistics, newspapers and libraries, international intergovernmental organizations, and more.

Space Research Institute

http://arc.iki.rssi.ru/Welcome.html

For a change of pace, browse through this home page of Russia's Space Research Institute for information on its Environment Monitoring Information Systems, the IKI Satellite Situation Center, and its Data Archive.

Worldwatch Institute

http://www.worldwatch.org

The Worldwatch Institute, dedicated to fostering the evolution of an environmentally sustainable society, presents this site with access to *World Watch Magazine* and *State of the World 2000.* Click on Alerts and Press Briefings for discussions of current problems.

We highly recommend that you review our Web site for expanded information and our other product lines. We are continually updating and adding links to our Web site in order to offer you the most usable and useful information that will support and expand the value of your Annual Editions. You can reach us at: *http://www.dushkin.com/annualeditions/*.

UNIT 1

The Global Environment: An Emerging World View

Unit Selections

1. **Ideas Matter: A Political History of the Twentieth-Century Environment**, J. R. McNeill
2. **Environmental Surprises: Planning for the Unexpected**, Chris Bright
3. **The Paradox of Global Environmentalism**, Ramachandra Guha
4. **Harnessing Corporate Power to Heal the Planet**, L. Hunter Lovins and Amory B. Lovins

Key Points to Consider

- How are the histories of technology, energy, population, and economics linked with the history of environmental change? What are the implications of these linkages for future environmental impact?

- Why are environmental changes so difficult to predict, and how can human planning systems develop mechanisms to adjust to continued environmental change?

- Explain the "paradox of global environmentalism." Is the author of this article too critical of the global environmental movement simply because it is centered in high-consumption societies? Defend your answer.

- What is meant by the term "natural capitalism" and how does the application of principles of this form of economics promise a new creation of appropriate environmental strategies? How do the concepts of natural capitalism fit into the present methods for dealing with the world's resources and systems?

 **Links: www.dushkin.com/online/**
These sites are annotated in the World Wide Web pages.

Earth Science Enterprise
http://www.earth.nasa.gov

National Geographic Society
http://www.nationalgeographic.com

Santa Fe Institute
http://acoma.santafe.edu

United Nations
http://www.unsystem.org

United Nations Environment Programme (UNEP)
http://www.unep.ch

More than three decades after the celebration of the first Earth Day in 1970, public apprehension over the environmental future of the planet has reached levels unprecedented even during the late 1960s and early 1970s "Age of Aquarius." No longer are those concerned about the environment dismissed as "ecofreaks" and "tree-huggers." Many serious scientists have joined the rising clamor for environmental protection, as have the more traditional environmentally conscious public interest groups. There are a number of reasons for this increased environmental awareness. Some of these reasons arise from environmental events; it is, for example, becoming increasingly difficult to deny the effects of global warming. But more arise simply from the process of globalization: the increasing unity of the world's economic, social, and information systems. Hailed by many as the salvation of the future, globalization has done little to make the world a better or safer place. Diseases once confined to specific regions now have the capacity for widespread dissemination. Increasing human mobility has allowed human-caused disruptions to political, cultural, and economic systems also to spread, and acts of terrorism now take place in locations once thought safe from such manifestations of hatred and despair. On the more positive side, the expansion of global information systems has fostered a maturation of concepts about the global nature of environmental processes.

Much of what has been learned through this increased information flow, particularly by American observers, has been of the environmentally ravaged world behind the old Iron Curtain—a chilling forecast of what other industrialized regions as well as the developing countries can become in the near future unless strict international environmental measures are put in place. For perhaps the first time ever, countries are beginning to recognize that environmental problems have no boundaries and that international cooperation is the only way to solve them.

The subtitle of this first unit, "An Emerging World View," is an optimistic assessment of the future: a future in which less money is spent on defense and more on environmental protection and cleanup—a new world order in which political influence might be based more upon leadership in environmental and economic issues than upon military might. It is probably far too early to make such optimistic predictions, to conclude that the world's nations—developed and underdeveloped—will begin to recognize that Earth's environment is a single unit. Neither have those nations shown any tendency to recognize humankind as a single unit in a world where what harms one harms all. The recent emergence of wide-scale terrorism as an instrument of political and social policy is evidence of such a failure of recognition. Nevertheless, there is growing international realization—aided by the information superhighway—that we are all, as environmental activists have been saying for decades, inhabitants of Spaceship Earth and will survive or succumb together.

The articles selected for this unit have been chosen to illustrate the increasingly global perspective on environmental problems and the degree to which their solutions must be linked to political, economic, and social problems and solutions. In the lead piece of the unit, "Ideas Matter: A Political History of the Twentieth-Century Environment," environmental historian J. R. McNeill explains how "the grand social and ideological systems" that people develop to define themselves have unexpected consequences for environmental systems. McNeill views the histories of technology, energy, population, and economic growth as all being interconnected and linked to human impact on the environment as well. The second article in this section also deals with the unanticipated. In "Environmental Surprises: Planning for the Unexpected," Chris Bright of the Worldwatch Institute describes how environmental futurists are discovering new ways to identify environmental problems and to plan solutions for them around the complex web of politics, economies, and societies. He notes that the process of environmental decline need not be slow and gradual but may come abruptly. This means that planning for change has to be capable of adjusting quickly to new environmental conditions. And since environmental conditions are never stable, Bright concludes, "plan to keep on planning."

The third selection in the unit is also directed toward the interconnectedness of environmental systems, toward new concepts and ways of thinking about the environment and human impact upon it, and toward the unexpected consequences of human/environment relationships. "The Paradox of Global Environmentalism," by Indian historian and anthropologist Ramachandra Guha, suggests that "green missionaries" may be more dangerous than either their economic or religious counterparts in terms of environmental impact. The paradox cited by Guha is that of well-meaning environmentalists in the highly developed regions who want nature preserved in less developed countries without regard for the needs of the people living there. In addition, few of the environmentalists from the developed world are willing to abandon their high-consumption lifestyles but expect sacrifices from others. A direct counterpoint to Guha's article is the fourth selection in the opening section. In "Harnessing Corporate Power to Heal the Planet," L. Hunter Lovins and Amory B. Lovins of the Rocky Mountain Institute, a nonprofit environmental policy center, introduce the concept of "natural capitalism," a new form of economics that takes into account the value of natural resources and, more important, the ecosystem services such as water and nutrient recycling, atmospheric and ecological stability, and biodiversity. Traditional capitalistic accounting has dealt with both resources and services as "free" and, hence, have paid little attention to their depletion or deterioration. The Lovinses suggest that treating the environment as something of tangible monetary value can increase resource productivity, eliminate the concept of waste, create better environmental services, and promote investment in natural capital—in other words, natural capitalism could result in the reversal of the worldwide deterioration of the ecosystem.

Ideas Matter: A Political History of the Twentieth-Century Environment

"The grand social and ideological systems that people construct for themselves invariably carry large consequences, for the environment no less than for more strictly human affairs. Among the swirl of ideas, policies, and political structures of the twentieth century, the most ecologically influential were the growth imperative and the (not unrelated) security anxiety that together dominated policy around the world.... By 1970, however, something new was afoot."

J. R. McNeill

The twentieth century has no rivals when it comes to human impact on the environment. Why this is so has much to do with the interconnected histories of technology, energy, population, and economic growth. But their trajectories, and the fate of the environment, have also been bound to the realms of ideas and politics, in ways both obvious and subtle.

What people think has affected the environment because to some extent it has shaped their behavior. And of course, the changing environment has played a part in affecting what people have thought. Here there are two related points. First, what people thought specifically about the environment, nature, and life mattered only marginally before 1970. Second, at all times, but more so before 1970, other kinds of ideas governed the human behavior that most affected the environment.

Ideological lock-in

Big ideas somehow succeed in molding the behavior of millions. They are usually about economics and politics. Ideas, like genetic mutations and technologies, are hatched all the time, but most quickly disappear for want of followers. Ruthless selection is always at work, but, again like mutations and technologies, the notion of increasing returns to scale often applies. When an idea becomes successful, it easily becomes even more

successful: it gets entrenched in social and political systems, which assist in its further spread. It then prevails even beyond the times and places where it is advantageous to its followers. Technology historians refer to analogous situations as "technological lock-in." For example, the narrow-gauge railway track adopted in the nineteenth century, once it became the standard, could not be replaced even after it prevented improvements that would allow for faster trains. Too much was already invested in the old ways. Ideological lock-in, the staying power of orthodox ideas, works the same way. Big ideas all became orthodoxies, enmeshed in social and political systems, and difficult to dislodge even if they became costly.

At the outset of the century the ideas with mass followings remained the great religions. Their doctrines include various injunctions about nature. The God of the ancient Hebrews enjoins believers to "Be fruitful and multiply, and fill the earth and subdue it" (Genesis 1:26–29). This and other biblical passages inspired an argument to the effect that Christianity, or the Judeo-Christian tradition, uniquely encouraged environmental despoliation. But the record of environmental ruin around the world, even among followers of Buddhism, Taoism, and Hinduism (seen in this argument as creeds more reverent of nature), suggests this is not so: either other religious traditions similarly encouraged predatory conduct, or religions did not notably constrain behavior with respect to the natural world.

A variation on the Judeo-Christian theme is the notion that Western humanism, rationalism, or the Scientific Revolution uniquely licensed environmental mayhem by depriving nature of its sacred character. While the lucubrations of Erasmus, Descartes, and Francis Bacon probably did not filter into the calculations of peasants, fishermen, or most landowners in the twentieth century or before, something can be said for this proposition. Western science helped recast environments everywhere indirectly, by fomenting technological change. Sir Isaac Newton said that if he had seen further than others it was because he stood on the shoulders of giants. Scientists of the twentieth century, such as Haber and Midgley, whose work proved enormously consequential in ecological terms, stood on the shoulders of giants of scientific method who held the notion that science's job was to unlock the secrets of nature and to deploy scientific knowledge in the service of human health and wealth.

This persuasive and pervasive idea legitimated all manner of environmental manipulation wherever modern science took hold. Applied science brought, for example, the chemical industry, which came of age in the mid- to late nineteenth century. By 1990 it had generated some 80,000 new compounds that found routine use, and inevitably also found their way into ecosystems un-adapted to them. A small proportion of these, even at tiny concentrations, proved disruptive, poisoning birds and fish, damaging genes, and causing other usually unwelcome effects. Some entered ecosystems at high concentrations; while the world's chemical industry in 1930 produced only a million tons of organic chemicals, by 1999 the total had grown a thousandfold. Slowly but surely the chemical industry came to play a part in ecology, introducing new selective criteria in biological evolution, namely compatibility with existing chemicals present in the environment. This development and others like it were an accidental result of rigorous scientific work over a century or more. In science more than religion, ideas from earlier eras exerted an impact on environmental history in the twentieth century.

Nationalism and nature

Modern political ideas also shaped environmental history. Nationalism, born of the French Revolution, proved an enormously successful idea in the twentieth century. It traveled across cultures and continents better than any other European idea, appearing in several guises. It powerfully affected environmental change, although in no single consistent direction.

In some contexts, nationalism served as a spur to landscape preservation. As Europe industrialized quickly after 1880, nostalgic notions of German, Swiss, or English countryside acquired special patriotic overtones. In 1926 British architect Patrick Abercrombie could write that "the greatest historical monument that we possess, the most essential thing that is England, is the countryside, the market town, the village, the hedgerow, the lanes, the copses, the streams and farmsteads." The Swiss waxed sentimental and patriotic about their mountains and farms, resisting railroads near the Matterhorn and other threatening manifestations of "Americanism." Germans honed such forms of nationalism to a fine edge, alloyed them

with idyllic romanticism, and organized countless countryside-preservation societies. Such ideas added a current to Nazism. Himmler's SS (the Nazi special forces) dreamed of converting Poland into a landscape redolent of German tribal origins, with plenty of primeval forest to reflect the peculiarly German love of nature.

Nationalism's preservationist, arcadian component lost out to a rival one that emphasized power and wealth.

Similar equations of national identity with rural righteousness, the sanctity of (our) land, and nature preservation cropped up wherever cities and industrialization spread. Russian populism before 1917; Russian (not Soviet) nationalism after 1917; western Canada's Social Credit movement; D. H. Lawrence's nature worship; the back-to-nature romanticism of best-selling and Nobel prize-winning Norwegian novelist Knut Hamsun; the intellectual hodgepodge underlying Mediterranean fascism and Japanese militarism; Mao's peasant populism; and all manner of back-to-the-land, antimodern currents—all these reflected political and cultural revulsion at urban and industrial transformations. In the Mediterranean, they provoked some small-scale reforestation schemes, including projects that won Mussolini's favor because he thought they would make Italy colder and thereby make Italians more warlike.

The SS did not carry out its plans for Poland; in general, nationalism's preservationist, arcadian component lost out to a rival one that emphasized power and wealth, and therefore favored industrialization and frontier settlement, regardless of ecological implications. The nationalism unleashed in the Mexican Revolution (1910–1920), for instance, quickly abandoned peasant causes in favor of accelerated industrialization. Argentina and Brazil pursued the same vision, without the revolutions, after 1930. In Japan, nationalism and industrialization were yoked together from the Meiji restoration to World War II (1868–1945), and in a more subdued and less militaristic way, after 1945 as well. The vast changes in land use and pollution patterns brought on by industrialization were in part a consequence of nationalisms.

So were the changes provoked by efforts to populate "empty" frontiers. States earned popular support for steps to settle (and establish firm sovereignty over) the Canadian Arctic, Soviet Siberia, the Australian Outback, Brazilian (not to mention Peruvian and Ecuadorian) Amazonia, and the outer islands of Indonesia. Settling and defending such areas involved considerable environmental change: deforestation in some cases, oil infrastructure in others, and road building in nearly all. It is also often involved displacing, resettling, or even killing off unassimilated indigenous populations whose loyalty to the states within whose borders they lived was doubtful.

Nationalism lurked behind other population policies too, notably pronatalism. Many twentieth-century states sought security in numbers, especially in Europe, where birth rates were sagging. Hypernationalist regimes in particular tried to boost birth rates, in France after the humiliation at the hands of Prussia in 1871, in fascist Italy, and in Nazi Germany. The most successful was Romania, under Nicolae Ceausescu (1918–1989). In 1965 he set a growth target of 30 million Romanians by the year 2000, banned all forms of birth control including abortion, and subjected women of childbearing age to police surveillance to make sure they were not shirking their reproductive duties. At the time, abortions outnumbered live births 4 to 1 in Romania. After 1966, Romanian maternity wards were deluged, sometimes obliged to wedge two delivering mothers into a single bed. Ceausescu temporarily reversed the demographic transition and doubled the birth rate, all for the greater glory of Romania. Other embattled societies, such as Stalin's Soviet Union and Iran after the 1979 revolution also sought to maximize population to safeguard the nation. Nationalism, in its myriad forms and through the multiple policies it inspired, was a crucially important idea because of its effects on the environment—especially when its adherents gave this connection no thought.

Karl Marx endorsed the French socialists who urged that the "exploitation of man by man" give way to the "exploitation of nature by man."

Communism, another European idea that traveled well, was in some respects the highest form of nationalism. Its political success in Russia and China, in Cuba and Vietnam, depended as much on its promise of independence from foreign domination as on its promise of social justice. The same ambitions—economic development and state power—that inspired state-sponsored industrialization elsewhere drove the heroic sprints of communist five-year plans.

Correcting nature's "mistakes"

But communism had other components too. Deep in Marxism is the belief that nature exists to be harnessed by labor. Friedrich Engels believed, like today's most cheerful optimists, that "the productivity of the land can be infinitely increased by the application of capital, labour and science." Karl Marx endorsed the French socialists who urged that the "exploitation of man by man" give way to the "exploitation of nature by man."

Explicitly linking communist progress with environmental transformation, the wordsmith V. Zazurbin in 1926 addressed the Soviet Writers Congress: "Let the fragile green breast of Siberia be dressed in the cement armor of cities, armed with the stone muzzle of factory chimneys, and girded with iron belts of railroads. Let the taiga be burned and felled, let the steppes be trampled. Let this be, and so it will be inevitably. Only in cement and iron can the fraternal union of all peoples, the iron brotherhood of all mankind, be forged."

Communists, especially in the Soviet Union and Eastern Europe, liked things big. Ostensibly this was to realize economies of scale, but it became an ideology, a propaganda tactic, and eventually an end in itself. Gigantism most famously affected architecture and statuary but also industry, forestry, and agriculture. The Soviets typically built huge industrial complexes, like Norilsk and Magnitogorsk, concentrating pollution. When the Soviet Union faced a timber shortage during the course of its first five-year plan in 1929–1933, millions of prisoners and collective farmworkers were sent to the forests to cut trees as quickly as possible. The resulting deforestation and erosion put sandbars in the Volga, inhibiting traffic on the country's chief waterway. In collectivizing agriculture, the Soviet government created not merely huge farms but huge fields, stretching as far as the eye could see, far larger than necessary to realize efficiencies from mechanization. This maximized wind and water erosion. Gigantism, together with the Marxist enthusiasm for conquering nature, led to the slow death of the Aral Sea, the creation of the world's biggest artificial lake and the world's biggest dam, and countless efforts "to correct nature's mistakes" on the heroic scale.

Communism, at least after its initial consolidation in power, also resisted technological innovation. With fixed production quotas pegged to five-year plans, Soviet and Eastern European factory bosses could ill afford to experiment with new technologies. Subsidized energy prices helped ossify industry in the Soviet Union and Eastern Europe so that, for example, most steel mills in 1990 still used the open-hearth process, an obsolete nineteenth-century invention long since replaced in Japan, South Korea, and the West. The political system stymied decarbonization and dematerialization, leaving the communist world with an energy-guzzling pollution-intensive coke-town economy to the end—a fact that helped bring on that end in the Soviet sphere.

The growth fetish

Communism aspired to become the universal creed of the twentieth century, but a more flexible and seductive religion succeeded where communism failed: the quest for economic growth. Capitalists, nationalists—indeed almost everyone, communists included—worshipped at this same altar because economic growth disguised a multitude of sins. Indonesians and Japanese tolerated endless corruption as long as economic growth lasted. Russians and Eastern Europeans put up with clumsy surveillance states. Americans and Brazilians accepted vast social inequalities. Social, moral, and ecological ills were sustained in the interest of economic growth; indeed, adherents to the faith proposed that only more growth could resolve such ills. Economic growth became the indispensable ideology of the state nearly everywhere. How?

This state religion had deep roots in earlier centuries, at least in imperial China and mercantilist Europe. But it succeeded only after the Great Depression of the 1930s. Like an exotic intruder invading disturbed ecosystems, the growth fetish colonized ideological fields around the world after the dislocations of the Depression: it was the intellectual equivalent of the European rabbit. After the Depression, economic rationality trumped all other concerns except security. Those who promised to deliver the holy grail became high priests.

These were economists, mostly Anglo-American economists. They helped win World War II by reflating and managing the American and British economies. The international dominance of the United States after 1945 assured wide acceptance of American ideas, especially in economics, where American success was most conspicuous. Meanwhile the Soviet Union proselytized within its geopolitical sphere, offering a version of the growth fetish administered by engineers more than by economists.

American economists cheerfully accepted credit for ending the Depression and managing the war economies. Between 1935 and 1970 they acquired enormous prestige and power because—or so it seemed—they could manipulate demand through minor adjustments in fiscal and monetary policy so as to minimize unemployment, avoid slumps, and assure perpetual economic growth. They seized every chance to spread the gospel: infiltrating the corridors of power and the groves of academe, providing expert advice at home and abroad, training legions of acolytes from around the world, and writing columns for popular magazines. Their priesthood tolerated many sects, but agreed on fundamentals. Their ideas fit so well with contemporary social and political conditions that in many societies they locked in as orthodoxy. All this mattered because economists thought, wrote, and prescribed as if nature did not.

If Judeo-Christian monotheism took nature out of religion, Anglo-American economists (after about 1880) took nature out of economics.

This was peculiar. Earlier economists, most notably the Reverend Thomas Malthus (1766–1834) and W. S. Jevons (1835–1882), tried hard to take nature into account. But with industrialization, urbanization, and the rise of the service sector, economic theory by 1935 to 1960 became a bloodless abstraction in which nature figured, if at all, as a storehouse of resources waiting to be used. Nature did not evolve, nor did it adjust when

tweaked. Economics, once the dismal science, became the jolly science. One American economist in 1984 cheerfully forecast 7 billion years of economic growth—only the extinction of the sun could cloud the horizon. Nobel Prize winners such as Robert Solow could claim, without risk to their reputations, that "the world can, in effect, get along without natural resources." These were extreme statements but essentially canonical views. If Judeo-Christian monotheism took nature out of religion, Anglo-American economists (after about 1880) took nature out of economics.

The growth fetish, while on balance quite useful in a world with empty land, shoals of undisturbed fish, vast forests, and a robust ozone shield, helped create a more crowded and stressed one. Despite the disappearance of ecological buffers and mounting real costs, ideological lock-in reigned in both capitalist and communist circles. No reputable sect among economists could account for depreciating natural assets. The true heretics, economists who challenged the fundamental goal of growth and sought to recognize value in ecosystem services, remained beyond the pale to the end of the century. Economic thought did not adjust to the changed conditions it helped create; it thereby continued to legitimate, and indeed indirectly to cause, massive and rapid ecological change. The overarching priority of economic growth was easily the most important idea of the twentieth century.

From about 1880 to 1970, the intellectual world was aligned so as to deny the massive environmental changes afoot. While economists ignored nature, ecologists pretended humankind did not exist. Rather than sully their science with the uncertainties of human affairs, they sought out pristine patches in which to monitor energy flows and population dynamics. Consequently they had no political, economic—or ecological—impact.

Thinking environmentally

In contrast to the big ideas of the twentieth century, explicitly environmental thought mattered little before 1970. Acute observers, such as Aldo Leopold (1887–1948) in the United States, remarked on changes to forests, wildlife, soils, and biogeochemical flows. Fears of global resource exhaustion, although almost always mistaken, provoked laments and warnings. But the audience was small and the practical results few. Environmental thinking appealed only to a narrow slice of society. Small nature conservation societies arose almost everywhere in the Western world by 1910. Nature preserves and national parks, mostly isolated from economic use, emerged after 1870, first in Australia and North America, where plenty of open space was available after the near elimination of aboriginal and Amerindian peoples. These efforts inspired widespread imitation, but in most countries preserves and parks had to be small and had to accommodate existing economic activity. These developments scarcely slowed the momentum of environmental change. The ideas, however sound and elegantly put, did not mesh with the times. That began to change in the 1960s.

The 1960s were turbulent times. From Mexico to Indonesia and from China to the United States, received wisdom and constituted authority came under fierce attack. Of the many ideas and movements nurtured in these heated conditions, two continue to have an important impact: women's equality and environmentalism. The success of environmentalism (loosely defined as the view that humankind ought to seek peaceful co-existence with, rather than mastery of, nature) depended on many things. In the industrial world, pollution loads and dangerous chemicals had built up quickly in the preceding decades. Wealth had accumulated (and diffused through Fordism) to the point where most citizens could afford to worry about matters beyond money. In a sense, the economic growth of the industrial countries from 1945 to 1973 provoked its own antithesis in environmentalism.

Successful ideas require great communicators to bring about wide conversion. The single most effective catalyst for environmentalism was an American aquatic zoologist with a sharp pen, Rachel Carson (1907–1964). In 1962 her salvo against indiscriminate use of pesticides, *Silent Spring*, appeared. She compared the agrochemical companies to the Renaissance Borgias with their penchant for poisoning. This earned her denunciations from chemical manufacturers and the Department of Agriculture as a hysterical and unscientific woman. The resulting hue and cry brought her, and her detractors, onto national television in 1963. But her scientific information, mainly concerning the deleterious effects of DDT and other insecticides on bird life, was mostly sound and her message successful. After serialization in *The New Yorker*, her book became a bestseller in several languages. President John F. Kennedy, against the wishes of the Department of Agriculture, convened a government panel to look into pesticide problems, and its findings harmonized with Carson's. Eventually she had elementary schools named for her and her face graced postage stamps.

In another age Rachel Carson's ideas might have been ignored. Instead she, and hundreds like her, inspired followers and imitators. Millions now found the pollution they had known most of their lives to be unnecessary and intolerable. Earth Day in 1970 mobilized some 20 million Americans in demonstrations against assaults on nature. By the 1980s, anxieties about tropical deforestation, climate change, and the thinning ozone shield added a fillip (and a new focus) to environmentalism. Earth Day in 1990 attracted 200 million participants in 140 countries. American popular music—a global influence—added the environment to its repertoire of subjects. Mainstream religious leaders, from the Dalai Lama to the Greek Orthodox patriarch (of Istanbul), embraced aspects of environmentalism, as did some fundamentalist religions groups. Big science and its government funders converted too. The United Nations launched its Man and the Biosphere research program in 1971, and by 1990 most rich countries had global-change science programs. Taken together, by 1998 these amounted to the largest research program in world history.

Between 1960 and 1990 a remarkable and potentially earth-shattering (earth-healing?) shift took place. For millions of people, swamps long suited only for draining had become wetlands worth conserving. Wolves graduated from varmints to noble savages. Nuclear energy, once expected to fuel a cornucopian future, became politically unacceptable. Pollution no longer signified industrial wealth but became a crime against nature and society. People held such views with varying emphases and degrees of commitment. Movements based on them became schismatic in the extreme, but all shared a common perceptual shift. The package of ideas proved highly successful, to the point where by the late 1980s, oil companies and chemical firms caved in and instructed their public relations staff to construct new "green" identities. While the sincerity of their conversions remained open to doubt, their fig leaves showed that in the realm of ideology, environmentalism had arrived.

By far the most important political forces for environmental change were inadvertent and unwitting.

This extraordinary intellectual and cultural shift started in the rich countries but emerged almost everywhere. Environmentalism had many faces, each with its own issues and agendas. Where it was once systematically repressed—in some countries of the Soviet bloc ecological data were state secrets—environmentalism soon helped topple regimes. In countries as poor as India, vigorous environmental groups emerged by 1973 and coalesced by the 1980s. In poor countries environmentalism normally was entwined in social struggles over water, fish, or wood and had little to do with nature for nature's sake. The full meaning of this new current will take decades, conceivably centuries, to reveal itself.

Security anxiety...

By far the most important political forces for environmental change were inadvertent and unwitting. Explicit, conscious environmental politics, while growing in impact after 1970, still operated in the shadow cast by conventional politics. This was true on both the international and national scales.

At the international level, the dominant characteristic of the twentieth-century system was its highly agitated state. By the standards of prior centuries, the big economies and populous countries conducted their business with war very much on their minds, especially from about 1910 to 1991. War efforts in the two world wars were all-consuming. Security anxiety between the wars, especially during the long cold war, was high given the perceived costs of unpreparedness; states and societies had strong incentives to maximize their military strength, to industrialize (and militarize) their economies, and after 1945, to develop nuclear weapons. The international system, in Darwinian language, selected rigorously against ecological prudence in favor of policies dictated by short-term security considerations.

Security anxiety had countless environmental ramifications. In France after the defeat of 1870, the army won the power to preserve public and private forests in northeastern France, using them in a reorganized frontier defense system designed to channel German invaders along narrow, well-fortified corridors. (The next German invasion, in 1914, came through Belgium.) Many tense borders became de facto nature preserves because ordinary human activities were prohibited (for example, the border between Bulgaria and Greece, the demilitarized zone between North and South Korea, and the border between Iran and the Soviet Union). But other border regions became targets for intensive settlement intended, among other goals, to secure sovereignty, and consequently witnessed wide deforestation, as in Brazilian and Ecuadorian Amazonia. Many states built road and rail systems with geopolitical priorities in mind, such as imperial Russia's Trans-Siberian Railroad, Hitler's autobahns, America's interstate system, and the Karakoram Highway between Pakistan and China. Such major transport systems inevitably affected land-use patterns.

The largest environmental effect of security anxiety occurred with the construction of military-industrial complexes. After World War I, aside from plenty of young men, clearly the main ingredient of military power was heavy industry. Horses and heroism were obsolete. All the great twentieth-century powers adopted policies to encourage the production of munitions, ships, trucks, aircraft—and nuclear weapons.

... and environmental liberties

No component of military-industrial complexes enjoyed greater subsidy, protection from public scrutiny, and latitude in its environmental impact than the nuclear weapons business. At least nine countries built nuclear weapons, although only seven admitted doing so (the United States, Britain, France, the Soviet Union, China, India, and Pakistan). Israel and South Africa developed nuclear weapons while pretending they had not (a pretense South Africa abandoned by publicly dismantling its nuclear weapons program after 1994).

The American weapons complex involved some 3,000 sites. The United States built tens of thousands of nuclear warheads and tested more than a thousand of them. The jewel in this crown was the Hanford Engineering Works, a sprawling bomb factory on the Columbia River in the bone-dry expanse of south-central Washington state. It opened during World War II and built the bomb that destroyed Nagasaki. Over the next 50 years, Hanford released billions of gallons of radioactive wastes into the Columbia and accidentally leaked some more into groundwater. In 1949, shortly after the Soviet Union had exploded its first atomic bomb, the Americans conducted a secret experiment at Hanford. The fallout detected from the Soviet test had prompted questions about how quickly the Soviet Union could process plutonium. In response, American officials decided to use "green" uranium, less than 20 days out of the reactor, to test their hypotheses about Soviet activities. The Green Run, as it was known to those in on the secret, released nearly 8,000 curies of iodine-131, dousing the downwind region with radiation levels between 80 and 1,000 times the limit then thought tolerable. The local populace learned of these events in

1986, when Hanford became the first of the American nuclear weapons complexes to release documents concerning the environmental effects of weapons production. The Green Run shows the environmental liberties the Americans took under the influence of cold war security anxiety.

But that was just the tip of the iceberg. More environmentally serious were the wastes, which in the heat of the cold war were left for the future to worry about. A half century of weapons production around the United States left a big mess, including tens of millions of cubic meters of long-lived nuclear waste. Partial cleanup is projected to take 75 years and cost $100 billion to $1 trillion, the largest environmental remediation project in history. Full cleanup is impossible. More than half a ton of plutonium is buried around Hanford alone.

The Soviet government was more cavalier. Its nuclear program began with Stalin, who wanted atomic weapons as fast as possible, whatever the human and environmental cost. As it happened, the Soviet command economy was rather good at such things: a large nuclear weapons complex arose from nothing in only a few years. The Soviet Union built approximately 45,000 warheads and carried out about 715 nuclear tests between 1949 and 1991, mostly at Semipalatinsk (in what is now Kazakhstan) and on the Arctic island of Novaya Zemlya. The government used nuclear explosions to create reservoirs and canals and to open mine shafts. In 1972 and 1984 it detonated three nuclear weapons in an attempt to loosen ores from which phosphate (for fertilizer) was derived. The Soviet government dumped much of its nuclear wastes at sea, mostly in the Arctic Ocean, some of it in shallow water. It also scuttled defunct nuclear submarines at sea.

The Soviet Union had only one center for reprocessing used nuclear fuel: the Mayak complex in the upper Ob River basin in western Siberia, now the most radioactive place on earth. It accumulated 26 metric tons of plutonium, 50 times Hanford's total. From 1948 to 1956, the Mayak complex dumped radioactive waste into the Techa River, an Ob tributary and the sole source of drinking water for 10,000 to 20,000 people. After 1952, storage tanks held some of Mayak's most dangerous wastes, but in 1957 one exploded, raining 20 million curies down onto the neighborhood—about 40 percent of the level of radiation released by the 1986 nuclear reactor accident at Chernobyl. After 1958, liquid wastes were stored in Lake Karachay. In 1967 a drought exposed the lakebed's radioactive sediments to the steppe winds, sprinkling dangerous dust with 3,000 times the radioactivity released at Hiroshima over an area the size of Belgium and onto a half million unsuspecting people. By the 1980s, anyone standing at the lakeshore for an hour received a lethal dose of radiation (600 roentgens/hr). A former chairman of the Soviet Union's Supreme Soviet's Subcommittee on Nuclear Safety, Alexander Penyagin, likened the situation at Mayak to 100 Chernobyls. No one knew the extent of contamination in the former Soviet Union because the nuclear complex was so large and so secret. Much of the complex was shut down in the last years of the Soviet Union, but the radioactive waste remains and Russia cannot afford much in the way of cleanup.

War's other collateral damage

Much less has been done in war than in the name of war. The twentieth century did not lack for prolonged combat, but most of the environmental changes wrought in combat proved fleeting. Bombers flattened most of Berlin and Tokyo near the end of World War II, but both cities sprang up again within a decade or two. American bombers put some 20 million craters in Vietnam, but vegetation covered most of these wounds, while some eventually served as fishponds. In the war between Japan and China (1937–1945), Chinese Nationalists, hoping to forestall a Japanese advance, destroyed the dikes holding the Huanghe (Yellow River) in 1938. Probably the single most environmentally damaging act of war, it drowned several hundred thousand Chinese (and many thousand Japanese), destroyed millions of hectares of cropland in three provinces, and flooded 11 cities and 4,000 villages. But the labor of surviving Chinese made good the devastation in a few years. The intense combat on the Western front and at Gallipoli during World War I and the scorched-earth policies of the German-Soviet struggle during World War II brought correspondingly intense environmental devastation. But patient labor and the processes of nature have hidden these scars and assimilated into the surrounding countryside the sites of even the most ferocious battles—except where a conscious effort has been made to preserve the battlefields as memorials. In the 1991 Persian Gulf War, Iraqi forces ignited huge oil fires that darkened the skies, and spilled further oil into the shallow and biologically rich Persian Gulf. The atmospheric pall dissipated in a few months when the burning wells were capped, but marine life took (and will take) years to recover. The Gulf War may prove an exception to the rule about the fleeting nature of environmental damage from combat.

While environments governed by irrigation works such as China's were the most vulnerable to war's destruction, deforestation took (and will take) longer to heal. Dryland agriculture recovered quickly from war, on average in about three years. Pasture and grassland often took a little longer, perhaps 10 years. But forests would take a century or three. For centuries warfare had featured forest destruction as policy. Caesar burned Gallic woods. In the twentieth century the prominence of guerrilla tactics meant that war played an unusually large role in deforestation. Many wars of colonial resistance in Africa and Southeast Asia involved guerrilla campaigns. During the cold war, many of the proxy wars fought in Africa, Asia, and Central America did too. Guerrillas had to hide, and forests provided ideal cover; hence antiguerrilla forces destroyed forest. At times guerrillas did too, often as acts of arson aimed at occupying powers or forces of constituted order.

Twentieth-century technology made forest destruction much easier than in Caesar's (or William Tecumseh Sherman's) day. The French pioneered incendiary bombing of forests in the Rif War, an up rising of Moroccan Berbers against Spanish and French colonial power in the 1920s. Napalm debuted in World War II in flame-throwers and proved its effectiveness against forest cover in the Greek Civil War before becoming a major weapon in the American arsenal in Vietnam. The British inaugurated the use of chemical defoliants in the Malaya insurgency in the 1950s. The Americans used them on a large scale (for example, Agent Orange) in Vietnam. The Soviet-Afghan War begun in 1979 witnessed the use of a variety of high-tech defoliants. These and a hundred cases like them constitute some of the more durable ecological effects of combat.

Outside of combat, war efforts had other ecological impacts. European wheat demand in World War I led to the plowing up of about 6 million hectares of grasslands on the American High Plains and in Canada's prairie provinces. This helped prepare the way for the dust bowl of the 1930s. The British war effort in World War II consumed about half of Britain's forests. To build Liberty Ships in 11 days, as Americans did in Portland, Oregon, during that war, took a lot of electricity, justifying additional hydroelectric dams on the Columbia River, where two large dams had already been built in the late 1930s. Frantic drives to raise production of food, fuel, minerals, and other resources surely led to sharp ecological disruptions in every combatant nation, as did crash road- and railroad-building efforts. More recently, belligerents in the civil wars that raged in Cambodia and eastern Burma financed their campaigns by contracting with Thai logging companies to strip forest areas under their control.

By suppressing normal economic activity, war temporarily reduced some ordinary environmental pressures. Despite depth charges and oil spills in the U-boat campaigns, World War II brought back halcyon days for North Atlantic fish populations, because fishing fleets sat out the war. Industrial emissions slackened because of coal shortages and factory destruction, at least in Europe and Japan. Iraqi land mines in the Kuwaiti desert kept people out and allowed a resurgence of animal and plant life in the 1990s. Combat had its impacts on the environment, occasionally acute but usually fleeting. More serious changes arose from the desperate business of preparing and mobilizing for industrial warfare.

The impact of political change

Although war was the most dramatic, some of the twentieth century's other major political forces—imperialism, decolonization, and democratization—also had a lasting impact on the earth's environment.

As the twentieth century began, Russia, Japan, the United States, and especially the Western European powers had embarked on imperial expansions. This often involved the displacement of existing populations, as in South Africa and Algeria. Colonial powers reoriented local economies toward mining and logging, and toward export monocultures of cotton, tea, peanuts, or sisal. Normally these changes were imposed with no thought to environmental consequences: the only goals were to make money for the state and for entrepreneurs, and to assure the mother country ready access to strategic materials. By the 1940s the French and British at least claimed to have local interests at heart when converting as much as possible of Mali to cotton or of Tanganyika to peanut production. But through ecological ignorance they nonetheless brought salinization in the Niger bend region of Mali and turned marginal land into useless hardpan in central Tanganyika.

Decolonization surprisingly changed little of this. Newly independent regimes often continued the economic policies of their predecessors. Big prestige projects carried on the tradition of colonial environmental manipulation in places such as Ghana, Sudan, and India. Financially weak regimes in Indonesia, Papua New Guinea, and Ivory Coast often sold off timber and minerals as fast as possible, regardless of environmental impacts. Many rulers arrived in office by way of a coup and saw fit to cash in before the next colonel or sergeant followed suit. The decolonization of Soviet Central Asia brought no changes in the water regime that was strangling the Aral Sea. In environmental matters, as in so many respects, independence often proved no more than a change in flags.

Democratization was another matter. A global wind of democratization blew across Greece and Iberia in the 1970s, much of Latin America and East Asia in the 1980s, and parts of Eastern Europe and Africa in the 1990s. In some cases, environmental protests helped in modest ways to undermine the legitimacy of authoritarian (for example, Chile) and communist (Poland) regimes. Such regimes had encouraged pollution-intensive economies and ecologically heedless resource extraction in their quests to build state power and maximize economic growth. They normally had kept ecological information under tight control. Democratization broke the hold these regimes enjoyed over information, and brought to light many environmental problems. Those caused by foreigners, the military, or specific factories were often addressed and sometimes resolved. Those brought about by the consumption patterns of ordinary citizens often grew worse under democracy, when, for instance, Eastern Europe and Russia dropped subsidized public transport in favor of private cars. Moreover the media spotlight shone only on certain kinds of environmental problems, usually those inspiring maximum dread such as industrial accidents and nuclear issues. Slow-moving crises such as soil erosion or biodiversity loss remained hidden in the shadows, uncompelling to the media and the public and entirely irrelevant to politicians attuned to the next election. Thus democracy tended to generate its own characteristic environment.

The turn to environmentalism

Almost all the environmental changes generated by imperialism, decolonization, democratization, and to a lesser degree, war were inadvertent effects of politics and policies designed for other ends. In contrast, the politics and policies in which environmental considerations formed a conscious element had modest effects.

Environmental politics and policies, as such, began only in the 1960s. Before that, local, national, and (on a very limited scale) international laws and treaties regulated some aspects of pollution, land use, fishing, and other issues. Smoke nuisance ordinances go back at least 700 years. Britain established a regulatory agency for a specific source of pollution, the Alkali Inspectorate, in 1865. But all this was uncoordinated—specific policies and laws for very specific instances. On the international scale, neighboring countries occasionally had agreed to restrain fishing or water use. A multilateral agreement in 1911

checked the exploitation, in the Bering Sea's Pribilof Islands, of fur seals that Russian, Japanese, Canadian, and American sealers had hunted nearly to extinction between 1865 and 1900. In the aftermath of World War II, international institutions sprang up, including some concerned with the environment such as the International Union for the Conservation of Nature. Others regulated the environment without making it their explicit focus, such as the World Health Organization. But no coordinated policies or political currents dealt with the environment as such. This changed in the 1960s, a direct result of the tumult in the world of ideas.

Two general phases are discernible in the environmental politics and policies of the late twentieth century. The first began in the mid-1960s and lasted until the late 1970s. In this phase, environmental movements and, in some cases, political parties, sprang up in the rich countries. New Zealand's Values Party, born in the late 1960s, was the first explicitly green party but far from the most successful: it splintered after some 15 years on the edges of New Zealand politics. Environmental movements focused mainly on pollution issues but also on fears of resource exhaustion, spurred by the actions of the Organization of Petroleum Exporting Countries in 1973. Governments responded by creating new agencies charged with protecting the environment as a whole. Sweden (1967) and the United States (1970) led the way. International regimes of cooperation remained very weak, despite the efforts made after the first international conference on the environment in Stockholm (1972). That led to the United Nations Environment Program, headquartered in Nairobi.

In the second phase, beginning around 1980, poorer countries established their own environmental protection agencies, often given the status of ministry. In many cases, such as Nigeria or Russia, environmental laws and policy existed only on paper. In some cases, for example, Angola or Afghanistan, ongoing wars meant no environmental politics or policy existed even on paper. But in India, Brazil, Kenya, and elsewhere, grassroots environmental movements germinated and, whether through civil disobedience or official channels, began to affect national politics. India boasted hundreds of environmental organizations by the 1980s, ranging from scientific research institutes that served as watchdogs—such as New Delhi's Center for Science and Environment—to coalitions composed mainly of peasant women, such as the Chipko movement that sought to check logging in the Himalaya. These movements were often led by women whose lives were most affected by fuelwood shortage (because fuelwood gathering almost everywhere was women's and children's work), soil erosion (where women worked the land, as in much of Africa and the Indian Himalaya), and water pollution (because women fetched water and were responsible for children's health). Ordinarily these grassroots environmental movements were embedded in peasant protest or some other social struggle. When strong enough, they won some concessions from governments; when not, they solidified anti-environmental attitudes in the corridors of power, inadvertently inviting elites to equate environmentalism with subversion and treason.

In the rich countries in this second phase, new concerns added new dimensions to environmental politics: tropical forests, climate change, ozone depletion. In the United States an ideological crusade to roll back environmental regulation boomeranged, as provocative statements by President Ronald Reagan's officials served as recruiting devices for environmental pressure groups. Leadership in terms of innovative institutions and planning passed to northern European countries, notably the Netherlands, and to Japan. Green parties entered politics, and in some cases (such as Germany in 1983) parliaments as well. In 1998 the German Green Party took part in a coalition government, and its members held some important ministries. The Europeans pioneered a consensual politics of environmental moderation, based on corporatist traditions in which government, business, and organized labor hammered out agreements after prolonged bargaining. The Dutch in particular, beginning in 1989, arrived at an integrated national environmental plan, designed to harness the power of the influential ministries and special interests resistant to ecological prudence, such as agribusiness.

Cooperation and confrontation

The second phase featured unprecedented efforts at international cooperation. Regional and global problems, such as acid rain or ozone depletion, required new institutions, agreements, and regimes of restraint. In 1987 the United States Congress helped browbeat the World Bank into environmental awareness. That same year also saw the Brundtland Report. The fruit of four years of UN-sponsored inquiry into the relationship between environment and economic development, the report offered intellectual underpinnings for environmental planning, for regimes of restraint, and for the ambition of ecologically sustainable development. Also in 1987, the Montreal Protocol to protect the ozone layer and subsequent accords showed what good science and diplomacy could do. Thousands of international environmental agreements were reached from the mid-1960s onward and many had real effects. Optimistic observers saw in this a nascent "global governance regime" that could address the world's cross-border environmental problems.

The impact of all this, from 1967 on, was considerable in the rich countries. The technically and politically easiest environmental problems were in fact significantly reduced. Industrial wastewater was cleaned up, with benefits to the Rhine, the North American Great Lakes, and elsewhere. Sulfur dioxide emissions waned. Leaded gasoline vanished into history. Municipal sewage treatment improved. In general, problems that arose from a single institution or point source were addressed with some success. Initially at least, local solutions such as taller smokestacks merely shunted ill effects elsewhere. Sometimes more systematic solutions succeeded in their specific task but at the same time deepened other problems. Scrubbers used to control particulate emissions from smokestacks worsened acid rain. Most truculent of all were those problems that derived from citizen behavior or from diffuse sources. Nitrous oxides from vehicle exhausts and toxic farm runoff, for instance, continued to mount in North America and Europe.

Moreover, in most rich countries some powerful industries resisted environmental regulation by launching endless lawsuits or controlling the decisive ministries. This helped prevent serious reform in transport, energy, and agribusiness, the myriad environmental impacts of which scarcely abated. The United States automobile industry fought successfully to hamstring fuel-efficiency standards. California agribusiness kept getting water at dirt-cheap prices. The coal industry in Germany retained its giant subsidies. More often than not, major decisions affecting the environment remained the province of powerful ministries—trade, finance, industry, agriculture—rather than of environmental agencies.

Environmental politics ran up against the limits of the possible at the international level too. Although the United States became more amenable to international agreements after the late 1980s, it still made clear at the 1992 UN Conference on Environment and Development in Rio de Janeiro that American "lifestyles" were not negotiable. Other countries matched this stance. Japan proved recalcitrant on whaling prohibitions (as did Norway) and the trade in elephant ivory. Saudi Arabia and other oil producers fought against agreements on carbon emissions. Brazil insisted on its right to develop Amazonia as it wished, regardless of the implications of burning the world's largest rainforest. India and China declined to join the Montreal Protocol and subsequent accords on ozone-destroying chloro-fluorocarbons, and in general adamantly refused to compromise their industrial ambitions in the interest of their, or the global, environment. Mexico and many other countries resisted pressures to harmonize environmental laws with those of richer nations: countries with more relaxed laws (or enforcement) found multinational firms more eager to invest in new steel mills and chemical plants. While many fault lines and alliances existed in this late-century world of international environmental politics, the main one divided rich from poor. Called, with dubious geography, a North-South confrontation, it crystallized at Rio in 1992 and particularly bedeviled climate-change negotiations, which had achieved only toothless accords up to 1999.

In short, both before and after 1970, for good and for ill, real environmental policy, both on the international and national levels, was made inadvertently, as side effects of conventional politics and policies. Britain managed to reduce its sulfur emissions after 1985 because Prime Minister Margaret Thatcher scuttled the coal industry in her quest to break the political power of trade unionism. Farm subsidies, especially in Japan and Europe, helped create and maintain a chemical-intensive agriculture and dense populations of pigs and cattle, with deleterious consequences. Soviet and Chinese policies reduced the mobility of Central Asian nomads, aggravating overgrazing and desertification. China's collectivization and Cultural Revolution in the 1960s destroyed village-level constraints on marriage and fertility, provoking a gargantuan baby boom that lay behind many aspects of China's manifold environmental crisis in the 1980s and beyond. China's collectivization also helped inspire Tanzania's "villagization" scheme of the 1970s, the largest resettlement plan in African history, and one that led to deep environmental problems. Even in the age of environmental politics and overt environmental policy, real environmental policy almost always derived from other concerns, and traditional politics exercised stronger influence over environmental history.

Human history and unintended consequences

The grand social and ideological systems that people construct for themselves invariably carry large consequences, for the environment no less than for more strictly human affairs. Among the swirl of ideas, policies, and political structures of the twentieth century, the most ecologically influential were the growth imperative and the (not unrelated) security anxiety that together dominated policy around the world. Both were venerable features of the intellectual and political landscape, and both solidified their hold on imaginations and institutions in the twentieth century. Both, but particularly the growth imperative, meshed well with the simultaneous trends and trajectories in population, technology, energy, and economic integration. Indeed, successful ideas and policies had to mesh with these trends.

Domestic politics in open societies proved mildly more responsive to environmental problems that annoyed citizens than did more authoritarian societies, especially after 1970, but there were clear limits to the ecological prudence that citizens wanted. Regardless of political system, policymakers at all levels responded more readily to clear and present dangers (and opportunities) than to more subtle and gradual worries about the environment. The prospect of economic depression or military defeat commanded attention that pollution, deforestation, or climate change could not. More jobs, higher tax revenues, and stronger militaries all appealed, with an immediate lure that cleaner air or diversified ecosystems could not match.

By 1970, however, something new was afoot. The interlocked, mutually supporting (and co-evolving) social, ideological, political, economic, and technological systems that we conveniently call industrial society spawned movements that cast doubt on the propriety and prudence of business as usual. Some of these movements demanded the antithesis of industrial society, denouncing technology, wealth, and large-scale organization. Others called for yet more and better technology and or-

ganization, and more wealth for those who had least, as solutions to environmental problems. To date these new movements exercise only modest influence over the course of events, but they are still young. When Zhou Enlai, longtime prime minister of Mao's China, was asked about the significance of the French Revolution some 180 years after the event, he replied that it was still too early to tell. So it is, after only 35 years, with modern environmentalism.

Environmental change of the scale, intensity, and variety witnessed in the twentieth century required multiple, mutually reinforcing causes. The most important immediate cause was the enormous surge of economic activity. Behind that lay the long booms in energy use and population. Economic growth had the environmental implications that it had because of the technological, ideological, and political histories of the twentieth century. All these histories mutually affected one another; they also determined, and in some measure were themselves determined by, environmental history.

Few people paused to contemplate these complexities. In the struggles for survival and power, in the hurly-burly of getting and spending, few citizens and fewer rulers spared a thought for the ecological impacts of their behavior or ideas. Even after 1970, when environmental awareness had hurriedly dawned, easy fables of good and evil dominated public and political discourse. In this context, environmental outcomes continued, as before, to derive primarily from unintended consequences. Many specific outcomes were in a sense accidental. But the general trend of increasing human impact and influence was no accident. It was, while unintended, strongly determined by the trajectories of human history.

J. R. McNEILL is a professor at Georgetown University, where he has taught environmental and international history since 1985. His books include *Something New Under the Sun: An Environmental History of the Twentieth-Century World* (New York: Norton, 2000), from which this article is excerpted with permission, and *The Mountains of the Mediterranean World: An Environmental History* (New York: Cambridge University Press, 1992). Copyright 2000 by J. R. McNeill.

From *Current History,* November 2000, pp. 371-382, originally excerpted from "An Environmental History of the Twentieth-Century World" (New York-Norton 2000). © 2000 by Current History, Inc. Reprinted by permission.

Environmental Surprises: Planning for the Unexpected

Environmental futurists are finding new ways to anticipate the effects of complex trends.

By Chris Bright

The process of environmental decline usually seems gradual and predictable. We are comforted by the thought that even if we have not turned the trends around there will be time for our children to rise to the challenge.

But this way of thinking is like sleepwalking. To understand why, you have to look at decline close up. Here is how it has happened in one small country, with big implications.

The Honduran Predicament

In the early 1970s, Honduras was caught up in a drive to build agricultural exports. Landowners in the south increased their production of cattle, sugarcane, and cotton. This more-intensive farming reduced the soil's water absorbency, so more and more rain ran off the fields and less remained to evaporate back into the air. The drier air reduced cloud cover and rainfall. The region grew a lot warmer.

As the land became less productive, people began to leave. Many moved north to work on newly developed plantations or to carve their own small farms out of the area's rain forests. Much of this northern agriculture was devoted to export crops, too, primarily bananas, melons, and pineapples.

But it is difficult to mass-produce big, succulent fruits near rain forests—even badly fragmented rain forests—because there are so many insects and fungi around to eat them. So the plantations came to rely heavily on pesticides. From 1989 to 1991, Honduran pesticide imports increased more than fivefold, to about 8,000 tons.

The steaming, ragged forest was a perfect habitat for malaria mosquitoes. Around the plantations, the insecticide drizzle suppressed them for a time, but they eventually acquired resistance to a whole spectrum of chemicals. As a result, the mosquitoes were basically released from human control. When their populations bounced back, they encountered a landscape stocked with their favorite prey: people. And since these people were from an area where malaria infection had become rare, their immunity to the disease was low. Malaria rapidly reasserted itself: From 1987 to 1993, the number of cases in Honduras jumped from 20,000 to an estimated 90,000.

The situation was brought to light in 1993 by a group of researchers concerned about the public health implications of environmental decline. But their primary interest was not in what had already happened—it was in what might happen next. Some very nasty surprises might be tangled up somewhere in this web of pressures. They argued, for example, that deforestation and changing patterns of disease had made the country vulnerable to climate change.

They were right. In October 1998, Hurricane Mitch slammed into the Gulf Coast of Central America and stalled there for four days. Nightmarish mudslides obliterated entire villages; half the population of Honduras was displaced, and the country lost 95% of its agricultural production.

Mitch was the fourth-strongest hurricane to enter the Caribbean in the twentieth century, but much of the damage was caused by deforestation: If forests had been gripping the soil on those hills, fewer villages would have been buried in mudslides. And in the chaos and filth of Mitch's wake there followed tens of thousands of additional cases of malaria, cholera, and dengue fever.

Complexities of Change

It is hard to shake the feeling that "normal change"—even change for the worse—should not happen this way. In the first place, too many trends are spiking. Instead of gradual change, the picture is full of *discontinuities*—very rapid shifts that are much harder to anticipate. There is a rapid warming in the south, then an abrupt expansion in deforestation in the north, as plantations are developed. Then malaria infections jump. Then the mudslides, in addition to killing thousands of people, cause a huge increase in the rate of topsoil loss.

There also seem to be too many overlapping pressures—too many

synergisms. The mudslides were not the work of Mitch alone; they were caused by Mitch plus the social conditions that encouraged the farming of upland forests. The malaria emerged not just from the mosquitoes, but from the movement of a low-immunity population into a mosquito-infested area, and from heavy pesticide use.

The future of a trend—any trend—depends on the behavior of the entire system in which it is embedded.

Discontinuities and synergisms frequently catch us by surprise. They tend to subvert our sense of the world because we so often assume that a trend can be understood in isolation. It is tempting, for example, to believe that a smooth line on a graph can be used to see into the future: All you have to do is extend the line. But the future of a trend—any trend—depends on the behavior of the entire system in which it is embedded. When we isolate a phenomenon in

order to study it, we may actually be preventing ourselves from knowing the most important things about it.

Such a fragmented form of inquiry is becoming increasingly dangerous—and not just because we might miss problems in small, poor countries like Honduras. After all, there is nothing special about the pressures in the Honduran predicament. Deforestation, climate change, chemical contamination, and many other forms of environmental corrosion are at work on a global scale. Each has engendered its own minor research industry. But even as the publications pile up, we may actually be missing the biggest problem of all: What might the inevitable convergence of these forces do?

"When one problem combines with another problem, the outcome may be not a double problem, but a super-problem," writes Norman Myers, an Oxford-based ecologist who is one of the most active pioneers in the field of environmental surprise. We have hardly begun to identify those potential super-problems, but in the planet's increasingly stressed natural systems, the possibility of rapid, unexpected change is

Three Types of Environmental Surprise

Type of Surprise	Example
A **discontinuity** is an abrupt shift in a trend or a previously stable state. The abruptness is not necessarily apparent on a human scale; what counts is the time frame of the processes involved.	Overfishing has pushed some fish species into a population crash rather than a gradual decline. As recently as the 1970s, for example, the white abalone occurred along the coast of northern Mexico and southern California at densities of up to 10,000 per hectare. Today its total population is probably no more than a few dozen, and its extinction is imminent.
A **synergism** is a change in which several phenomena combine to produce an effect that is greater than would have been expected from adding up their effects taken separately.	The monstrous 1998 flood of China's Yangtze River did $30 billion in damage, displaced 223 million people, and killed another 3,700. The damage was a synergism caused not just by heavy rains, but by extremely dense settlement of the floodplain and by deforestation—the Yangtze basin has lost 85% of its forest cover.
An **unnoticed trend**, even if it produces no discontinuities or synergisms, may still do a surprising amount of damage before it is discovered.	In the United States, where natural areas are monitored with much greater attention than in most parts of the world, aggressive nonnative weeds may still have to be in the country for 30 years or have spread to thousands of hectares before they are even discovered. In the United States and elsewhere, such weeds displace native plants, upset fire and water cycles, and do billions of dollars in agricultural damage every year.

Sources: TREE (Trends in Ecology and Evolution); U.S. Congress, Office of Technology Assessment; Worldwatch Institute.

pervasive and growing. Important theaters of surprise include coral reefs, the atmosphere, and an ecosystem I will discuss in detail: tropical rain forests.

Fires in the Forest

Eight thousand years ago, before people began to clear land on a broad scale, forests covered more than 6 billion hectares (14.8 billion acres), or around 40% of the planet's land surface. Today, the earth's natural forests (as opposed to tree plantations) amount to 3.6 billion hectares (8.9 billion acres) at most. Every year, at least another 14 million hectares are lost. Among the many thousands of species that are believed to go extinct every year, the majority are forest creatures, primarily tropical insects, who have lost their habitat.

Currently, well over 90% of forest loss is occurring in the tropics—on a scale so vast that it might appear to have exceeded its capacity to surprise us. In 1997 and 1998, fires set to clear land in Amazonia claimed more than 5.2 million hectares of Brazilian forest, brush, and savanna—an area nearly 1.5 times the size of Taiwan. In Indonesia, some 2 million hectares of forest were torched during 1997 and 1998.

All this is certainly news, but if you are interested in conservation, it is the kind of dreadful news you have come to expect. And yet our expectations may not be an adequate guide to the sequel, assuming the destruction continues at its current pace. A substantial portion of the damage is hidden—it does not show up in the conventional analysis. But once you take the full extent of the damage into account, you can begin to make out some of the surprises it is likely to trigger.

Consider, for example, the destruction of Amazonia. Over half the world's remaining tropical rain forest lies within the Amazon basin, where more forest is being lost than anywhere else on earth. Deforestation statistics for the area are in-

tended primarily to track the conversion of forest into farms and ranches. Typically, the process begins with the construction of a road, which opens up a new tract of forest to settlement.

In June 1997, for example, some 6 million hectares of forest were officially released for settlement along a major new highway, BR-174, which runs from Manaus, in central Amazonia, over 1,000 kilometers (620 miles) north to Venezuela. Ranchers and subsistence farmers clear cut patches of forest along the road and burnt the slash during the July–November dry season. (The farmers generally have few other options: Brazil has large numbers of poor, land-hungry people, and the plots they cut from the forest lose their fertility rapidly, so there is a constant demand for fresh soil.)

But the damage to the forest generally extends much farther than the areas that are "deforested" in this conventional sense, because of the way fire works in Amazonia. In the past, major fires have not been a frequent enough occurrence to promote any kind of adaptive "fire proofing" in the region's dominant tree species. Some temperate-zone and northern trees, by contrast, are "fire-adapted" in one way or another—they may have especially thick bark, for example, or the ability to resprout after burning. The lack of such adaptations in Amazonian trees means that even a small fire can begin to unravel the forest.

During the burning season, the flames often escape the cuts and sneak into neighboring forest. Even in intact forest, there will be patches of forest floor that are dry enough to allow a small "surface fire" to feed on the dead leaves. Surface fires do not climb trees and become crown fires. They crackle along the forest floor in patches of flame, going out at night when the temperature drops and rekindling the next day. They are fatal to most of the smaller trees they touch. Overall, an initial surface fire may kill 10% of the living forest biomass.

The damage may not seem dramatic, but another tract of forest may already be doomed by an incipient positive feedback loop of fire and drying. After a surface fire, the amount of shade is reduced from about 90% to 60%, and the dead and injured trees rain debris down on the floor. So a year or two later, the next fire in that spot finds more tinder and a warmer, drier floor. Some 40% of forest biomass may die in the second fire. At this point, the forest integrity is seriously damaged; grasses and vines invade and contribute to the accumulation of combustible material. The next dry season may eliminate the forest entirely. Once the original forest is gone, the scrubby second growth or pasture that replaces it will almost certainly burn too frequently to allow the forest to restore itself.

In a recent survey, researchers cross-checked satellite maps with field observations and concluded that conventional deforestation estimates for Brazilian Amazonia were missing some 1 to 1.5 million hectares of severe forest damage done by logging every year. Surface fire damage is harder to quantify, but the same researchers did a fire survey and found that the amount of standing forest that had suffered a surface fire in 1994 and 1995 was 1.5 times the area fully deforested in those years. Overall, they suggested, the area of Amazonian forest attacked by surface fire every year may be roughly equivalent to the area deforested outright. And in some parts of the basin, the extent of this cryptic damage is so great that the conventional measurements may no longer be all that useful. In one region, around Paragominas in eastern Amazonia, the researchers found that, although 62% of the land was classified as forested, only about one-tenth of this consisted of undisturbed forest.

Other Tropical Surprises

As the Amazonian forest dwindles, a surprising second-order effect may emerge as the hydrological cycle changes. Because trees exhale so much water vapor, a forest to some degree creates its own climate. Much of this water vapor condenses below the canopy and drips back into the soil; some of it rises higher before falling back in as rain. Researchers estimate that most of the Amazon's rainfall comes from water vapor exhaled by the forest. Widespread deforestation will therefore tend to make the region substantially drier, and that will accelerate the feedback loop created by the fires.

Other kinds of surprises are lurking in tropical forests as well. As developing countries industrialize, some forest maladies better known in the industrial world are likely to appear in these countries, too. Acid rain, for example, is already reported to be affecting the forests of southern China. In parts of South Asia, Indonesia, South America, and West Africa, this form of pollution is bound to increase as industrialization proceeds and cities enlarge. The soil in these areas tends to be fairly acidic already, which would make it incapable of buffering large doses of additional acid. At least in some of these places, acid-induced decline may therefore be much more abrupt than in the temperate zone.

Other development pressures may be unique to the tropics. Increased hunting pressure, for instance, has reduced animal populations in a good number of tropical forests. Many forest-dwelling peoples have armed themselves with shotguns and rifles, which are far more lethal than traditional weapons. And logging is bringing additional hunters into forest interiors. Hunting often supplements the logging: It feeds the loggers, and the surplus is sold as bush meat in towns and cities farther from the frontier. Such hunting is typically very indiscriminate; almost any creature of any size is potential game. Hundreds—even thousands—of animals may fall prey to a single camp.

In the Republic of Congo, for example, the annual take in a single camp of 648 people was found to be 8,251 animals, amounting to 124 tons of meat. In tropical forests the world over, mammals, birds, and reptiles play critical roles in pollinating trees, dispersing their seeds, and eating other creatures that prey on the seeds. If hunting continues at its present rate, some tropical tree species are liable to disappear along with the animals themselves.

Nature has no reset button. Environmental corrosion is not just killing off individual species—it is setting off system-level changes that are, for all practical purposes, irreversible.

One last forest surprise: Recent research suggests that the Central African bush meat trade may have sparked the AIDS epidemic. It is likely that the original host of the HIV virus was a population of chimpanzees in Cameroon and Gabon. Chimps in that area are commonly hunted for their meat; now, apparently, one of their diseases is hunting us.

An Agenda for the Unexpected

Human pressures on the earth's natural systems have reached a point at which they are more and more likely to engender problems that we are less and less likely to anticipate. Dealing with this predicament is obviously going to require more than simply reacting to problems as they appear. We need to forge a new ethic for managing our relationship with nature—one that emphasizes minimal interference in the lives of wild beings and in the broad natural processes that sustain all living things. Such an ethic might begin with three basic principles.

First, nature is a system of unfathomable complexity. Our predominant response to that complexity has been specialization, in both the sciences and public policy. Learning a lot about a little is a form of progress, but it comes at a cost. Experience is seductive: It is easy for specialists to get into the habit of thinking that they understand all the consequences of a plan. But in a complex, highly stressed system, the biggest consequences may not emerge where the experts are in the habit of looking. This inherent unpredictability condemns us to some degree of error, so it is important to err on the side of minimal disruption whenever possible.

Second, nature gives away nothing for free. You cannot get an appreciable quantity of anything out of nature without sacrificing something in the process. Even sustainable resource management is a trade-off—it's simply one we regard as acceptable. In our dealings with nature, as with any other sort of transaction, we need to know the full cost of the goods before deciding whether they are worth the price, or whether there is a better way to pay for them.

Third, nature has no reset button. Environmental corrosion is not just killing off individual species—it is setting off system-level changes that are, for all practical purposes, irreversible. For example, even if all the world's coral reef species were miraculously to survive the impending bout of rapid climate change, that does not mean that our descendants will be able to reconstruct reef communities. The near impossibility of restoring complex systems to some previous state is another strong argument for minimal disruption.

These are basic features of the natural world: We will never understand it completely, it will not do our bidding for free, and we cannot put it back the way it was. A policy ethic sensitive to these facts of life might emphasize the following themes.

•**Monoculture technologies are brittle, so plan for diversity**. Huge, uniform sectors generally exhibit a kind of superficial efficiency because

An Alternative View: Privatization Proponents See Positive Trends

If you're tired of hearing only bad news about the environment, then *Earth Report 2000* may be the book for you. Published by the Competitive Enterprise Institute (CEI), a think tank in Washington, D.C., *Earth Report* lives up to its billing as a rebuttal to environmental doomsayers:

"For years, The Worldwatch Institute and the environmental establishment have generated publications trumpeting the inevitable destruction of the world's forests, rivers, wetlands, and wildlife. Each year finds the predictions more and more dire, yet somehow the catastrophe never seems to materialize," write CEI researchers.

In contrast, *Earth Report 2000* presents a drumbeat of optimism, arguing that deforestation, declining world fisheries, global warming, population stress, and the loss of biodiversity are simply not the mega-tragedies described by alarmists, but rather problems that can be solved through human ingenuity, privatization of environmental resources, and competition unfettered by government interference.

In a chapter called "Fishing for Solutions," for example, CEI research associate Michael De Alessi cites the fast-growing aquaculture industry as a positive response to the stagnation of the world's ocean fish catch. Private ownership is key, he notes, because resource entrepreneurs have a stronger incentive to maintain environmental quality than do government bureaucrats. However, in applying the privatization scenario to the waning oyster beds of the Chesapeake Bay, it isn't clear whether De Alessi would include among the property rights of oystermen the power to restrict landowners' rights to use fertilizer and pesticides on farms in the surrounding Maryland and Delaware watershed.

The emphasis on privatization in *Earth Report 2000* tends to reject the environmentalists' concept of the *commons*, in which citizens who do not own ecosystems are recognized as having a stake in environmental quality and resource use. According to CEI researchers, government environmental regulations such as the Clean Air Act

have been at best ineffective, at worst restrictive of private enterprise, and too costly. It should be noted, however, that, while *Earth Report 2000* criticizes government subsidies for renewable energy research in solar, wind power, and biomass, it does not mention long-standing federal subsidies that benefit the grazing, timber, and mining industries.

Earth Report 2000 and Worldwatch's *State of the World 2000* represent nothing less than a clash of world views—over science, government environmental policy, economics, and politics. Reading and considering both books may not settle the debate, but it's a lively place to start.

—*Dan Johnson*

Source: *Earth Report 2000: Revisiting the True State of the Planet* by Ronald Bailey et al. McGraw-Hill. 2000. 362 pages. Paperback. Available from the Futurist Bookstore for $19.95 ($17.95 for Society members), cat no. B-2329.

they generate economies of scale. You can see this in fossil-fuel-based power grids, megadam projects, and even in woodpulp plantations. But because they are beholden to vast quantities of invested capital—both financial and political—industrial monocultures are extremely difficult to reform when their hidden costs begin to catch up with them. More diverse technologies are liable to be more adaptable because their investors are not all "betting" on exactly the same future. And whether the goal is the production of irrigation water, paper, or electricity, a more-adaptable system is likely to be more durable over the long term.

•**Direct opposition to a natural force is generally counterproductive, so plan to work with nature**. Heavy-handed approaches sometimes exacerbate a problem, as when intensive pesticide use causes a population explosion of resistant pests.

And some successes are often worse than outright failures. Dams and levees, for instance, may end up controlling a flood-prone river—and largely killing the riverine ecosystem in the process. Sound management often tends to be more oblique than direct. Restoration of water-absorbing floodplain ecosystems can make for more-effective flood control than dams. Cropping systems that mimic natural floral diversity make it harder for any particular pest to dominate a field.

•**You can never have just one effect, so plan to have several**. Thinking through the likely "ripple effects" of a plan will help locate not just the risks, but additional opportunities. Encouraging organic agriculture in the U.S. Midwest, for example, could help reduce nutrient leakage into the Mississippi River. That, in turn, could ease the stress on reefs and other marine communities

in the Gulf of Mexico. Marine conservation may therefore "overlap" with agricultural reform; it might even be possible to extend this overlap to include reform of the North American diet.

Environmental policy is full of such latent positive synergisms. In many countries, for instance, there may be a powerful overlap between the need for meaningful employment and the need to replace the "throwaway" economy with one that emphasizes durable goods.

•**Solutions are almost never permanent, so plan to keep on planning**. In the 1950s, organochlorine pesticides were hailed as a permanent "fix" for insect pest problems; given the pervasive ecological damage that these chemicals are now known to cause, the idea of a permanent chemical solution to anything may seem rather naïve today. But because our relationship with nature is in a constant state of flux, even realistic fixes will need regular revision. The 1986 Montreal Protocol (calling for a phase-out of ozone-depleting chemicals) is not a permanent patch for the ozone layer, in part because climate change will probably exacerbate ozone loss. The Green Revolution is not a permanent answer to world hunger, in part because conventional agriculture is overtaxing aquifers.

The growing strain on the earth's natural systems will probably force an increase in the tempo of policy revision—so it makes sense to take full advantage of the powerful new information and communications technologies. Because of their ability to bring together enormous quantities of data from different areas and disciplines, such technologies could help counter the blinkering effects of specialization.

• **None of us may find the answer alone, but together we probably can**. In social as well as natural systems, there is a potent class of properties that exists only on the system level—properties that cannot be directly attributed to any particular component. In a political system, for example, institutional pluralism can create a public space that no single institution could have created alone. One of the most important policy activities may therefore be to encourage innovation outside policy institutions.

Policy may need to become increasingly a matter of creating not so much solutions per se as the conditions from which solutions can arise. In the face of the unexpected, our best hopes may lie in our collective imagination.

Chris Bright is senior editor of *World Watch* magazine and a research associate focusing on biodiversity issues at the Worldwatch Institute, 1776 Massachusetts Avenue, N.W., Washington, D.C. 20036. Telephone 1-202-452-1999; fax 1-202-296-7365; e-mail cbright@ worldwatch.org; Web site www.worldwatch. org.

This article draws on his chapter, "Anticipating Environmental Surprise," in *State of the World 2000* by Lester Brown et al. W.W. Norton. 2000. 276 pages. Paperback. Available from the Futurist Bookstore for $19.95 ($17.95 for Society members), cat. no. B-2138.

Originally published in the July/August, 2000 issue of *The Futurist*, pp. 41–47. © 2000 by the World Future Society, 7910 Woodmont Avenue, Suite 450, Bethesda, MD 20814. Tel. 301/656-8274; fax 301/951-0394; http://www.wfs.org. Reprinted by permission.

The Paradox of Global Environmentalism

"Wilderness lovers like to speak of the equal rights of all species to exist. This ethical cloaking cannot hide the truth that green missionaries are possibly more dangerous, and certainly more hypocritical, than their economic or religious counterparts."

Ramachandra Guha

The central paradox of global environmentalism is that the people who are the most vocal in defense of nature are the people who most actively destroy it. As biologists have repeatedly reminded us, the present epoch is witness to an unprecedented attack on species and habitats. The most vital as well as the most glamorous of these species and habitats are found in the poorer countries of the South, such as Brazil, Ecuador, Kenya, Tanzania, Indonesia, and India. However, the movement for their conservation is fueled principally by processes originating in the richer countries of the North, such as Norway, Australia, Germany and, preeminently, the United States.

The American wilderness movement has a history that extends back more than a century. Its two most influential and venerated figures have been John Muir (1838–1914), who founded the Sierra Club, and Aldo Leopold (1887–1948), who co-founded the Wilderness Society. Muir and Leopold advanced both scientific and ethical reasons for protecting endangered species and ecosystems. They, and their colleagues, helped inspire the creation of the National Park Service, which in turn put in place perhaps the world's best managed system of protected areas.

Until the middle decades of this century, wilderness protection in the United States was the preoccupation of precocious pioneers, whose shouts of alarm sometimes led to changes in public policy. However, when environmentalism emerged as a popular movement in the 1960s and 1970s, it principally focused on two concerns: the threats to human health posed by pollution, and the threats to wild species and wild habitats posed by economic expansion. The latter concern became, in fact, the defining motif of the movement. The dominance of wilderness protection in American environmentalism has promoted an essentially negative agenda: the protection of parks and their animals by freeing them of human habitation and productive activities. As the historian Samuel Hays points out, "natural environments which formerly had been looked upon as 'useless,' waiting only to be developed, now came to be thought of as 'useful' for filling human wants and needs. They played no less a significant role in the advanced consumer society than did such material goods as hi-fi sets or indoor gardens."[1] While saving these islands of bio-diversity, American environmentalists have paid scant attention to what was happening outside them. This was especially apparent in their indifference to America's growing consumption of energy and materials.

The growing popular interest in the wild and the beautiful has thus not merely accepted the parameters of the affluent society but tends to see nature itself as merely one more good to be consumed. The uncertain commitment of most nature lovers to a more comprehensive environmental ideology is illustrated by the puzzle that they are willing to drive thousands of miles, using scarce oil and polluting the atmosphere, to visit national parks and sanctuaries—thus using anti-ecological means to marvel at the beauty of forests, swamps, or mountains protected as specimens of a "pristine" and "untouched" nature.

Consuming nature abroad

Crucially, the most gorgeous examples of pristine nature are located outside the United States (and outside Europe as well). The most charismatic mammals—the tiger and the elephant, the rhinoceros and the lion—are found in Asia and Africa; the most charismatic habitats, such as the rainforest, in Latin America. In the decades after World War II, and more so since the 1970s, the gaze of the North Atlantic wilderness lover has increasingly turned outward. What his or her homeland offered was not quite as exotic or attractive as what might be found overseas. And the appeal of foreign species was enhanced by new technologies, such as satellite television, which brought the beauties of the tiger or the rainforest into the living room. Meanwhile, air travel had become cheaper, more extensive, and more reliable; within days of reading about a tiger or watching it on your screen, you could be with it in its own wild habitat.

In response to a growing global market for nature tourism and driven also by strong domestic pressures, many nations in the developing South have undertaken ambitious programs to conserve and demarcate habitats and species for strict protection. For instance, when India became independent in 1947, it had less than a half-dozen wildlife reserves; it now has more than 400 parks and sanctuaries, covering 4.3 percent of the country (there are proposals to double this area). A similar expansion of territory under wilderness conservation can be observed in other Asian and African countries too. These parks are governed by two axioms: that wilderness has to be big, continuous wilderness and that all human intervention is bad for the retention of diversity. These axioms have led to the constitution of numerous very large sanctuaries, with a total ban on human ingress in their "core" areas. In the process, hundreds of thousands of Indian villagers have been uprooted from their homes, and millions more have had their access to fuel, fodder, and small timber restricted or cut off.

Five major groups fuel the movement for wildlife conservation in the South. The first are the city-dwellers and foreign tourists who merely season their lives, a week at a time, with the wild. Their motive is straightforward: pleasure and fun. The second group consists of ruling elites who view the protection of particular species (for example, the tiger in India) as central to the retention or enhancement of national prestige. The third group is composed of international conservation organizations such as the World Conservation Union (IUCN) and the World Wildlife Fund, whose missions are "educating" people and politicians about the virtues of biological conservation. A fourth group consists of functionaries of the state forest or wildlife service mandated by law to physically control the parks. While some officials are genuinely inspired by a love of nature, the majority—at least in Asia and Africa—are motivated merely by the power and spin-off benefits (overseas trips, for example) that come with the job. The final group are biologists, who believe in wilderness and species preservation for the sake of "science."

These five groups are united in their hostility to the farmers, herders, swidden cultivators, and hunters who have lived in the "wild" from well before it became a "park" or "sanctuary." They see these human communities as having a destructive effect on the environment, their forms of livelihood aiding the disappearance of species and contributing to soil erosion, habitat simplification, and worse. Often their feelings are expressed in strongly pejorative language. Touring Africa in 1957, one prominent member of the Sierra Club sharply attacked the Masai for grazing cattle in African sanctuaries. He held the

Masai to be illustrative of a larger trend, wherein "increasing population and increasing land use," rather than industrial exploitation, constituted the main threat to the world's wilderness areas. The Masai and "their herds of economically worthless cattle," he remarked, "have already overgrazed and laid waste to much of the 23,000 square miles of Tanganyika they control, and as they move into the Serengeti, they bring the desert with them, and the wilderness and wildlife must bow before their herds."[2]

Thirty years later, the World Wildlife Fund initiated a campaign to save the Madagascar rainforest, the home of the ring-tailed lemur, the Madagascar serpent eagle, and other endangered species. The group's fund-raising posters boasted spectacular sketches of the lemur and the eagle and of the half-ton elephant bird that once lived on the island but is now extinct. Man "is a relative newcomer to Madagascar," noted the accompanying text, "but even with the most basic of tools—axes and fire—he has brought devastation to the habitats and resources he depends on." The posters also had a picture of a muddy river with the caption: "Slash-and-burn agriculture has brought devastation to the forest, and in its wake, erosion of the topsoil."

Environmental imperialism

This poster succinctly summed up the conservationist position with regard to the tropical rainforest. This holds that the enemy of the environment is the hunter and farmer living in the forest, who is too short-sighted for his, and our, good. This belief (or prejudice) has informed the many projects, spread across the globe, to constitute nature parks by evicting the original human inhabitants of these areas, with scant regard for their past or future. All this is done in the name of the global heritage of biological diversity. Cynics might conclude, however, that tribal people in the Madagascar or Amazon forest are expected to move out only so that residents of London or New York can have the comfort of knowing that the lemur or toucan has been saved for posterity—evidence of which is then provided for them by way of the wildlife documentary they can watch on their television screens.

Raymond Bonner's remarkable 1993 book on African conservation, *At the Hand of Man: Peril and Hope for Africa's Wildlife*, laid bare the imperialism, unconscious and explicit, of Northern wilderness lovers and biologists working on that luckless continent. Bonner remarks that:

> Africans [have been] ignored, overwhelmed, manipulated and outmaneuvered—by a conservation crusade led, orchestrated, and dominated by white Westerners.... As many Africans see it, white people are making rules to protect animals that white people want to see in parks that white people visit. Why should Africans support these programs?... Africans do not use the parks and they do not receive any significant benefits from them. Yet they are paying the costs. There are indirect economic costs—government revenues that go to parks instead of schools. And there are direct

personal costs [that is, from the ban on hunting and fuel collecting, or through physical displacement].

A Zambian biologist, E. N. Chidumayo, echoes Bonner's argument: "The only thing that is African about most conventional conservation policies is that they are practiced on African land."[3]

Bonner's book focuses on the elephant, one of approximately six animals that have come to acquire "totemic" status among Western wilderness lovers. Animal totems existed in most premodern societies, but as the Norwegian scholar Arne Kalland points out, in the past the injunction not to kill the totemic species applied only to members of the group. Hindus do not ask others to worship the cow, but those who love and cherish the elephant, seal, whale, or tiger try to impose a worldwide prohibition on its killing. No one, they say, anywhere, anytime, shall be allowed to harm the animal they hold sacred even if (as with the elephant and several species of whale) scientific evidence has established that small-scale hunting will not endanger its viable populations and will, in fact, save human lives put at risk by the expansion, after total protection, of the *lebensraum* of the totemic animal. The new totemists also insist that their species is the only true inhabitant of the ocean or forest, and ask that human beings who have lived in the same terrain (and with these animals) for many generations be sent elsewhere.[4]

The Northern wilderness lover has largely been insensitive to the needs and aspirations of human communities that live in or around habitats they wish to "preserve for posterity."

Throughout Asia and Africa, the management of parks has sharply posited the interests of poor villagers who have traditionally lived in them against those of wilderness lovers and urban pleasure seekers who wish to keep parks "free of human interference"—free, that is, of humans other than themselves. This conflict has led to violent clashes between local people and government officials. At present, the majority of wildlife conservationists, domestic or foreign, seem to believe that species and habitat protection can succeed only through a punitive guns-and-guards approach. However, some Southern scientists have called for a more inclusively democratic approach to conservation, whereby tribal people and peasants can be involved in management and decision making and can be fairly compensated for the loss of their homes and livelihood.[5]

Environmaterialists?

The Northern wilderness lover has largely been insensitive to the needs and aspirations of human communities that live in or around habitats they wish to "preserve for posterity." At the same time, he or she has also been insensitive to the deep asymmetries in global consumption, to the fact that it is precisely the self-confessed environmentalist who practices a lifestyle that lays an unbearable burden on the finite natural resources of the earth.

The United States and the countries of Western Europe consume a share of the world's resources radically out of proportion to their percentage of the world's population. A recent study by the Wuppertal Institute for Climate, Environment, and Energy, in Wuppertal, Germany, notes that the North lays excessive claim to the South's "environmental space." The way the global economy is currently structured, it argues, "the North gains cheap access to cheap raw materials and hinders access to markets for processed products from those countries; it imposes a system [the World Trade Organization] that favors the strong; it makes use of large areas of land in the South, tolerating soil degradation, damage to regional ecosystems, and disruption of local self-reliance; it exports toxic waste; [and] it claims patent rights to utilization of biodiversity in tropical regions...."

Seen "against the backdrop of a divided world," says the report, "the excessive use of nature and its resources in the North is a principal block to greater justice in the world.... A retreat of the rich from overconsumption is thus a necessary first step towards allowing space for improvement of the lives of an increasing number of people."

The problem thus identified, the report itemizes, in meticulous detail, how Germany can take the lead in reorienting its economy and society toward a more sustainable path. It begins with an extended treatment of overconsumption, the excessive use of the global commons by the West over the past 200 years, and the terrestrial consequences of profligate lifestyles—namely soil erosion, forest depletion, biodiversity loss, and air and water pollution. It then outlines a long-range plan for reducing the "throughput" of nature in the economy and cutting down on emissions.[6]

Consider, conversely, the approach to global environmental problems advocated by a man regarded as the "dean" of tropical biology, the American scientist Daniel Janzen. In an editorial written for the October 1988 issue of the journal *Conservation Biology*, Janzen asked his fellow biologists—professors as well as graduate students—to devote 20 percent of their funds and time to tropical conservation. He calculated that the $500 million and the 20,000 man-years thus generated would be enough to "solve virtually all neotropical conservation problems." "What can academics and researcher committees do?" asks Janzen. He offers this answer: "Significant input can be anything from voluntary secretarial work for a fund-raising drive to a megalomaniacal effort to bootstrap an entire tropical country into a permanent conservation ecosystem." Janzen assumes that money plus biologists will suffice to solve "virtually all neotropical conservation problems," although some of us think that a more effective solution would be for biologists to throw themselves into a megalomaniacal effort to bootstrap but one temperate country—Janzen's own—into living off its own resources.

Wilderness lovers like to speak of the equal rights of all species to exist. This ethical cloaking cannot hide the truth that green missionaries are possibly more dangerous, and certainly more hypocritical, than their economic or religious counterparts. The globalizing advertiser and banker works for a world in which everyone, regardless of class or color, is in an economic sense an American or Japanese—driving a car, drinking a Pepsi, owning a refrigerator and a washing machine. The missionary, having discovered Christ or Allah, wants all pagans or kaffirs also to share in the discovery. The conservationist wants to "protect the tiger or whale for posterity," yet expects other people to make the sacrifice, expects indigenous tribal people or fisherfolk to vacate the forest or the ocean so that he may enjoy his own brief holiday in communion with nature. But few among these lovers of nature scrutinize their own lifestyle, their own heavy reliance on nonrenewable resources, and the ecological footprint their consumption patterns leave on the soil, forest, waters, and air of lands other than their own.

Notes

1. Samuel Hayes, "From Conservation to Environment: Environmental Politics in the United States since World War Two," *Environmental Review*, vol. 6, no. 1, 1982, p. 21f. See also Hayes's *Beauty, Health and Permanence: The American Environmental Movement, 1955–85* (Cambridge: Cambridge University Press, 1987).
2. Lee Merriam Talbot, "Wilderness Overseas," *Sierra Club Bulletin*, vol. 42, no. 6 (1957).
3. E. N. Chidumayo, "Realities for Aspiring Young African Conservationists," in Dale Lewis and Nick Carter, eds., *Voices from Africa: Local Perspectives on Conservation* (Washington, D.C.: World Wildlife Fund, 1993), p. 49.
4. Arne Kalland, "Seals, Whales and Elephants: Totem Animals and the Anti-Use Campaigns," in *Proceedings of the Conference on Responsible Wildlife Management* (Brussels: European Bureau for Conservation and Development, 1994). See also Kalland's "Management by Totemization: Whale Symbolism and the Anti-Whaling Campaign," *Arctic*, vol. 46, no. 2 (1993).
5. For thoughtful suggestions as to how the interests of wild species and those of poor humans might be made more compatible, see M. Gadgil and P. R. S. Rao, "A System of Positive Incentives to Conserve Biodiversity," *Economic and Political Weekly* (Mumbai, India), August 6, 1994; See also Ashish Kothari, Saloni Suri, and Neena Singh, "Conservation in India: A New Direction," *Economic and Political Weekly*, October 28, 1995.
6. Wolfgang Sachs, Reinhard Loske, Manfred Linz, et al., *Greening the North: A Post-Industrial Blueprint for Ecology and Equity* (London: Zed Books, 1998).

RAMACHANDRA GUHA is a historian and anthropologist living in Bangalore. He has taught at Yale University, the Indian Institute of Science, and the University of California at Berkeley. His books include a history of the Chipko movement, *The Unquiet Woods*, 2d ed. (Berkeley: University of California Press, 2000), and *Environmentalism: A Global History* (New York: Longman, 2000).

Reprinted from *Current History*, November 2000, pp. 367-370. © 2000 by Current History, Inc. Reprinted by permission.

HEALTH OF THE EARTH

Harnessing Corporate Power to Heal the Planet

Pioneering companies in sectors ranging from wire to plastic films and planned residential communities have already demonstrated that today's environmental challenges hold many profit-enhancing opportunities.

L. Hunter Lovins and Amory B. Lovins

The late twentieth century witnessed two great intellectual shifts: the fall of communism, with the apparent triumph of market economics; and the appearance, in a rapidly growing number of businesses, of the end of the war against the earth, and the emergence of a new form of economics we call natural capitalism.

This term implies that capitalism as practiced is an aberration, not because it is capitalist but because it is defying its own logic. It does not value, but rather liquidates, the most important form of capital: *natural* capital, in other words the natural resources and, more importantly, the ecosystem services upon which all life depends.

Deficient logic of this sort can't be corrected simply by placing a monetary value on natural capital. Many key ecosystem services have no known substitutes at any price. For example, the $200 million Biosphere II project, despite a great deal of impressive science, was unable to provide breathable air for eight people. Biosphere I, our planet, performs this task daily at no charge for six billion of us.

Ecosystem services give us tens of trillions of dollars' worth of benefits each year, or more than the global economy. But none of this is reflected on anyone's balance sheets.

The best technologies cannot substitute for water and nutrient cycling, atmospheric and ecological stability, pollination and biodiversity, topsoil and biological productivity, and the process of assimilating and detoxifying society's wastes. With the human race increasing by 8,700 people every hour, more people are chasing after fewer resources. The limits to economic growth are coming to be set by scarcities of natural capital.

Sometimes the value of ecosystem services becomes apparent only when they are lost. In China's Yangtze basin in 1998, for example, deforested watersheds fostered flooding that killed 3,700 people, dislocated 223 million, and inundated 60 million acres of cropland. That $30 billion disaster forced a logging moratorium and a $12 billion crash program of reforestation.

This is not to say we're running out of such commodities as copper and oil. Even with recent fluctuations, prices for almost all commodities are near record lows and will fall for some time, in part because of improvements in extraction technologies. But in many instances these technologies impose environmental costs that further degrade the ability of living systems to sustain a growing human population.

Before the Industrial Revolution, from which capitalism emerged, it was inconceivable that people could work more productively. Nonetheless, textile mills introduced in the late 1700s enabled one Lancashire spinner to do the work previously done by 200 weavers, and the mills were only one of many technologies that increased the productivity of workers.

Profit-maximizing capitalists economized on their scarcest production factor: skilled people. They substituted seemingly abundant resources and the ability of the planet to absorb their pollution to enable people to do more work.

COURTESY OF SOUTHWIRE CORPORATION

Southwire Corporation, which nearly went bankrupt due to severe competition in the wire industry, survived by cutting its energy use per a pound of wire in half as it also enhanced the efficiency of all aspects of operations. Today the company is thriving.

Given today's patterns of scarcity and abundance, that same business logic dictates using more people and more brains to wring 4, 10, or even 100 times as much benefit from each unit of energy, water, materials, or anything else borrowed from the planet. Success at this will be the basis of competitiveness in the decades to come and will be the hallmark of the next industrial revolution.

The first of four principles of natural capitalism is to increase resource efficiency radically. This not only increases profits, but also solves most of the environmental dilemmas facing the world today. It greatly slows resource depletion at one end of the economic process and discharge of pollution (resources out of place) at the other end. It creates profits by reducing the costs of both resources and pollution. And it also buys time, forestalling the threatened collapse of natural systems.

That time should then be used to implement the other three principles of natural capitalism. These are: (2) eliminate the concept of waste by redesigning the economy based on models that close the loops of mate-

rials flows; (3) shift the focus of the economy from processing materials and making things to creating service and flow; and (4) reverse the destruction of the planet now under way by instituting programs of restoration that invest in natural capital.

By applying the four principles of natural capitalism, businesses can behave as if ecosystem services were properly valued and begin to reverse the loss of such services even as they increase profits.

Just the energy thrown away by U.S. power stations as waste heat equals the total energy used by Japan.

---

1. Increase resource productivity

It is relatively easy to profit by using resources more efficiently because they are used incredibly wastefully now. The stuff that drives the industrial metabolism of the United States

currently amounts to, for each American, more than 20 times your body weight every day, or more than a million pounds per year.

Globally, this is a flow of half a trillion tons per year. But only about 1 percent of all the materials mobilized in the economy is ever embodied in a product that endures six months after sale. Cutting such waste represents a vast business opportunity.

Nowhere are the opportunities for savings easier to see than in energy. The United States has already cut its annual energy bills by $150 billion relative to what they would have been if savings had not been implemented since the first oil shock in 1973. However, we still waste $300 billion worth of energy each year in an economy whose total energy bill is more than $500 billion ($516 billion in 1995). Just the energy thrown away by U.S. power stations as waste heat, for example, equals the total energy used by Japan.

Fortunately, we already have ample examples of companies that have shown how to improve energy efficiency (i.e., reduce waste) and increase profits.

Southwire Corporation—an energy-intensive maker of cable, rod, and wire—halved its energy per pound of product in six years. The savings roughly equaled the company's profits during that period, and company officials estimated that the energy-efficiency effort probably secured 4,000 jobs at 10 plants in six states that were jeopardized by competitive market forces. The company then went on to save even more energy, achieving two-year paybacks despite all the earlier energy-efficiency improvements.

Dow's Louisiana division (now Dow Louisiana Operations) implemented over 900 worker-suggested energy-saving projects during the period 1981 to 1993, with average annual returns on investment of over 200 percent. Both returns and savings *rose* in later years, even after the accumulated annual savings from the projects had passed $100 million, because the engineers were learning faster than they were using up the cheapest opportunities.

State-of-the-shelf technologies can make old buildings three- to fourfold more energy-efficient and new buildings nearer 10-fold—and cheaper to build. Examples include large and small buildings in climates ranging from well below freezing to sweltering, both types of which can be kept comfortable with no heating or cooling systems. Industries can achieve profitable savings in motor systems, process designs, and materials productivity. Rocky Mountain Institute's Hypercar design synthesis for automobiles and other road vehicles will produce huge energy and materials savings—and spell the end of the car, oil, steel, aluminum, coal, and electricity industries as we know them [see "The Car of the Future?" THE WORLD & I, August 1996, p. 148].

How can such savings be captured? An international company recently redesigned a standard industrial pumping loop slated for installation in its new Shanghai fac-

tory. The original, supposedly optimized, design needed 95 horsepower for pumping. Dutch engineer Jan Schilham made two embarrassingly simple design changes that cut that 95 hp to only 7 hp—a 92 percent reduction. The redesigned system cost less to build and worked better in all respects.

First, Schilham chose big pipes and small pumps rather than small pipes and big pumps. The friction in a pipe falls inversely as nearly the fifth power of its diameter. In considering how big to make the pipes, normal engineering practice balances the capital cost of the pipe against the ongoing energy costs of pumping fluid through the pipe.

But this textbook optimization ignores the capital cost of the pumping equipment—the pump, motor, variable-speed electronic control, and electrical supply—that must be big enough to fight the pipe friction. Ignoring the potential equipment saving, and optimizing one component (the pipe) in isolation, "pessimizes" the system. Optimizing the *whole system* instead, and counting savings in total capital cost as well as in energy cost, makes it clear that, within a critical range, as pipe size increases the capital cost falls more rapidly for equipment than it rises for the much fatter pipe. The whole system therefore costs less but works better.

Schilham's second innovation was to lay out the pipes first, then the equipment. The normal sequence is the opposite: install the equipment in traditional positions (far apart, at the wrong height, facing the wrong way, with other stuff in between), then tell the pipe fitter to hook it all up. The resulting long, crooked pipes have about three to six times as much friction as short, straight pipes. Using short, straight pipes to minimize friction cuts both capital and operating costs. In this case, it also saved 70 kilowatts of heat loss, because straight pipes are easier to insulate.

This matters because pumping is a major user of electricity worldwide. Optimizing a whole pumping

system, at the level of a whole building or a whole factory, can typically yield energy savings of 3- to 10-fold and cost less to operate. But more importantly, the thought process of whole systems thinking applies to almost every technical system that uses resources.

Consider real estate development. Typical tract home developments drain storm water in expensive underground sewers. Village Homes, an early solar housing development in Davis, California, instead installed natural drainage swales. This saved $800 per house and provided more green space. The company then used the saved money to pay for edible landscaping that provided shade, nutrition, beauty, community focus, and crop revenues that paid the homeowners' assessments and paid for a community center. The people-centered site planning (narrower, tree-lined streets, with the housing fronting on the greenways) saved more land and more money. It also cooled off the microclimate, yielding better comfort at lower cost, and it created safe and child-friendly neighborhoods that cut crime 90 percent compared with neighboring subdivisions. Real estate brokers once described the project as weird. It is now the most desirable real estate in town, with market values $11 per square foot over average.

The same approach can diminish the risk of climate change. DuPont is proposing to reduce its greenhouse gas emissions 65 percent from 1990 levels by 2010. In addition, by 2010 DuPont aims to derive a tenth of its energy and a quarter of its raw materials from renewables. It is making these changes in the name of increasing shareholder value. In a similar vein, ST Microelectronics, a manufacturer of microchips, has set a goal of zero new emissions while it implements a 20-fold production increase.

Many executives are realizing that protecting the climate is not costly but profitable, because saving fuel costs less than buying fuel. Using energy in a way that saves money is therefore an important way to strengthen the bottom line and the whole economy, while also resolving the climate problem.

This is why the European Union has already adopted at least a four-fold ("Factor Four") gain in resource productivity as the new basis for sustainable development policy and practice. Some countries, like the Netherlands and Austria, have declared this a national goal. Environment ministers from the OECD (Organization for Economic Cooperation and Development), the government of Sweden, and distinguished industrial and academic leaders in Europe, Japan, and elsewhere have gone even further, adopting "Factor Ten" improvements as their goal. The World Business Council for Sustainable Development and the UN Environment Programme have called for "Factor Twenty." There is growing evidence that such ambitious goals are feasible and achievable in the marketplace. They may, in fact, offer even greater profits.

2. Eliminate the concept of waste

Resource efficiency is natural capitalism's cornerstone, but only its beginning.

Natural capitalism would eliminate the entire concept of waste by adopting biological patterns, processes, and often materials. This implies eliminating any industrial output that represents a disposal cost rather than a salable product.

Architect Bill McDonough tells the story of being asked by the Steelcase subsidiary DesignTex to design a "green" textile for upholstering office chairs. The fabric it was to replace used such toxic chemicals to treat and dye the cloth that the Swiss government had declared its edge trimmings a hazardous waste. McDonough's team eventually found a chemical firm that would let them explore its textile chemistry in detail.

They screened more than 8,000 chemicals, rejecting any that were toxic, built up in food chains, or caused cancer, mutations, birth defects, or endocrine disruption. The 38 that passed could make all colors. The cloth would look better, feel better in the hand, and last longer, because the natural fibers wouldn't be damaged by harsh chemicals. Fewer and cheaper feedstocks, as well as no health and safety concerns, meant that production cost less. The new fabric was beautiful and won design awards.

The Swiss environmental inspectors who tested the new plant thought their equipment was malfunctioning when the effluent water proved cleaner than the Swiss drinking water input: the cloth itself was acting as a filter. More important, the redesign of the process "took the filters out of the pipes and put them where they belong, in the designers' heads."

Professor Hanns Fischer noticed that the University of Zurich's basic chemistry lab course was turning pure, simple reagents into mainly hazardous wastes, incurring costs at both ends. The students were also learning once-through, linear thinking. So in some of the lessons, the students turned the toxic wastes back into pure, simple reagents. Students volunteered vacation time to repurify the wastes, because it was so much fun. Demand for wastes soon outstripped supply. Waste production declined 99 percent, costs fell by about $20,000 a year, and the students learned the closed-loop thinking that must ultimately save the chemical industry.

This is an archetype for the emerging world where environmental regulation will be an anachronism. In that biological world, the design lessons of nature will improve business—as well as health, housing, mobility, community, and national security. Such a world emerges from the cybernetics of not inflicting on others any emission to which you wouldn't expose yourself: How clean a car would you buy if its exhaust pipe, instead of being aimed at pedestrians, fed directly into the passenger compartment?

How clean would a city or factory make the water it discharges if its intake pipes were downstream of its outlets? We all live downwind, downstream.

3. Create service and flow

A further key element of natural capitalism is to shift the structure of the economy from focusing on the production and sale of things to focusing on providing the customer a continuous flow of service and value.

This change in the business model provides incentives for a continuous improvement in the elimination of waste, because it structures the relationships so that *the provider and customer both make money by finding more efficient solutions that benefit both.* That contrasts sharply with the conventional sale or leasing of physical goods in which the vendor wants to provide more things more often—increasing waste—and at a higher price, while the customer has the opposite interests.

For example, Schindler leases vertical transportation services instead of selling elevators, Electrolux/Sweden leases the performance of professional floor-cleaning and commercial food-service equipment rather than the equipment itself, and Dow and Safety-Kleen lease dissolving services rather than selling solvents. Both customer and provider profit from minimizing the flow of energy and materials.

Carrier, the world's largest manufacturer of air conditioners, is experimenting with leases of comfort instead of sales of air conditioners. Making the equipment more efficient or more durable will give Carrier greater profits and its customers better comfort at lower cost. So too, however, will making the building itself more efficient, so that less cooling yields the same comfort. Carrier is therefore starting to team up with other firms that can improve lighting, glazings, and other building systems. Providing a more systemic solution, creating a relationship that continually aligns interests, is obviously better for customers, shareholders and the earth than selling air conditioners.

A striking example is emerging at Atlanta carpet maker Interface. Most broadloom carpet is replaced every decade because it develops worn spots. An office is shut down, furniture removed, and carpet torn up and sent to landfill. (The millions of tons deposited each year will last up to 20,000 years.) New carpet is laid down, the office restored, operations resumed, and workers get sick from the carpet-glue fumes.

Instead, Interface prefers to lease floor-covering services. People want to walk on and look at carpet, not own it. They can obtain these services at much lower cost if Interface owns the carpet and remains responsible for keeping it clean and fresh. For a monthly fee, Interface will visit regularly and replace the 10–20 percent of the carpet tiles that show 80–90 percent of the wear. This reduces the mass flow of carpet to landfill by about 80 percent and provides better service at lower cost. It also increases net employment, eliminates the disruption (worn tiles are seldom under furniture), and turns a capital expenditure into an operating lease.

Interface's latest technical innovation goes further. Other manufacturers say that they recycle carpet. Actually they downcycle it—reusing it in lower-grade products. In contrast, Interface's new Solenium product provides floor covering that is almost completely remanufacturable into identical carpet. This will cut their net flow of materials and energy it takes to make them by 97 percent.

It will also provide better service, because the new floor covering, which may be leased or sold, is non-toxic, virtually stainproof, easy to clean with water, four times as durable, one-third less materials-intensive, renewably produced, and otherwise superior in every respect.

Interface's first four years on this systematic quest to turn avoided waste into profit returned doubled revenues, tripled operating profits, and nearly doubled employment. Its latest $250 million revenue came with no increase in energy or materials inputs, from mining internal waste.

Whole-system solutions create more life, more value, and ultimately more profits.

Or consider the Films division of DuPont. Once failing, it now leads its 59-firm market because it makes its films thinner, stronger, and better matched to customers' needs. This enables it to produce higher-value products using fewer materials. It also recycles used film, closing the materials loops, getting it back from customers with a process now coming to be known as "reverse logistics," a new topic of study in business schools. Jack Krol, past chairman of DuPont, has remarked that he sees no end to DuPont's ability to profit in this way.

4. Invest in natural capital

The fourth principle of natural capitalism is to invest to reverse the worldwide destruction of the ecosystem.

If natural capital is the most important, valuable, and indispensable form of capital, a true capitalist will restore it where degraded and sustain it where healthy—the better to create wealth and sustain life. Once toxicity and waste are designed out of industry, then forestry, farming, and fishing must be redesigned to be restorative to natural ecosystems. This will be especially important as the primary inputs to industry come to be grown, not mined, and living nanotechnologies replace vast industries.

This will place a premium on understanding biological models and on using nature as model and mentor rather than as a nuisance to be evaded. The incentive will derive not just from the goodwill of corporations, but from the scent of real profits and the promise of long-term corporate survival. No doubt some managers (the commercial about the company that is not interested in e-business comes to mind) will lack the willingness to tackle natural capitalism, and their companies will likely fossilize. Meanwhile, more visionary, adventurous managers will lead the wave of companies that embrace the new competitive grounds set by natural capitalism.

Catching up with centuries of deferred but unbooked planetary maintenance might sound expensive. But whole-system solutions create more life, more value, and ultimately more profits. Production is automatically carried out; people need only create hospitable conditions and do no harm. In this exciting sphere of innovation lie such opportunities as these:

•Dr. Allan Savory, cofounder of the Albuquerque-based Center for Holistic Management, has redesigned ranching to mimic the migration of large herds of native grazers that coevolved with grasslands. This can greatly improve the carrying capacity of even degraded rangelands, which turn out to have been not overgrazed but undergrazed, out of ignorance of how brittle ecosystems evolved.

•The California Rice Industry Association partnered with environmental groups to switch from burning rice straw to flooding the rice fields after harvest. They now flood 30 percent of California's rice acreage, from which they can harvest a more profitable mix of wildfowl, high-silica straw, groundwater recharge, and other benefits, with rice as a by-product.

•Dr. John Todd of Ocean Arks International and Living Technologies, based in Burlington, Vermont, builds biological "Living Machines" that turn sewage into clean water—plus valuable flowers, a tourist venue, and other by-products—with no toxicity, no odor, and reduced capital costs. Such "bioneers" are using living organisms to "bioremediate" toxic pollutants into forms that are harmless or salable or both.

These practices adopt the design experience of nearly four billion years of evolutionary testing in which products that failed were recalled by the Manufacturer. Though many details of such nature-mimicking practices are still evolving, the broad contours of the lessons they teach are already clear [see "The Living Building," THE WORLD & I, October 1999, p. 160].

Some of the most exciting developments are modeled on nature's low-temperature, low-pressure assembly techniques, whose products rival anything man-made. Janine Benyus' book *Biomimicry* points out that spiders make silk as strong as Kevlar—but much tougher—from digested crickets and flies, without needing boiling sulfuric acid and high-pressure extruders.

The abalone makes an inner shell twice as tough as ceramics, and diatoms make seawater into glass; neither need furnaces. Trees turn air, sunlight, and soil into cellulose, a sugar stiffer and stronger than nylon. We may never be as skillful as spiders, abalone, diatoms, or trees, but such benign natural chemistry may be a better model than industrialism's primitive approach of "heat, beat, and treat."

Beyond profits: What's in it for us?

Natural capitalism implemented in a company creates an extraordinary outpouring of energy, initiative, and enthusiasm at all levels. It removes the contradiction between what people do at work and what they want for their families when they go home. This makes natural-capitalist firms some of the most exciting places in the world to work.

Civilization in the twenty-first century is imperiled by the dissolution of civil societies into lawlessness and despair; weakened life-support systems; and the dwindling public purse needed to address these problems and reduce human suffering. These three threats share a common cause—waste.

Natural capitalism implemented in a company creates an extraordinary outpouring of energy, initiative, and enthusiasm at all levels.

The leaders in waste reduction will be in the corporate sector. But there remains a vital role for governments and for civil society. It is important to remember markets' purposes and limitations. Markets make a splendid servant but a bad master and a worse religion. Markets produce value, but only communities and families produce values. A society that substitutes markets for politics, ethics, or faith is dangerously adrift. Commerce can create a durable system of production and consumption by properly applying sound market principles. Yet not all value is monetized; not every priceless thing is priced. Nor is accumulating money the same thing as creating wealth or improving people. Many of the best things in life are not the business of business. And as the Russians are finding under "gangster capitalism," unless democratic institutions establish and maintain a level playing field, only the most ruthless can conduct business.

One of government's most powerful tools is tax policy. Such taxes as FICA and other penalties on employment that grew out of the first industrial revolution encourage companies to use more resources and fewer people. Gradual and fair tax shifting and desubsidization can provide more of what we want—jobs and income—and less of what we don't want: environmental and social damage.

But government's power is limited. Today over half the world's 100 largest economic entities are not countries but companies. Corporations may be the only institution with the resources, agility, organization, and motivation to tackle the toughest problems.

Firms that pursue the four principles of natural capitalism—profiting from advanced resource productivity, closing materials loops and eliminating waste, providing their customers with efficient solutions, and reinvesting in natural capital—will gain a commanding competitive advantage. They'll be behaving as if natural and human capital were properly valued. And as Ed Woollard, former chairman of DuPont, once remarked, companies that don't take these principles seriously won't be a problem, because they won't be around.

Perhaps the only problem with capitalism—a system of wealth creation built on the productive flow and expansion of all forms of capital—is that it is only now beginning to be tried.

L. Hunter Lovins and Amory B. Lovins are co-CEOs of the Rocky Mountain Institute (RMI), a nonprofit resource policy center they cofounded in 1982 in Snowmass, Colorado. The Lovinses are co-authors along with Paul Hawken of the book Natural Capitalism: Creating the Next Industrial Revolution *(Little, Brown, 1999), available from RMI at orders@rmi.org.*

UNIT 2
The World's Population: People and Hunger

Unit Selections

5. **The Population Surprise**, Max Singer
6. **Population Control Today—and Tomorrow?** Jacqueline R. Kasun
7. **Population and Consumption: What We Know, What We Need to Know**, Robert W. Kates
8. **Feeding the World in the New Millennium: Issues for the New U.S. Administration**, Per Pinstrup-Andersen
9. **Escaping Hunger, Escaping Excess**, Gary Gardner and Brian Halweil
10. **Growing More Food With Less Water**, Sandra Postel

Key Points to Consider

- What is meant by the term "demographic transition" and how does it relate to the long-term potential for population change in the world's industrialized and developing nations?

- Assess the criticisms that some detractors have leveled against Planned Parenthood and its founder. Are these criticisms valid and is the message of Planned Parenthood therefore discredited?

- Describe the IPAT formula for calculating environmental impact and tell why it may no longer be appropriate to use when attempting to describe the role of population growth, affluence, and technology in environmental deterioration.

- Why should policy makers in the more developed countries of the world become more aware of the true dimensions of the world's food problem? How can increased awareness of food scarcity and misallocation lead to solutions for both food production and environmental protection?

- Explain the differences between "malnutrition" in a country like the United States and a nation in sub-Saharan Africa. To what extent is the world's food problem more one of distribution and availability of food than of total supply?

 **Links: www.dushkin.com/online/**
These sites are annotated in the World Wide Web pages.

The Hunger Project
http://www.thp.org

Penn Library Resources
http://www.library.upenn.edu/resources/websitest.html

World Health Organization
http://www.who.int

WWW Virtual Library: Demography & Population Studies
http://demography.anu.edu.au/VirtualLibrary/

One of the greatest setbacks on the road to the development of more stable and sensible population policies came about as a result of inaccurate population growth projections made in the late 1960s and early 1970s. The world was in for a population explosion, the experts told us back then. But shortly after the publication of the heralded works *The Population Bomb* (Paul Ehrlich, 1975) and *Limits to Growth* (D. H. Meadows et al., 1974), the growth rate of the world's population began to decline slightly. There was no cause and effect relationship at work here. The decline in growth was simply demographic transition at work, a process in which declining population growth tends to accompany increasing levels of economic development. Since the alarming predictions did not come to pass, the world began to relax a little. However, two facts still remain—population growth in biological systems must be limited by available resources, and the availability of Earth's resources is finite.

Consider the following: In developing countries, high and growing rural population densities have forced the use of increasingly marginal farmland that was once considered to be too steep, too dry, too wet, too sterile, or too far from market for efficient agricultural use. Farming this land damages soil and watershed systems, creates deforestation problems, and adds relatively little to total food production. In the more developed world, farmers also have been driven—usually by market forces—to farm more marginal lands and to rely more on environmentally harmful farming methods that utilize high levels of agricultural chemicals (such as pesticides and artificial fertilizers). These chemicals create hazards for all life and rob the soil of its natural ability to renew itself. The increased demand for food production has also created an increase in the use of precious groundwater reserves for irrigation purposes, depleting those reserves beyond their natural capacity to recharge and creating the potential for once-fertile farmland and grazing land to be transformed into desert. The continued demand for higher production levels also contributes to a soil erosion problem that has reached alarming proportions in all agricultural areas of the world, whether hihh or low on the scale of economic development. The need to increase the food supply and its consequent effects on the agricultural environment are not the only results of continued population growth. For industrialists, the larger market creates an almost irresistible temptation to accelerate production, requiring the use of more marginal resources and resulting in the destruction of more fragile ecological systems, particularly in the tropics. For consumers, the increased demand for products means increased competition for scarce resources, driving up the cost of those resources until only the wealthiest can afford what our grandfathers would have viewed as an adequate standard of living.

The articles selected for this second unit all relate, in one way or another, to the theory and reality of population growth (and its relationship to food supply). In the first selection, "The Population Surprise," Max Singer notes that the pattern of population dynamics in the more developed nations of the world, what we call "the demographic transition," indicates that when population growth begins to slow down—as it has worldwide over the last few decades—the decline in growth does not stop at the replacement rate but often continues beneath that point: in other words, populations begin to decline. In "Population Control Today—and Tomorrow?" economist Jacqueline R. Kasun takes a sharply critical look at past population control efforts, with a particularly pointed criticism of Planned Parenthood and its founder, Margaret Sanger. Kasun claims that, while population control has achieved success in reducing growth rates significantly, the funding of programs for further population control are still increasing. More important, she notes, money spent on population control in developing nations is often money that is not spent on basic medicines. The unit's third article moves from a discussion of population dynamics to one of how to analyze the relationship between population growth and environmental degradation. In "Population and Consumption: What We Know, What We Need to Know," the author notes that growth in population, affluence, and technology are all contributors to deteriorating, environments.

The remaining articles in the unit address the issue of food consumption and supply—the other component of any discussion of population. In "Feeding the World in the New Millennium," Per Pinstrup-Andersen, director-general of the International Food Policy Research Institute, lays out the world food problem in terms of what needs to be done by the administration in Washington, D.C. One out of every five persons in the developing world (Middle and South America, Africa, Asia) is hungry, notes the author, and that is something that policy makers in the developed world (North America and Europe) should be concerned about. Hunger produces illness, violence, and political instability, and it may even be a factor in the recent emergence of terrorism directed toward the United States. Food shortages also do not contribute to the levels of economic development that producers in Europe, Japan, Canada, and the United States would like to see in order to enhance their markets. Hunger also produces environmental degradation. Doing something about it makes good economic and political sense. It is also the morally right thing to do. In the section's next piece, Gary Gardner and Brian Halweil of the Worldwatch Institute discuss the issue of inequitable food distribution and consumption in more concrete terms. In "Escaping Hunger, Escaping Excess," Gardner and Halweil note that it is not just the scarcity of food in poor countries that should be viewed with concern but the conspicuous overconsumption in the richer countries. Malnourishment can just as easily be brought on by too much consumption as by too little, and both forms of malnutrition are as costly for national economic activities as they are for personal health. Finally, in the section's concluding article, Sandra Postel of the Global Water Policy Project discusses one of the most critical problems facing food producers in developing and developed countries alike: the stresses placed on the world's freshwater supply by the demands of irrigation agriculture, a necessary component of increasing global food production. Postel describes the inefficiencies of existing irrigation systems and calls for the development of a sustainable base of irrigation agriculture if both food and environmental problems are to be solved.

All the authors of the selections in this unit make it clear that the global environment is being stressed by population growth and that more people means more pressure and more poverty. But while it should be evident that we can no longer afford to permit the unplanned and unchecked growth of the planet's dominant species, it should also be apparent that doomsday predictions of population and food imbalances are less commonplace now than they were just a few years ago.

The Population Surprise

*The old assumptions about world population trends need to be rethought.
One thing is clear: in the next century the world is in for some rapid downsizing*

by Max Singer

FIFTY years from now the world's population will be declining, with no end in sight. Unless people's values change greatly, several centuries from now there could be fewer people living in the entire world than live in the United States today. The big surprise of the past twenty years is that in not one country did fertility stop falling when it reached the replacement rate—2.1 children per woman. In Italy, for example, the rate has fallen to 1.2. In Western Europe as a whole and in Japan it is down to 1.5. The evidence now indicates that within fifty years or so world population will peak at about eight billion before starting a fairly rapid decline.

Because in the past two centuries world population has increased from one billion to nearly six billion, many people still fear that it will keep "exploding" until there are too many people for the earth to support. But that is like fearing that your baby will grow to 1,000 pounds because its weight doubles three times in its first seven years. World population was growing by two percent a year in the 1960s; the rate is now down to one percent a year, and if the patterns of the past century don't change radically, it will head into negative numbers. This view is coming to be widely accepted among population experts, even as the public continues to focus on the threat of uncontrolled population growth.

As long ago as September of 1974 *Scientific American* published a special issue on population that described what demographers had begun calling the "demographic transition" from traditional high rates of birth and death to the low ones of modern society. The experts believed that

birth and death rates would be more or less equal in the future, as they had been in the past, keeping total population stable after a level of 10–12 billion people was reached during the transition.

Developments over the past twenty years show that the experts were right in thinking that population won't keep going up forever. They were wrong in thinking that after it stops going up, it will stay level. The experts' assumption that population would stabilize because birth rates would stop falling once they matched the new low death rates has not been borne out by experience. Evidence from more than fifty countries demonstrates what should be unsurprising: in a modern society the death rate doesn't determine the birth rate. If in the long run birth rates worldwide do not conveniently match death rates, then population must either rise or fall, depending on whether birth or death rates are higher. Which can we expect?

The rapid increase in population during the past two centuries has been the result of lower death rates, which have produced an increase in worldwide life expectancy from about thirty to about sixty-two. (Since the maximum—if we do not change fundamental human physiology—is about eighty-five, the world has already gone three fifths as far as it can in increasing life expectancy.) For a while the result was a young population with more mothers in each generation, and fewer deaths than births. But even during this population explosion the average number of children born to each woman—the fertility rate—has been falling in modernizing societies. The prediction that world population will

soon begin to decline is based on almost universal human behavior. In the United States fertility has been falling for 200 years (except for the blip of the Baby Boom), but partly because of immigration it has stayed only slightly below replacement level for twenty-five years.

Obviously, if for many generations the birth rate averages fewer than 2.1 children per woman, population must eventually stop growing. Recently the United Nations Population Division estimated that 44 percent of the world's people live in countries where the fertility rate has already fallen below the replacement rate, and fertility is falling fast almost everywhere else. In Sweden and Italy fertility has been below replacement level for so long that the population has become old enough to have more deaths than births. Declines in fertility will eventually increase the average age in the world, and will cause a decline in world population forty to fifty years from now.

Because in a modern society the death rate and the fertility rate are largely independent of each other, world population need not be stable. World population can be stable only if fertility rates around the world average out to 2.1 children per woman. But why should they average 2.1, rather than 2.4, or 1.8, or some other number? If there is nothing to keep each country exactly at 2.1, then there is nothing to ensure that the overall average will be exactly 2.1.

The point is that the number of children born depends on families' choices about how many children they want to raise. And when a family is deciding whether to have another child, it is usually thinking about things other than the national or the world population. Who would know or care if world population were to drop from, say, 5.85 billion to 5.81 billion? Population change is too slow and remote for people to feel in their lives—even if the total population were to double or halve in only a century (as a mere 0.7 percent increase or decrease each year would do). Whether world population is increasing or decreasing doesn't necessarily affect the decisions that determine whether it will increase or decrease in the future. As the systems people would say, there is no feedback loop.

WHAT does affect fertility is modernity. In almost every country where people have moved from traditional ways of life to modern ones, they are choosing to have too few children to replace themselves. This is true in Western and in Eastern countries, in Catholic and in secular societies. And it is true in the richest parts of the richest countries. The only exceptions seem to be some small religious communities. We can't be sure what will happen in Muslim countries, because few of them have become modern yet, but so far it looks as if their fertility rates will respond to modernity as others' have.

Nobody can say whether world population will ever dwindle to very low numbers; that depends on what values people hold in the future. After the approaching peak, as long as people continue to prefer saving effort and money by having fewer children, population will continue to decline. (This does not imply that the decision to have fewer children is selfish; it may, for example, be motivated by a desire to do more for each child.)

Some people may have values significantly different from those of the rest of the world, and therefore different fertility rates. If such people live in a particular country or population group, their values can produce marked changes in the size of that country or group, even as world population changes only slowly. For example, the U.S. population, because of immigration and a fertility rate that is only slightly below replacement level, is likely to grow from 4.5 percent of the world today to 10 percent of a smaller world over the next

two or three centuries. Much bigger changes in share are possible for smaller groups if they can maintain their difference from the average for a long period of time. (To illustrate: Korea's population could grow from one percent of the world to 10 percent in a single lifetime if it were to increase by two percent a year while the rest of the world population declined by one percent a year.)

World population won't stop declining until human values change. But human values may well change—values, not biological imperatives, are the unfathomable variable in population predictions. It is quite possible that in a century or two or three, when just about the whole world is at least as modern as Western Europe is today, people will start to value children more highly than they do now in modern societies. If they do, and fertility rates start to climb, fertility is no more likely to stop climbing at an average rate of 2.1 children per woman than it was to stop falling at 2.1 on the way down.

In only the past twenty years or so world fertility has dropped by 1.5 births per woman. Such a degree of change, were it to occur again, would be enough to turn a long-term increase in world population of one percent a year into a long-term decrease of one percent a year. Presumably fertility could someday increase just as quickly as it has declined in recent decades, although such a rapid change will be less likely once the world has completed the transition to modernity. If fertility rises only to 2.8, just 33 percent over the replacement rate, world population will eventually grow by one percent a year again—doubling in seventy years and multiplying by twenty in only three centuries.

The decline in fertility that began in some countries, including the United States, in the past century is taking a long time to reduce world population because when it started, fertility was very much higher than replacement level. In addition, because a preference for fewer children is associated with modern societies, in which high living standards make time valuable and children financially unproductive and expensive to care for and educate, the trend toward lower fertility couldn't spread throughout the world until economic development had spread. But once the whole

world has become modern, with fertility everywhere in the neighborhood of replacement level, new social values might spread worldwide in a few decades. Fashions in families might keep changing, so that world fertility bounced above and below replacement rate. If each bounce took only a few decades or generations, world population would stay within a reasonably narrow range—although probably with a long-term trend in one direction or the other.

The values that influence decisions about having children seem, however, to change slowly and to be very widespread. If the average fertility rate were to take a long time to move from well below to well above replacement rate and back again, trends in world population could go a long way before they reversed themselves. The result would be big swings in world population—perhaps down to one or two billion and then up to 20 or 40 billion.

Whether population swings are short and narrow or long and wide, the average level of world population after several cycles will probably have either an upward or a downward trend overall. Just as averaging across the globe need not result in exactly 2.1 children per woman, averaging across the centuries need not result in zero growth rather than a slowly increasing or slowly decreasing world population. But the long-term trend is less important than the effects of the peaks and troughs. The troughs could be so low that human beings become scarcer than they were in ancient times. The peaks might cause harm from some kinds of shortages.

One implication is that not even very large losses from disease or war can affect the world population in the long run nearly as much as changes in human values do. What we have learned from the dramatic changes of the past few centuries is that regardless of the size of the world population at any time, people's personal decisions about how many children they want can make the world population go anywhere— to zero or to 100 billion or more.

Max Singer was a founder of the Hudson Institute. He is a co-author, with Aaron Wildavsky, of *The Real World Order* (1996).

Population Control Today—and Tomorrow?

by Jacqueline R. Kasun

The success of the population control movement over the past four decades has been nothing less than astonishing. Places like Bangladesh and Kenya are awash in condoms (even though basic medicines are scarce), and population is actually falling in some countries and heading in that direction in many others.

Yet the movement is astonishing in another way, too: Despite its success, it is expanding at a breakneck pace in terms of both funding and programs.

One of the least-remarked events of the year 2000 was the announcement by the UN Population Division that "in the next 50 years, the populations of most developed countries are projected to become smaller and older as a result of low fertility and increased longevity.... Population decline is inevitable in the absence of replacement migration."

The division reported that 44 percent of the world's population lives in countries where birthrates are too low to prevent population decline. If present trends continue, there will be 100 million fewer people in Europe and 21 million fewer in Japan 50 years from now. The birthrate in the United States has fallen from 24.3 per thousand population in 1950–55 to 14.6 in 1998, a trend that is likely to continue for some time because the female population of reproductive age will decline by several million during the next decade (unless offset by immigration). Also, the U.S. death rate has been rising slightly but perceptibly, because the population is aging. (The death rate for a fixed group of people is still 100 percent sooner or later, despite rising life expectancy, and older populations have higher death rates, other factors being equal.)

The UN Population Division predicts that world fertility will continue to decline from its present average of less than three children per woman (the one-child family is now typical in Europe and Japan) while the death rate rises. Thus, the proportion of people over 60 will rise to exceed the proportion of people under 15, for the first time in history.

Nevertheless, groups supporting population control continue to press for more funding for their programs both at home and abroad. Population Action International, for example, reported in June 2000 that "the Clinton administration intends to fight for additional funds" and that "Hollywood celebrities mingled with top policymakers and international family planning advocates on... World Health Day... to show the... administration's support for population assistance."

ROOTS OF THE MOVEMENT

The quest of those in power to control population is at least as old as the Exodus story of Pharaoh killing Hebrew baby boys. In our time, the movement has received stimuli from both eugenics and environmental worries.

Eugenics was a rather popular cause in the first half of the twentieth century in the United States and England. "More children from the fit, less from the unfit," Margaret Sanger, the founder of Planned Parenthood, wrote in her popular magazine, *Birth Control Review*, in 1919. Thirty-one states passed compulsory sterilization laws in the first half of the century.

Early this year, the Virginia State legislature expressed its "regrets" to Raymond Ludlow for forcibly sterilizing him at the age of 16 in 1941 for repeatedly running away from home. Ludlow, one of thousands sterilized by force across the country, subsequently served as a radioman in the Army, earning a Bronze Star, a Purple Heart, and a Prisoner of War Medal.

> **To reduce the U.S. birthrate, Planned Parenthood proposed ideas like putting "fertility control agents" in the water supply and encouraging homosexuality.**

At the close of World War II, Guy Irving Burch, the founder of the Population Reference Bureau, submitted his plan to solve all world problems through compulsory sterilization of "all persons who are inadequate, either biologically or socially," as he wrote in *Population*

Roads to Peace or War. Although Congress did not endorse Burch's plan, his bureau subsequently received millions of dollars in government grants and contracts for "population education" and other activities.

THE 'EXPLOSION'

New concerns emerged in the postwar years. A sudden spurt in population growth occurred in the 1960s as antibiotics and improvements in sanitation sharply reduced death rates. (Birthrates had been declining throughout the century as women joined the workforce and curtailed childbearing.) As death rates plunged below the falling birthrates, world population grew at an unprecedented pace. The response was intense.

Population Negation

- Birthrates have been declining precipitously around the world.

- Forty-four percent of the world's people live in nations whose population has shrunk or at least stalled in its growth.

- If present trends continue, Europe's population will fall by 100 million and Japan's by 21 million in the next 50 years.

- Yet funding and programs for population control are increasing.

- The population-control market is saturated, with a surfeit of contraceptives in many developing countries that otherwise lack basic medicines.

- Many Third World countries are suffering from the cultural seeds planted by the family planning movement, especially promiscuity, which spreads sexually transmitted diseases.

Congress held hearings. President Johnson recommended legislation, which Congress passed in 1965 and '67,

providing for the world's largest program of publicly financed birth control, targeted both at home and abroad.

In 1970, President Nixon appointed the Commission on Population and the American Future, under the chairmanship of John D. Rockefeller III, founder of the Population Council. That same year, Planned Parenthood published a list of "proposed measures to reduce U.S. fertility," among them putting "fertility control agents" in the water supply, encouraging homosexuality, imposing a "substantial" marriage tax, discouraging home ownership, requiring permits for couples to have children, making abortion compulsory, and mandating sterilization of all women who had borne two children. The United Nations proclaimed a World Population Year in 1974.

In a document that remained classified from 1974 to 1980, the U.S. State Department warned that "mandatory population control measures" might be necessary to bring about a "two-child family on the average" throughout the world by the year 2000. By 1975, the U.S. Agency for International Development (AID) was the world's chief player in world population control, spending more money on it than did all other countries combined.

In 1978, AID officials initiated and Congress enacted Section 104(d) of the new foreign aid legislation, which stipulated that "all... activities proposed for financing... shall be designed to build motivation for smaller families." The World Bank also began to impose population control requirements on its lending, as did other international institutions and countries. Henceforth, developing countries seeking international aid would be required to give evidence of their "commitment" to the "control of population growth."

PROMISED CALAMITIES

The justifications were a long, varied list of calamities that would ensue in the absence of swift, stern action. Starvation was looming, according to experts such as biologist Paul Ehrlich of Stanford University. The Sierra Club published his book *The Population Bomb,* which

became required reading in many high schools and colleges.

The House Select Committee on Population announced in 1978 that the "major biological systems that humanity depends upon... are being strained by rapid population growth... [and] in some cases, they are... losing productive capacity." Created by the Smithsonian Institution at about the same time, a traveling exhibit for schoolchildren called "Population: The Problem Is Us" featured a picture of a dead rat on a dinner plate as an example of "future food sources."

The Carter administration's Council on Environmental Quality and State Department together warned that "the staggering growth of human population... [was creating]... possibilities of... permanent damage to the planet's resource base." Robert McNamara, then director of the World Bank, warned in 1977 that continued population growth would cause "poverty, hunger, stress, crowding, and frustration" that would threaten social, economic, and military stability.

Sen. (later Vice President) Al Gore warned in his 1992 book *Earth in the Balance* of the approach of an "environmental holocaust without precedent," like a "black hole" caused by "expansion beyond the environment's carrying capacity." To stave off this catastrophe, he wrote, "the first strategic goal should be the stabilizing of world population."

Herman Daly, a World Bank economist, proposed in his 1990 book *For the Common Good* that, as a step toward the "sustainable society," births be limited by a government-operated licensing system. The UN Population Fund (UNFPA, not to be confused with the UN Population Division, a statistical agency) gave millions of dollars a year starting in 1979 to China's population control program, which featured forced abortion.

The funding increased along with the pressure. By 1994, federal and state governments were spending more than $2 billion a year directly on domestic and foreign population control. (Probably much more was being spent due to population-control requirements attached to

Population Control Pillar

Margaret Sanger, who founded the Planned parenthood Federation of America in 1942, is often viewed as the patron saint of the modern population control movement.

Her critics, citing numerous references in her writings, denounce her as a white supremacist, Nazi sympathizer, and an advocate of free sex.

Her supporters dismiss the criticism, saying that the references are confined to a small number of sources and are often taken out of context. Esther Katz, editor and director of the Margaret Sanger Papers Project, has said, "As a historian, I take issue with [such] gross misuse of historical sources to support those views."

One thing that critics often say about Sanger is that she viewed blacks as inferior and wanted to use birth control and abortion to reduce their numbers. They cite Sanger's quotation: "We don't want word to get out that we want to exterminate the Negro population."

Supporters say the full context of the quote proves Sanger did not want to eliminate blacks. In a letter to philanthropist Clarence Gamble in 1939 about her "Negro Project," she said: "The minister's work is also important and also he should be trained, perhaps by the [Birth Control] Federation [of America] as to our ideals and the goal that we hope to reach. We do not want word to go out that we want to exterminate the Negro population, and the minister is the man who can straighten out that idea if it ever occurs to any of their more rebellious members."

Alexander Sanger, president of Planned Parenthood of New York City and Sanger's grandson, says Sanger was committed to helping all women "regardless of race or nationality." He highlights her slogan "Let every child be a wanted child."

But Sanger's extensive written comments over several decades continue to make life difficult for population control advocates who would otherwise like to unreservedly embrace her. For example:

- In her 1922 *Pivot of Civilization*, she clearly called for the sterilization of "genetically inferior races," the elimination of "human weeds" and the cessation of charity.

- In the same book, she advocated the segration of "morons, misfits, and maladjusted."

- The *Birth Control Review*, founded by Sanger in 1917, sounded eugenics themes for decades and categorized blacks, southern Europeans, and other immigrants as mentally inferior, calling them a nuisance and a menace to society.

—**The Editor**

many other programs in the $12–15 billion a year U.S. foreign aid budget.)

Eventually, unmistakable signs of population-control market saturation became evident around the world. In 1994, at the International Conference on Population and Development in Cairo, Margaret Ogola, a Kenyan pediatrician, reported that clinics in her country had an abundance of every kind of contraceptive but lacked the "simplest medicines" to treat common childhood diseases. Similar reports came from other places. In Bangladesh, newspapers reported that unwanted birth control pills were piling up in warehouses.

In addition, many Third World countries have been put off by the cultural appurtenances of the family-planning movement. A sticking point in Cairo, for example, was the insistence by the United Nations that countries provide "sexual health care" for adolescents without their parents' supervision or knowledge, as is done in the United States. Gadul Haqq Ali Gadul Haqq, the grand imam sheikh of al-Azar University, one of many critics of this policy, said, "Islam can by no means agree to give young generations full freedom to do what they like." Other countries also objected, but International Planned Parenthood and other agencies funded by the United States have continued to promote sexual freedom.

CONDOM FAILURE

Stephen Karanja, a Kenyan gynecologist, visited the United States in 2000 to report on what he said were the devastating effects of U.S. population programs in his country. Under the pretext of preventing AIDS, he said, foreign-paid family-planning workers promote promiscuity by indiscriminately distributing condoms and are taking over the healthcare system to perform sterilizations.

"Over and over, we have seen it in Africa—condoms do not stop HIV/AIDS," he said. "In the last two years in Kenya, more than 100 million condoms have been used [while] the number of HIV/AIDS people doubled. Stopping HIV/AIDS is a behavior thing. It is a thing to do with not having sexual activity outside of marriage. We do not need the African family to be attacked."

As to the reputed economic benefits of lowering fertility, several countries said in Cairo that they had reduced or eliminated population growth without improving their economies. But those nations with free economies—even those as densely populated as South Korea—reported not only sturdy growth in wealth but no problems of overpopulation.

Shortly before the Cairo conference, economists at the IMF listed the causes of Africa's severe economic problems. They blamed excessive government spending, high taxes on farmers, inflation, restrictions on trade (a Zambian representative in Cairo said that "trade barriers by developed countries cost developing countries 10 times as much in lost trade as they receive in development assistance"), too much government ownership, overregulation of private economic activity, and government creation of "powerful vested interests." There was no mention of overpopulation—not surprisingly, since Africa has fewer than one-fourth as many people per square mile as prosperous Europe.

Nevertheless, the Sierra Club announced in Cairo its support for increased "international population assistance" and a "sustainable population level within the carrying capacity" of the United States—with its "local activists" being the ones determining that "carrying capacity."

MORE MONEY, MORE COMMITMENT

After the conference, the Clinton administration and the population-control network redoubled their efforts. By 1998, world flow of international aid for population programs amounted to $2.06 billion, with another $9 billion in local funds being reported by the targeted countries themselves.

In 1998, two of the largest U.S. recipients of federal family-planning funds, Planned Parenthood Federation of America and its affiliate, the Alan Guttmacher Institute, received $122 million from the federal government and additional amounts from the states. Medicaid alone spent $449 million for family-planning services in 1998, up over $100 million from 1994.

"More children from the fit, less from the unfit," wrote Margaret Sanger, the founder of Planned Parenthood.

AID, which cites stabilizing population growth as one of its foreign policy goals, asked for "total funding of $542 million from all grant-funded accounts" for population control and $569 million for "protecting human health" in its 2001 budget request. The agency also asked for $254 million for work against AIDS.

In the meantime, failing economies continue to fail, and, as the high-level negotiations regarding "sustainable development" and "reproductive health" proceed, the evidence of the programs' innate tendencies toward coercion mounts. Paid by the head for recruiting women and men for sterilization and other birth control procedures, local family-planning workers press forward to meet their targets in Asia, Africa, and Latin America.

Yet birthrates continue to fall throughout the world as more women work outside their homes. World food availability rose to unprecedented levels, according to the UN Food and Agriculture Organization (FAO). World forest acreage remained at the same levels as in the 1950s, according to FAO data. Some 19,000 scientists have signed a petition stating that there is "no convincing scientific evidence" that human release of greenhouse gases will cause "disruption of Earth's climate" (www.sitewave.net/pproject/s33p427.htm).

What the future will bring is anyone's guess. Perhaps the new Bush administration will exercise its conservative sinews and stanch the flow of federal funds to population control groups. Perhaps a cultural backlash in the Third World will gain strength and slow the population-control juggernaut to a crawl. But it may very well be that current programs, propelled by political inertia and sluiced by already open funding spigots, will continue and even grow.

Jacqueline R. Kasun is an economist and the author of **The War Against Population: The Economics and Ideology of World Population Control** *(Ignatius, 1999).*

Population and Consumption

What We Know, What We Need to Know

by Robert W. Kates

Thirty years ago, as Earth Day dawned, three wise men recognized three proximate causes of environmental degradation yet spent half a decade or more arguing their relative importance. In this classic environmentalist feud between Barry Commoner on one side and Paul Ehrlich and John Holdren on the other, all three recognized that growth in population, affluence, and technology were jointly responsible for environmental problems, but they strongly differed about their relative importance. Commoner asserted that technology and the economic system that produced it were primarily responsible.[1] Ehrlich and Holdren asserted the importance of all three drivers: population, affluence, and technology. But given Ehrlich's writings on population,[2] the differences were often, albeit incorrectly, described as an argument over whether population or technology was responsible for the environmental crisis.

Now, 30 years later, a general consensus among scientists posits that growth in population, affluence, and technology are jointly responsible for environmental problems. This has become enshrined in a useful, albeit overly simplified, identity known as IPAT, first published by Ehrlich and Holdren in *Environment* in 1972[3] in response to the more limited version by Commoner that had appeared earlier in *Environment* and in his famous book *The Closing Circle*.[4] In this identity, various forms of environmental or resource impacts (I) equals population (P) times affluence (A) (usually income per capita) times the impacts per unit of income as determined by technology (T) and the institutions that use it. Academic debate has now shifted from the greater or lesser importance of each of these driving forces of environmental degradation or resource depletion to debate about their interaction and the ultimate forces that drive them.

However, in the wider global realm, the debate about who or what is responsible for environmental degradation lives on. Today, many Earth Days later, international debates over such major concerns as biodiversity, climate change, or sustainable development address the population and the affluence terms of Holdrens' and Ehrlich's identity, specifically focusing on the character of consumption that affluence permits. The concern with technology is more complicated because it is now widely recognized that while technology can be a problem, it can be a

solution as well. The development and use of more environmentally benign and friendly technologies in industrialized countries have slowed the growth of many of the most pernicious forms of pollution that originally drew Commoner's attention and still dominate Earth Day concerns.

A recent report from the National Research Council captures one view of the current public debate, and it begins as follows:

> *For over two decades, the same frustrating exchange has been repeated countless times in international policy circles. A government official or scientist from a wealthy country would make the following argument: The world is threatened with environmental disaster because of the depletion of natural resources (or climate change or the loss of biodiversity), and it cannot continue for long to support its rapidly growing population. To preserve the environment for future generations, we need to move quickly to control global population growth, and we must concentrate the effort on the world's poorer countries, where the vast majority of population growth is occurring.*

Government officials and scientists from low-income countries would typically respond:

> *If the world is facing environmental disaster, it is not the fault of the poor, who use few resources. The fault must lie with the world's wealthy countries, where people consume the great bulk of the world's natural resources and energy and cause the great bulk of its environmental degradation. We need to curtail overconsumption in the rich countries which use far more than their fair share, both to preserve the environment and to allow the poorest people on earth to achieve an acceptable standard of living.*[5]

It would be helpful, as in all such classic disputes, to begin by laying out what is known about the relative responsibilities of both population and consumption for the environmental crisis, and what might need to be known to address them. However, there is a profound asymmetry that must fuel the frustration of the developing countries' politicians and scientists: namely, how much people know about population and how little they know about consumption. Thus, this article begins by examining these differences in knowledge and action and concludes with the alternative actions needed to go from more to enough in both population and consumption.[6]

Population

What population is and how it grows is well understood even if all the forces driving it are not. Population begins with people and their key events of birth, death, and location. At the margins, there is some debate over when life begins and ends or whether residence is temporary or permanent, but little debate in between. Thus, change in the world's population or any place is the simple arithmetic of adding births, subtracting deaths, adding immigrants, and subtracting outmigrants. While whole subfields of demography are devoted to the arcane details of these additions and subtractions, the error in estimates of population for almost all places is probably within 20 percent and for countries with modern statistical services, under 3 percent—better estimates than for any other living things and for most other environmental concerns.

Current world population is more than six billion people, growing at a rate of 1.3 percent per year. The peak annual growth rate in all history—about 2.1 percent—occurred in the early 1960s, and the peak population increase of around 87 million per year occurred in the late 1980s. About 80 percent or 4.8 billion people live in the less developed areas of the world, with 1.2 billion living in industrialized countries. Population is now projected by the United Nations (UN) to be 8.9 billion in 2050, according to its medium fertility assumption, the one usually considered most likely, or as high as 10.6 billion or as low as 7.3 billion.[7]

A general description of how birth rates and death rates are changing over time is a process called the demographic transition.[8] It was first studied in the context of Europe, where in the space of two centuries, societies went from a condition of high births and high deaths to the current situation of low births and low deaths. In such a transition, deaths decline more rapidly than births, and in that gap, population grows rapidly but eventually stabilizes as the birth decline matches or even exceeds the death decline. Although the general description of the transition is widely accepted, much is debated about its cause and details.

The world is now in the midst of a global transition that, unlike the European transition, is much more rapid. Both births and deaths have dropped faster than experts expected and history foreshadowed. It took 100 years for deaths to drop in Europe compared to the drop in 30 years in the developing world. Three is the current global average births per woman of reproductive age. This number is more than halfway between the average of five children born to each woman at the post World War II peak of population growth and the average of 2.1 births required to achieve eventual zero population growth.[9] The death transition is more advanced, with life expectancy currently at 64 years. This represents three-quarters of the transition between a life expectancy of 40 years to one of 75 years. The current rates of decline in births outpace the estimates of the demographers, the UN having reduced its latest medium ex-

pectation of global population in 2050 to 8.9 billion, a reduction of almost 10 percent from its projection in 1994.

Demographers debate the causes of this rapid birth decline. But even with such differences, it is possible to break down the projected growth of the next century and to identify policies that would reduce projected populations even further. John Bongaarts of the Population Council has decomposed the projected developing country growth into three parts and, with his colleague Judith Bruce, has envisioned policies that would encourage further and more rapid decline.[10] The first part is unwanted fertility, making available the methods and materials for contraception to the 120 million married women (and the many more unmarried women) in developing countries who in survey research say they either want fewer children or want to space them better. A basic strategy for doing so links voluntary family planning with other reproductive and child health services.

Yet in many parts of the world, the desired number of children is too high for a stabilized population. Bongaarts would reduce this desire for large families by changing the costs and benefits of childrearing so that more parents would recognize the value of smaller families while simultaneously increasing their investment in children. A basic strategy for doing so accelerates three trends that have been shown to lead to lower desired family size: the survival of children, their education, and improvement in the economic, social, and legal status for girls and women.

However, even if fertility could immediately be brought down to the replacement level of two surviving children per woman, population growth would continue for many years in most developing countries because so many more young people of reproductive age exist. So Bongaarts would slow this momentum of population growth by increasing the age of childbearing, primarily by improving secondary education opportunity for girls and by addressing such neglected issues as adolescent sexuality and reproductive behavior.

How much further could population be reduced? Bongaarts provides the outer limits. The population of the developing world (using older projections) was expected to reach 10.2 billion by 2100. In theory, Bongaarts found that meeting the unmet need for contraception could reduce this total by about 2 billion. Bringing down desired family size to replacement fertility would reduce the population a billion more, with the remaining growth—from 4.5 billion today to 7.3 billion in 2100—due to population momentum. In practice, however, a recent U.S. National Academy of Sciences report concluded that a 10 percent reduction is both realistic and attainable and could lead to a lessening in projected population numbers by 2050 of upwards of a billion fewer people.[11]

Consumption

In contrast to population, where people and their births and deaths are relatively well-defined biological events, there is no consensus as to what consumption includes. Paul Stern of the National Research Council has described the different ways physics, economics, ecology, and sociology view consumption.[12] For physicists, matter and energy cannot be consumed, so consumption is conceived as transformations of matter and energy with increased entropy. For economists, consumption is spending on consumer goods and services and thus distinguished from their production and distribution. For ecologists, consumption is obtaining energy and nutrients by eating something else, mostly green plants or other consumers of green plants. And for some sociologists, consumption is a status symbol—keeping up with the Joneses—when individuals and households use their incomes to increase their social status through certain kinds of purchases. These differences are summarized in the box below.

In 1977, the councils of the Royal Society of London and the U.S. National Academy of Sciences issued a joint statement on consumption, having previously done so on population. They chose a variant of the physicist's definition:

> *Consumption is the human transformation of materials and energy. Consumption is of concern to the extent that it makes the transformed materials or energy less available for future use, or negatively impacts biophysical systems in such a way as to threaten human health, welfare, or other things people value.*[13]

On the one hand, this society/academy view is more holistic and fundamental than the other definitions; on the other hand, it is more focused, turning attention to the environmentally damaging. This article uses it as a working definition with one modification, the addition of information to energy and matter, thus completing the triad of the biophysical and ecological basics that support life.

In contrast to population, only limited data and concepts on the transformation of energy, materials, and information exist.[14] There is relatively good global knowledge of energy transformations due in part to the common units of conversion between different technologies. Between 1950 and today, global energy production and use increased more than fourfold.[15] For material transformations, there are no aggregate data in common units on a global basis, only for some specific classes of materials including materials for energy production, construction, industrial minerals and metals, agricultural crops, and water.[16] Calculations of material use by volume, mass, or value lead to different trends.

Trend data for per capita use of physical structure materials (construction and industrial minerals, metals, and forestry products) in the United States are relatively complete. They show an inverted S shaped (logistic) growth pattern: modest doubling between 1900 and the depression of the 1930s (from two to four metric tons), followed by a steep quintupling with economic recovery until the early 1970s (from two to eleven tons), followed by a leveling off since then with fluctuations related to economic downturns (see Figure 1).[17] An aggregate analysis of all current material production and consumption in the United States averages more than 60 kilos per person per day (excluding water). Most of this material flow is split between energy and related products (38 percent) and minerals for construction (37 percent), with the remainder as industrial minerals (5 percent), metals (2 percent), products of fields (12 percent), and forest (5 percent).[18]

A massive effort is under way to catalog biological (genetic) information and to sequence the genomes of microbes, worms, plants, mice, and people. In contrast to the molecular detail, the number and diversity of organisms is unknown, but a conservative estimate places the number of species on the order of 10 million, of which only one-tenth have been described.[19] Although there is much interest and many anecdotes, neither concepts nor data are available on most cultural information. For example, the number of languages in the world continues to decline while the number of messages expands exponentially.

What Is Consumption?

Physicist: "What happens when you transform matter/energy"

Ecologist: "What big fish do to little fish"

Economist: "What consumers do with their money"

Sociologist: "What you do to keep up with the Joneses"

Trends and projections in agriculture, energy, and economy can serve as surrogates for more detailed data on energy and material transformation.[20] From 1950 to the early 1990s, world population more than doubled (2.2 times), food as measured by grain production almost tripled (2.7 times), energy more than quadrupled (4.4 times), and the economy quintupled (5.1 times). This 43-year record is similar to a current 55-year projection (1995–2050) that assumes the continuation of current trends or, as some would note, "business as usual." In this 55-year projection, growth in half again of population (1.6 times) finds almost a doubling of agriculture (1.8 times), more than twice as much energy used (2.4 times), and a quadrupling of the economy (4.3 times).[21]

Thus, both history and future scenarios predict growth rates of consumption well beyond population. An attractive similarity exists between a demographic transition that moves over time from high births and high deaths to low births and low deaths with an energy, materials, and information transition. In this transition, societies will use increasing amounts of energy and materials as consumption increases, but over time the energy and materials input per unit of consumption decrease and information substitutes for more material and energy inputs.

Some encouraging signs surface for such a transition in both energy and materials, and these have been variously labeled as decarbonization and dematerialization.[22] For more than a century, the amount of carbon per unit of energy produced has been decreasing. Over a shorter period, the amount of energy used to produce a unit of production has also steadily declined. There is also evidence for dematerialization, using fewer materials for a unit of production, but only for industrialized countries and for some specific materials. Overall, improvements in technology

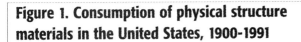

Figure 1. Consumption of physical structure materials in the United States, 1900-1991

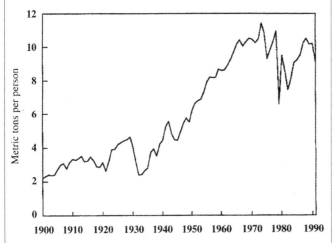

SOURCE: I. Wernick, "Consuming Materials: The American Way," *Technological Forecasting and Social Change*, 53 (1996): 114.

and substitution of information for energy and materials will continue to increase energy efficiency (including decarbonization) and dematerialization per unit of product or service. Thus, over time, less energy and materials will be needed to make specific things. At the same time, the demand for products and services continues to increase, and the overall consumption of energy and most materials more than offsets these efficiency and productivity gains.

What to Do about Consumption

While quantitative analysis of consumption is just beginning, three questions suggest a direction for reducing environmentally damaging and resource-depleting consumption. The first asks: *When is more too much for the life-support systems of the natural world and the social infrastructure of human society?* Not all the projected growth in consumption may be resource-depleting— "less available for future use"—or environmentally damaging in a way that "negatively impacts biophysical systems to threaten human health, welfare, or other things people value."[23] Yet almost any human-induced transformations turn out to be either or both resource-depleting or damaging to some valued environmental component. For example, a few years ago, a series of eight energy controversies in Maine were related to coal, nuclear, natural gas, hydroelectric, biomass, and wind generating sources, as well as to various energy policies. In all the controversies, competing sides, often more than two, emphasized environmental benefits to support their choice and attributed environmental damage to the other alternatives.

Despite this complexity, it is possible to rank energy sources by the varied and multiple risks they pose and, for those concerned, to choose which risks they wish to minimize and which they are more willing to accept. There is now almost 30 years of experience with the theory and methods of risk assessment and 10 years of experience with the identification and setting of en-

vironmental priorities. While there is still no readily accepted methodology for separating resource-depleting or environmentally damaging consumption from general consumption or for identifying harmful transformations from those that are benign, one can separate consumption into more or less damaging and depleting classes and *shift* consumption to the less harmful class. It is possible to *substitute* less damaging and depleting energy and materials for more damaging ones. There is growing experience with encouraging substitution and its difficulties: renewables for nonrenewables, toxics with fewer toxics, ozone-depleting chemicals for more benign substitutes, natural gas for coal, and so forth.

The second question, *Can we do more with less?*, addresses the supply side of consumption. Beyond substitution, shrinking the energy and material transformations required per unit of consumption is probably the most effective current means for reducing environmentally damaging consumption. In the 1997 book, *Stuff: The Secret Lives of Everyday Things*, John Ryan and Alan Durning of Northwest Environment Watch trace the complex origins, materials, production, and transport of such everyday things as coffee, newspapers, cars, and computers and highlight the complexity of reengineering such products and reorganizing their production and distribution.[24]

Yet there is growing experience with the three Rs of consumption shrinkage: reduce, recycle, reuse. These have now been strengthened by a growing science, technology, and practice of industrial ecology that seeks to learn from nature's ecology to reuse everything. These efforts will only increase the existing favorable trends in the efficiency of energy and material usage. Such a potential led the Intergovernmental Panel on Climate Change to conclude that it was possible, using current best practice technology, to reduce energy use by 30 percent in the short run and 50–60 percent in the long run.[25] Perhaps most important in the long run, but possibly least studied, is the potential for and value of substituting information for energy and materials. Energy and materials per unit of consumption are going down, in part because more and more consumption consists of information.

The third question addresses the demand side of consumption—*When is more enough?*[26] Is it possible to reduce consumption by more satisfaction with what people already have, by *satiation*, no more needing more because there is enough, and by *sublimation*, having more satisfaction with less to achieve some greater good? This is the least explored area of consumption and the most difficult. There are, of course, many signs of *satiation* for some goods. For example, people in the industrialized world no longer buy additional refrigerators (except in newly formed households) but only replace them. Moreover, the quality of refrigerators has so improved that a 20-year or more life span is commonplace. The financial pages include frequent stories of the plight of this industry or corporation whose markets are saturated and whose products no longer show the annual growth equated with profits and progress. Such enterprises are frequently viewed as failures of marketing or entrepreneurship rather than successes in meeting human needs sufficiently and efficiently. Is it possible to reverse such views, to create a standard of satiation, a satisfaction in a need well met?

Can people have more satisfaction with what they already have by using it more intensely and having the time to do so? Economist Juliet Schor tells of some overworked Americans who would willingly exchange time for money, time to spend with family and using what they already have, but who are constrained by an uncooperative employment structure.[27] Proposed U.S. legislation would permit the trading of overtime for such compensatory time off, a step in this direction. *Sublimation*, according to the dictionary, is the diversion of energy from an immediate goal to a higher social, moral, or aesthetic purpose. Can people be more satisfied with less satisfaction derived from the diversion of immediate consumption for the satisfaction of a smaller ecological footprint?[28] An emergent research field grapples with how to encourage consumer behavior that will lead to change in environmentally damaging consumption.[29]

A small but growing "simplicity" movement tries to fashion new images of "living the good life."[30] Such movements may never much reduce the burdens of consumption, but they facilitate by example and experiment other less-demanding alternatives. Peter Menzel's remarkable photo essay of the material goods of some 30 households from around the world is powerful testimony to the great variety and inequality of possessions amidst the existence of alternative life styles.[31] Can a standard of "more is enough" be linked to an ethic of "enough for all"? One of the great discoveries of childhood is that eating lunch does not feed the starving children of some far-off place. But increasingly, in sharing the global commons, people flirt with mechanisms that hint at such—a rationing system for the remaining chlorofluorocarbons, trading systems for reducing emissions, rewards for preserving species, or allowances for using available resources.

A recent compilation of essays, *Consuming Desires: Consumption, Culture, and the Pursuit of Happiness*,[32] explores many of these essential issues. These elegant essays by 14 well-known writers and academics ask the fundamental question of why more never seems to be enough and why satiation and sublimation are so difficult in a culture of consumption. Indeed, how is the culture of consumption different for mainstream America, women, inner-city children, South Asian immigrants, or newly industrializing countries?

Why We Know and Don't Know

In an imagined dialog between rich and poor countries, with each side listening carefully to the other, they might ask themselves just what they actually know about population and consumption. Struck with the asymmetry described above, they might then ask: "Why do we know so much more about population than consumption?"

The answer would be that population is simpler, easier to study, and a consensus exists about terms, trends, even policies. Consumption is harder, with no consensus as to what it is, and with few studies except in the fields of marketing and advertising. But the consensus that exists about population comes from substantial research and study, much of it funded by governments and groups in rich countries, whose asymmetric concern readily identifies the troubling fertility behavior of others and only reluctantly considers their own consumption behavior. So while consumption is harder, it is surely studied less (see Table 1).

The asymmetry of concern is not very flattering to people in developing countries. Anglo-Saxon tradition has a long history of dominant thought holding the poor responsible for their con-

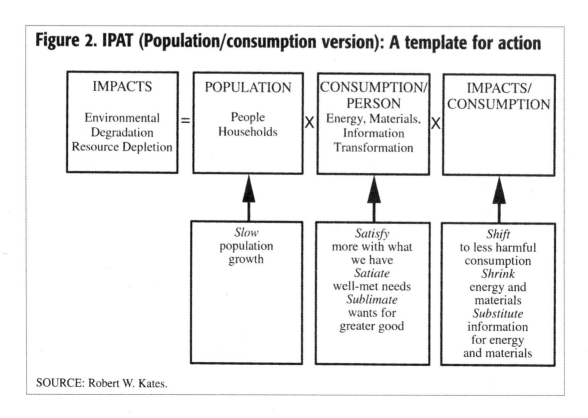

Figure 2. IPAT (Population/consumption version): A template for action

IMPACTS — Environmental Degradation Resource Depletion = POPULATION — People Households X CONSUMPTION/PERSON — Energy, Materials, Information Transformation X IMPACTS/CONSUMPTION

Slow population growth

Satisfy more with what we have *Satiate* well-met needs *Sublimate* wants for greater good

Shift to less harmful consumption *Shrink* energy and materials *Substitute* information for energy and materials

SOURCE: Robert W. Kates.

dition—they have too many children—and an even longer tradition of urban civilization feeling besieged by the barbarians at their gates. But whatever the origins of the asymmetry, its persistence does no one a service. Indeed, the stylized debate of population versus consumption reflects neither popular understanding nor scientific insight. Yet lurking somewhere beneath the surface concerns lies a deeper fear.

Table 1. A comparison of population and consumption	
Population	**Consumption**
Simpler, easier to study	More complex
Well-funded research	Unfunded, except marketing
Consensus terms, trends	Uncertain terms, trends
Consensus policies	Threatening policies

SOURCE: Robert W. Kates.

Consumption is more threatening, and despite the North–South rhetoric, it is threatening to all. In both rich and poor countries alike, making and selling things to each other, including unnecessary things, is the essence of the economic system. No longer challenged by socialism, global capitalism seems inherently based on growth—growth of both consumers and their consumption. To study consumption in this light is to risk concluding that a transition to sustainability might require profound changes in the making and selling of things and in the opportunities that this provides. To draw such conclusions, in the absence of convincing alternative visions, is fearful and to be avoided.

What We Need to Know and Do

In conclusion, returning to the 30-year-old IPAT identity—a variant of which might be called the Population/Consumption (PC) version—and restating that identity in terms of population and consumption, it would be: I = P*C/P*I/C, where I equals environmental degradation and/or resource depletion; P equals the number of people or households; and C equals the transformation of energy, materials, and information (see Figure 2).

With such an identity as a template, and with the goal of reducing environmentally degrading and resource-depleting influences, there are at least seven major directions for research and policy. To reduce the level of impacts per unit of consumption, it is necessary to separate out more damaging consumption and shift to less harmful forms, *shrink* the amounts of environmentally damaging energy and materials per unit of consumption, and *substitute* information for energy and materials. To reduce consumption per person or household, it is necessary to *satisfy* more with what is already had, *satiate* well-met consumption needs, and *sublimate* wants for a greater good. Finally, it is possible to *slow* population growth and then to *stabilize* population numbers as indicated above.

However, as with all versions of the IPAT identity, population and consumption in the PC version are only proximate driving forces, and the ultimate forces that drive consumption, the consuming desires, are poorly understood, as are many of the major interventions needed to reduce these proximate driving forces. People know most about slowing population growth, more about shrinking and substituting environmentally damaging consumption, much about shifting to less damaging consumption, and least about satisfaction, satiation, and sublimation. Thus the determinants of consumption and its alternative patterns have been identified as a key understudied topic for an emerging sustainability science by the recent U.S. National Academy of Science study.[33]

But people and society do not need to know more in order to act. They can readily begin to separate out the most serious problems of consumption, shrink its energy and material throughputs, substitute information for energy and materials, create a standard for satiation, sublimate the possession of things for that of the global commons, as well as slow and stabilize population. To go from more to enough is more than enough to do for 30 more Earth Days.

Robert W. Kates is an independent scholar in Trenton, Maine; a geographer; university professor emeritus at Brown University; and an executive editor of *Environment*. The research for "Population and Consumption: What We Know, What We Need to Know" was undertaken as a contribution to the recent National Academies/National Research Council report, *Our Common Journey: A Transition Toward Sustainability*. The author retains the copyright to this article. Kates can be reached at RR1, Box 169B, Trenton, ME 04605.

NOTES

1. B. Commoner, M. Corr, and P. Stamler, "The Causes of Pollution," *Environment*, April 1971, 2–19.

2. P. Ehrlich, *The Population Bomb* (New York: Ballantine, 1966).

3. P. Ehrlich and J. Holdren, "Review of The Closing Circle," *Environment*, April 1972, 24–39.

4. B. Commoner, *The Closing Circle* (New York: Knopf, 1971).

5. P. Stern, T. Dietz, V. Ruttan, R. H. Socolow, and J. L. Sweeney, eds., *Environmentally Significant Consumption: Research Direction* (Washington, D.C.: National Academy Press, 1997), 1.

6. This article draws in part upon a presentation for the 1997 De Lange-Woodlands Conference, an expanded version of which will appear as: R. W. Kates, "Population and Consumption: From More to Enough," in *In Sustainable Development: The Challenge of Transition*, J. Schmandt and C. H. Wards, eds. (Cambridge, U.K.: Cambridge University Press, forthcoming), 79–99.

7. United Nations, Population Division, *World Population Prospects: The 1998 Revision* (New York: United Nations, 1999).

8. K. Davis, "Population and Resources: Fact and Interpretation," K. Davis and M. S. Bernstam, eds., in *Resources, Environment and Population: Present Knowledge, Future Options*, supplement to *Population and Development Review*, 1990: 1–21.

9. Population Reference Bureau, *1997 World Population Data Sheet of the Population Reference Bureau* (Washington, D.C.: Population Reference Bureau, 1997).

10. J. Bongaarts, "Population Policy Options in the Developing World," *Science*, 263: (1994), 771–776; and J. Bongaarts and J. Bruce, "What Can Be Done to Address Population Growth?" (unpublished background paper for The Rockefeller Foundation, 1997).

11. National Research Council, Board on Sustainable Development, *Our Common Journey: A Transition Toward Sustainability* (Washington, D.C.: National Academy Press, 1999).

12. See Stern, et al., note 5 above.

13. Royal Society of London and the U.S. National Academy of Sciences, "Towards Sustainable Consumption," reprinted in *Population and Development Review*, 1977, 23 (3): 683–686.

14. For the available data and concepts, I have drawn heavily from J. H. Ausubel and H. D. Langford, eds., *Technological Trajectories and the Human Environment.* (Washington, D.C.: National Academy Press, 1997).

15. L. R. Brown, H. Kane, and D. M. Roodman, *Vital Signs 1994: The Trends That Are Shaping Our Future* (New York: W. W. Norton and Co., 1994).

16. World Resources Institute, United Nations Environment Programme, United Nations Development Programme, World Bank, *World Resources, 1996–97* (New York: Oxford University Press, 1996); and A. Gruebler, *Technology and Global Change* (Cambridge, Mass.: Cambridge University Press, 1998).

17. I. Wernick, "Consuming Materials: The American Way," *Technological Forecasting and Social Change*, 53 (1996): 111–122.

18. I. Wernick and J. H. Ausubel, "National Materials Flow and the Environment," *Annual Review of Energy and Environment*, 20 (1995): 463–492.

19. S. Pimm, G. Russell, J. Gittelman, and T. Brooks, "The Future of Biodiversity," *Science*, 269 (1995): 347–350.

20. Historic data from L. R. Brown, H. Kane, and D. M. Roodman, note 15 above.

21. One of several projections from P. Raskin, G. Gallopin, P. Gutman, A. Hammond, and R. Swart, *Bending the Curve: Toward Global Sustainability*, a report of the Global Scenario Group, Polestar Series, report no. 8 (Boston: Stockholm Environmental Institute, 1995).

22. N. Nakicénovíc, "Freeing Energy from Carbon," in *Technological Trajectories and the Human Environment*, eds., J. H. Ausubel and H. D. Langford. (Washington, D.C.: National Academy Press, 1997); I. Wernick, R. Herman, S. Govind, and J. H. Ausubel, "Materialization and Dematerialization: Measures and Trends," in J. H. Ausubel and H. D. Langford, eds., *Technological Trajectories and the Human Environment* (Washington, D.C.: National Academy Press, 1997), 135–156; and see A. Gruebler, note 16 above.

23. Royal Society of London and the U.S. National Academy of Science, note 13 above.

24. J. Ryan and A. Durning, *Stuff: The Secret Lives of Everyday Things* (Seattle, Wash.: Northwest Environment Watch, 1997).

25. R. T. Watson, M. C. Zinyowera, and R. H. Moss, eds., *Climate Change 1995: Impacts, Adaptations, and Mitigation of Climate Change—Scientific-Technical Analyses* (Cambridge, U.K.: Cambridge University Press, 1996).

26. A sampling of similar queries includes: A. Durning, *How Much Is Enough?* (New York: W. W. Norton and Co., 1992); Center for a New American Dream, *Enough!: A Quarterly Report on Consumption, Quality of Life and the Environment* (Burlington, Vt.: The Center for a New American Dream, 1997); and N. Myers, "Consumption in Relation to Population, Environment, and Development," *The Environmentalist*, 17 (1997): 33–44.

27. J. Schor, *The Overworked American* (New York: Basic Books, 1991).

28. A. Durning, *How Much Is Enough?: The Consumer Society and the Future of the Earth* (New York: W. W. Norton and Co., 1992); Center for a New American Dream, note 26 above; and M. Wackernagel and W. Ress, *Our Ecological Footprint: Reducing Human Impact on the Earth* (Philadelphia. Pa.: New Society Publishers, 1996).

29. W. Jager, M. van Asselt, J. Rotmans, C. Vlek, and P. Costerman Boodt, *Consumer Behavior: A Modeling Perspective in the Contest of Integrated Assessment of Global Change*, RIVM report no. 461502017 (Bilthoven, the Netherlands: National Institute for Public Health and the Environment, 1997); and P. Vellinga, S. de Bryn, R. Heintz, and P. Molder, eds., *Industrial Transformation: An Inventory of Research.* IHDP-IT no. 8 (Amsterdam, the Netherlands: Institute for Environmental Studies, 1997).

30. H. Nearing and S. Nearing. *The Good Life: Helen and Scott Nearing's Sixty Years of Self-Sufficient Living* (New York: Schocken, 1990); and D. Elgin, *Voluntary Simplicity: Toward a Way of Life That Is Outwardly Simple, Inwardly Rich* (New York: William Morrow, 1993).

31. P. Menzel, *Material World: A Global Family Portrait* (San Francisco: Sierra Club Books, 1994).

32. R. Rosenblatt, ed., *Consuming Desires: Consumption, Culture, and the Pursuit of Happiness* (Washington, D.C.: Island Press, 1999).

33. National Research Council, Board on Sustainable Development, *Our Common Journey: A Transition Toward Sustainability* (Washington, D.C.: National Academy Press, 1999).

From *Environment*, April 2000, pp. 10–19. Reprinted with permission of the Helen Dwight Reid Educational Foundation. Published by Heldref Publications, 1319 Eighteenth St., NW, Washington, DC 20036-1802. © 2000.

Feeding the World in the New Millennium

Issues for the New U.S. Administration

by Per Pinstrup-Andersen

One of every five people in the developing world is hungry. The new U.S. presidential administration must put solutions to this catastrophe high on its agenda—and encourage other nations to do likewise—for six reasons:

- Hunger spawns illness, instability, violent conflict, and refugees—problems that are seldom contained within national borders and often spill into the United States;

- poor and hungry people do not make good trading partners;

- developing countries offer the most promising future market for U.S. goods and services;

- hunger fuels environmental degradation as desperate people try to eke out a living on ever more marginal land and migrate to urban slums in search of livelihoods;

- environmental degradation in developing countries affects the United States; and

- it is the morally right thing to do.

Poverty and Hunger

About 1.2 billion people in developing countries—almost five times the U.S. population—live on $1 a day or less.[1] These people often cannot afford to buy all the food they need, although they may spend 50 to 70 percent of their incomes trying,[2] and many do not have access to land to produce food.

The number of hungry people has fallen since 1970, but 800 million people in developing countries (18 percent of the total) remain chronically undernourished.[3] Severe undernourishment is rare in industrialized countries, but many low-income people have difficulty meeting their food needs. The U.S. Department of Agriculture (USDA) considers 31 million people in the United States—about one in ten—"food insecure," i.e., unable to regularly afford an adequate diet.[4] The rising economic tide of the 1990s has left many boats stuck on the bottom.

The largest number of hungry people is in the Asia-Pacific region, especially South Asia (the Indian subcontinent). The other global hunger hot spot is sub-Saharan Africa, the only region in which the number of hungry people is expected to increase during the next 20 years. If the international community does not make significant policy changes, the developing world's hungry population will only fall to 650 million by 2015, with hunger even more concentrated in Africa and South Asia. This is far short of the 1996 World Food Summit goal, agreed to by the United States, to reduce hunger by half by 2015.

> Increasing demand for animal feedgrain means that by 2020 the demand for maize in developing countries will overtake the demand for wheat, as well as rice.

Malnutrition among preschool children is of particular concern. Each year, it contributes to 5 million deaths of children less than 5 years old in the developing world—this is ten times the number of people that die from cancer annually in the United States. Even when they reach their fifth birthday, malnourished children frequently suffer impaired physical and mental development. The silent scourge of malnutrition robs the human family of countless artists, scientists, community leaders, and productive workers. Currently, there are 150 million malnourished preschoolers in developing countries (27 percent of the total number of children less than five years old there). Malnourished mothers frequently have low-birth-weight babies who are vulnerable to malnutrition, in effect passing hunger across generations.[5]

The International Food Policy Research Institute (IFPRI) projects that by 2020, without any changes in national and international policies, the number of developing country malnourished preschoolers will still be 135 million (25 percent). The number will increase substantially in Africa, and 77 percent of all hungry preschoolers will live there and in South Asia.[6] Acute malnutrition is rare among children in the United States, but USDA estimates that 40 percent of people in the United States who are food-insecure—12 million—are children. The reduction in global child malnutrition between 1990 and 2020 will not reach 25 percent, even though the World Summit for Children in 1990 pledged to halve preschooler malnutrition by 2000.[7]

Hunger is not just a problem of consuming too little food. Diets may also lack vitamins and minerals. Roughly 75 percent of people in the developing world consume too little iron; one billion suffer from anemia as a result. Iron-deficiency anemia is also a problem in wealthier countries, including the United States. It can lead to mothers dying during childbirth, newborn deaths, poor health and development among surviving children, and limited learning and work capacity. In some developing countries, iron-deficiency anemia reduces national income by as much as 1 percent annually. Inadequate vitamin A intake among developing country children causes blindness, infections, and tens of thousands of deaths. Pregnant vitamin A–deficient women face increased risk of death in childbirth and mother-to-child HIV transmission.[8]

The Supply Side

For the past 25 years, total global food production has consistently been more than adequate to provide everyone with minimum calorie requirements if food were distributed according to needs. However, in recent years, food production has been stagnant or declining in countries the United Nations has designated as having a "low-income food deficit." Poor weather, economic difficulties, and violent conflict create short- and even long-term food shortages affecting millions of people.

IFPRI projects a gap between food production and demand in several parts of the world by 2020. Population growth, urbanization, changes in income levels, and associated changes in dietary preferences all affect demand. World population is projected to increase 25 percent to 7.5 billion in 2020. On average, 73 million people—equivalent to the population of the Philippines—will be added each year, virtually all in developing countries. Urban population in developing countries is expected to double. When people move to cities, they shift to foods that require less preparation time and to more meat, milk, fruit, and vegetables.

Average incomes in all developing regions are expected to increase through 2020, but income inequality is likely to persist within and between countries. Poverty is likely to remain entrenched in South Asia and Latin America and to increase considerably in Africa, where the average income per person will be less than $1 a day. Many millions of impoverished people will be unable to afford the food they need even if it is available in the marketplace.

A 58 percent increase in global meat demand is forecasted between 1995 and 2020, with almost all of it coming from the developing world. This livestock revolution is already under way and will double developing countries' feedgrain demand during the next two decades. Farmers will have to produce 40 percent more cereals in 2020, with 80 percent of additional output coming from yield increases rather than farmland expansion.

However, in both developed and developing countries, rates of increase in cereal yields are slowing as low cereal prices have reduced fertilizer use and levels of investment in agricultural research and technology are low. Poorly functioning markets and lack of infrastructure and credit also contribute. Africa is falling behind other regions.

IFPRI estimates that developing countries' cereal production will not keep pace with demand through 2020. Net cereal imports will increase by 80 percent between 1995 and 2020. With a 34 percent increase projected in net cereal exports, the United States will continue to capture a large market share, providing about 60 percent of the developing world's net cereal imports. However, competition will increase from Eastern Europe, the former Soviet Union, the European Union, and Australia.[9]

Constraints on Ending Hunger

Several factors could significantly influence the food outlook during the next few decades:

Trade Liberalization

Many developing countries have liberalized food and agricultural trade since the 1980s. Developed countries have not taken reciprocal measures, maintaining barriers to high value commodities from developing countries, such as beef, sugar, peanuts, dairy products, and processed goods.

Developing countries must be encouraged to participate effectively in upcoming global agricultural trade negotiations, pursuing better access to industrialized countries' markets. However, without appropriate domestic economic and agricultural policies, developing countries in general and poor people in particular will not fully capture potential benefits from trade liberalization. The distribution of benefits will be determined largely by distribution of productive assets, such as land, water, and credit. In addition, many developing countries lack the infrastructure and the administrative and technical capacity to comply with global trade rules. The African share of world agricultural trade continues to decline rapidly. The effect of current trade agreements is likely to be adverse for most African countries.

Low-income countries must try to strengthen their bargaining position and pursue changes in both domestic policies and international trade arrangements by:

- enacting domestic policy reforms to remove biases against small-scale farmers and poor people while facilitating access to benefits from more open trade;

- seeking elimination of industrial countries' export subsidies, taxes, and controls that exacerbate price fluctuations; and

- convincing donors to target adequate levels of food aid to poor groups in ways that do not displace domestic production.[10]

For their part, developed countries should provide technical assistance and financial support for poor countries' agriculture, as well as technical support to developing countries to create strong animal and plant health standards so that they can produce for developed country markets.

Decreasing Aid

Aid to agriculture and rural development shrunk by almost half between 1986 and 1997. The share of aid going to agriculture dropped from 25 to 14 percent. Aid to education has similarly declined, and overall development aid fell about 17 percent between 1992 and 1997.[11] Yet research has found that aid to developing country agriculture not only is effective in promoting sustainable development and poverty alleviation, but it also leads to increased export opportunities for developed countries, including increased agricultural exports as agricultural growth spurs more general economic growth and demand for food products.[12] In addition to reversing the aid decline, donors, including the United States, must rethink their 20-year emphasis on reducing government's economic role, which has contributed to developing countries' public disinvestment in agriculture.[13]

In the post–cold war era, the United States has fallen to last place among donors in terms of aid as a percentage of gross national product (less than 0.1 percent). In absolute terms, U.S. aid has consistently ranked second to Japan's.[14]

In the late 1990s, the United States expanded food aid after substantial reductions mid-decade.[15] Fluctuations stemmed from domestic market conditions rather than developing country needs, as the United States continues to tie food aid to U.S. farm products.

Conflict, Refugees, and Food Security

Since the end of the cold war, internal conflicts have proliferated in developing and transition countries, particularly in Africa. Fourteen million refugees have fled these struggles, which have displaced another 20 million to 30 million people within their own countries.[16] Uprooted people are vulnerable to malnutrition and disease and need humanitarian assistance to survive. Postconflict reconstruction takes years. Violent conflict not only causes hunger, but hunger often contributes to conflict, especially when resources are scarce and perceptions of economic injustice are widespread.[17]

Natural Resource Management

Soil fertility management. Policies and investments are needed to eradicate hunger and protect natural resources, thereby breaking the vicious cycle of poverty, low productivity, and environmental degradation.

Low soil fertility and lack of access to affordable fertilizers, along with past and current failures to replenish soil nutrients in many countries, must be rectified through efficient use of organic and inorganic fertilizers and improved soil management. Reduced chemical fertilizer use is warranted where heavy application is harming the environment. Nevertheless, it is critical to expand fertilizer use where soil fertility is low and a large share of the population is hungry, especially in Africa. This will help boost production and reduce the serious land degradation that affects 20 percent of African farmland.[18]

Pest management. Preharvest losses to pests (insects, animals, weeds, and plant diseases) reduce the potential value of farm output by 40 percent; postharvest losses cost another 10 percent. In developing countries, losses greatly exceed agricultural aid received.[19] Developing countries' share of the global pesticide market is expected to increase significantly during the early 21st century. Insecticides now used in developing countries are often older and acutely toxic and often banned in developed countries except for export.

Until recently, developing country governments and aid donors encouraged use of chemical pesticides. Now, consensus is emerging on the need for integrated pest management, emphasizing alternatives to synthetic chemicals and the use of chemicals only as a last resort. Alternatives include use of natural predators and biological pesticides as well as breeding pest-resistant crops.[20]

> The gap in average cereal yields is widening considerably within the developing world as sub-Saharan Africa lags further and further behind other developing regions.

Water. Globally, water supplies are sufficient to meet demand through 2020. But water is poorly distributed across countries, within countries, and between seasons. Competition is increasing among uses. Developing countries are projected to increase water withdrawals 43 percent between 1995 and 2020, doubling domestic and industrial uses at the expense of agriculture.

Policy reforms can save water, improve use efficiency, and boost crop output per unit of water while reducing the risk of armed conflict between countries sharing surface or ground water sources. These reforms should include establishing secure water rights, decentralizing and privatizing water management, and setting conservation incentives.[21]

Wild and marginal land. Poor people in developing countries tend to depend on annual crops (which generally degrade soils more than perennial crops) and on common property lands (which generally suffer greater degradation than privately managed land). They often cannot afford to invest in land improvements. Degradation and lack of access to high-quality land frequently push poor people to clear forests and pastures for cultivation, often at the expense of wildlife habitat, contributing to further degradation. Policies should raise the value of forests and pastures, offer incentives for sound management, and help create nonfarm employment opportunities.[22]

Broad-Based Agricultural Development Is Critical

Despite rapid urbanization, poverty remains overwhelmingly rural in developing countries and is likely to remain so for decades. Hence, agriculture is key to reducing poverty. Even when rural people are not farmers or farm workers, they work in jobs closely related to agriculture, such as employment in enterprises producing processed food, tools, household goods, or services for agriculture.[23] Research has shown that for every new dollar of farm income earned in developing countries, income in the economy as a whole rises by as much as $2.60 as increasing farm demand generates employment, income, and growth economywide.[24] Agricultural growth also helps meet rising food demand and creates incentives for sustainable management of the natural resource base necessary for agriculture.

Sound public policies are essential to guarantee that agricultural and rural development is broad-based, creating opportunities for small-scale farmers and other poor people. Markets have a critical role but by themselves cannot assure equity. Key public investments include

- assuring poor farmers access to yield-increasing crop varieties (including drought- and salt-tolerant and pest-resistant varieties), improved livestock, and other yield-increasing and environment-friendly technology;

- access to tools, fertilizer, pest management, and credit;

- extension services and technical assistance;

- improved rural infrastructure such as roads and effective markets;

- particular attention to the needs of women farmers, who grow much of the locally produced food in developing countries; and

- primary education, health care, clean water, safe sanitation, and good nutrition for all.

The policy atmosphere must promote poverty reduction, must not discriminate against agriculture, and should provide incentives for sound natural resource management, such as secure property rights for small-scale farmers. Policies and programs must engage low-income

people as active participants, not passive recipients; development efforts seldom succeed unless affected people have a sense of ownership. Unfortunately, public investment in agriculture is on the decline in developing countries. On average, these countries devote 7.5 percent of government spending to agriculture (and the figure is even lower in many African countries).[25]

Agricultural Research Is Essential

Public investment in agricultural research is crucial to food security. The private sector is unlikely to undertake research needed by small-scale farmers in developing countries—even though societal benefits may be extremely large—because it cannot expect sufficient gains to cover costs. Currently, low-income developing countries grossly underinvest in agricultural research: less than 0.5 percent of the value of agricultural production, compared with 2.0 percent in higher income countries.[26]

Research should focus on productivity gains on small farms, emphasizing staple food crops and livestock. More research must be directed to appropriate technology for sustainable intensification of agriculture in resource-poor areas, where many poor people live. All appropriate scientific tools and better utilization of indigenous knowledge should be mobilized to help small-scale farmers in developing countries. These tools include not only new technologies that rely on external inputs, but also agroecology, which focuses on locally available farm labor and organic material as well as improved knowledge and farm management. In addition to the strengthening of national agricultural research in developing countries, international agricultural research, particularly the work by the Consultative Group on International Agricultural Research (CGIAR), should be supported.

Developed countries stand to gain from support for agricultural research for developing countries. For example, high-yielding varieties of wheat and rice bred by the Future Harvest centers for use in developing countries are now widely planted in the United States as well as in the developing world.[27]

The Role of Modern Agricultural Biotechnology

Modern biotechnology[28] is not a silver bullet for ending hunger, but, used in conjunction with traditional and conventional agricultural research methods, it may be a powerful tool that should be made available to poor farmers and consumers. It has the potential to help enhance agricultural productivity in developing countries in ways that reduce hunger and poverty and promote sustainable natural resource use.

> Public investment in agricultural and rural development should focus particular attention on the needs of women farmers, who grow much of the locally produced food in developing countries.

Current applications of molecular biology–based science to agriculture are oriented toward industrial country farmers and commercial farmers in a few developing countries. The United States alone cultivates more than 70 percent of genetically modified (GM) crops.[29]

Strong opposition to GM food in the European Union has resulted in severe restrictions on modern agricultural biotechnology. Opposition stems from perceived lack of consumer benefits, uncertainty about possible negative health and environmental effects, and widespread sentiment that a few large corporations will be the main beneficiaries.

Consumers outnumber farmers by 20 to 1 in the European Union and spend only a tiny fraction of their incomes on food. U.S. agriculture employs 2.6 percent of the workforce, and people spend an average of 12.0 percent of their income on food.[30] These numbers contrast sharply with comparable developing country figures noted earlier.

Modern agricultural biotechnology offers many potential benefits to poor farmers and consumers in developing countries. It may help achieve the productivity gains needed to feed a growing global population, introduce resistance to pests without high-cost purchased inputs, heighten the tolerance of crops to adverse weather and soil conditions, offer more nutritious foods, and enhance products' durability during harvesting or shipping. Bioengineered products may reduce reliance on pesticides, lowering crop protection costs and benefiting the environment and public health. By increasing yields and lowering unit production costs, biotechnology could help reduce food prices, greatly benefiting poor consumers. Biotechnology-assisted research developed broader-leafed rice that denies weeds sunlight, increasing farm incomes in West Africa and reducing the time women farmers spend weeding, allowing more time for the child care essential for good nutrition. Development of cereal plants capable of capturing nitrogen from the air could contribute greatly to plant nutrition and soil health while helping small-scale farmers who cannot afford fertilizers. Biotechnology may offer cost-effective solutions to vitamin and mineral deficiencies, such as vitamin A- and iron-rich rice. By increasing productivity, agricultural biotechnology could help conserve wild and marginal land and biodiversity.

Public policy must guide research. In addition to increasing the public resources for agricultural research, including biotechnology research, the public sector can entice the private sector to develop technologies for poor people by offering to buy the exclusive rights and make technologies available to small-scale farmers.[31]

Before GM crops and foods are introduced, a country should have sound food safety and environmental regulations to assess the risks and opportunities involved. Health risks include the transfer of allergy-causing traits through genetic engineering. GM foods need to be tested for such transfers, and those with possible allergy risks should be labeled. Environmental risks requiring assessment include the spread of traits, such as herbicide resistance to unmodified plants (including weeds), the buildup of resistance among pests, and unintended harm to other species.

Recent mergers and acquisitions in the biotechnology industry may lead to reduced competition, monopoly or oligopoly profits, exploitation of small-scale farmers and consumers, and extraction of special favors from governments. Institutions to promote competition and an effective antitrust system must be established in developing countries.

The biggest risk is that modern agricultural biotechnology will bypass poor people in a kind of "scientific apartheid."[32] Opportunities for reducing poverty, food insecurity, child malnutrition, and natural resource degradation will be missed, and the productivity gap between developing and developed country agriculture will widen for the benefit of no one.

Recommendations

There is nothing inevitable about this rather pessimistic forecast regarding world hunger. It is possible to meet and even exceed the World Food Summit's goal. If the new U.S. administration takes the appropriate actions, world hunger could decrease significantly by 2020. Achieving this will require concerted and committed action by governments, citizen groups, and the international community to empower poor people; mobilize new technological developments—including those in biotechnology—to benefit poor and hungry people in developing countries; invest in the factors essential for agricultural growth, including agricultural research and human resource development; and harness the political will to adopt sound antipoverty, food security, and natural resource management policies. Failing to take these steps will mean continued low economic growth and rapidly increasing food insecurity and malnutrition in many low-income developing countries, environmental deterioration, forgone trading opportunities, widespread conflict, and an unstable world for all.

The United States should make the eradication of hunger the top priority of its relations with developing countries, as the Presidential Commission on World Hunger recommended 20 years ago. Leadership of global cooperation to end hunger requires new policies:

Domestic hunger. The United States must address domestic hunger with employment policies to ensure adequate incomes for everyone to meet their needs, along with a nutrition safety net. This safety net should include full funding of the Special Supplemental Nutrition Program for Women, Infants, and Children (WIC). Every dollar invested in WIC saves up to $3.50 on Medicaid and special education. All schools should provide breakfast and lunch, with free meals for low-income children. Food stamp benefit levels should increase by 10 percent, and eligibility restrictions should be loosened.

> In Pakistan, salinity and waterlogging problems in the main cereal production regions will limit crop yield growth between 1995 and 2020, while growth in population is expected to be rapid.

Increased aid. The United States spends a smaller percentage of national income on development assistance than any other Organisation for Economic Development and Cooperation (OECD) country. As the world's wealthiest country, the United States can afford to increase aid to developing countries. It would cost an additional $12 billion annually to bring the United States up to the industrialized country average of 0.24 percent of GNP devoted to aid, a figure equivalent to half of U.S. annual sales of sporting goods and one-quarter of U.S. spending on tobacco products. The increase comes to $44 per person per year. Such an expansion would leverage additional funds from other countries and accelerate progress toward ending hunger. Polls consistently show public support for aid aimed at reducing poverty and hunger.

Need-based aid. The United States should target aid resources based on need (i.e., to countries with high levels of poverty and hunger, particularly in South Asia and Africa). Aid should go to countries where government policies support poverty alleviation and sound natural resource management. Investment in building prosperity, peace, and stability in these countries promises the United

States the potential for future commercial relationships. It is a win-win proposition.

Improved aid. Qualitative improvements in aid are needed, such as giving priority to human capital development and high-impact interventions. These include aid to broad-based agricultural and rural development and national and international agricultural research (including increased support for CGIAR and international programs at Land Grant universities and colleges), clean water, safe sanitation, universal primary education, access to basic health care, access to credit, land reform, natural resource management, and democracy and popular participation in development. Resources should be redirected from higher income recipients and military aid to these priorities.

> Developed countries can benefit from supporting agricultural research for developing countries. For example, high-yielding wheat and rice varieties bred for use in developing countries are now also widely planted in the United States.

Social development. Research has shown that improvements in female education, food availability per person, health care, and women's social status all enhance child nutrition. Educating girls has an especially strong impact. Increased national incomes and democratic governance are also important.[33] Aid programs must include conflict prevention and resolution components. Postconflict assistance must focus on reconstruction and underlying social tensions.

Food aid. Food aid for humanitarian emergencies is essential. It should also be used to help ease transitions and dislocations caused by economic reforms and trade liberalization, for maternal and child health activities, for school meals and building, and to restore roads and irrigation systems through "food for work." The United States should continue to provide food aid through international organizations and nongovernmental organizations, allowing them to sell commodities to generate funds for high-priority development activities. The United States should significantly increase the use of food aid resources to procure food in recipient countries and other developing countries and for direct purchase of seeds, fertilizers, tools, and livestock for poor farmers.

Biotechnology. The United States should support agricultural biotechnology research oriented toward poor farmers and consumers in developing countries and assist developing countries in building capacity to enact and enforce food safety, biosafety, and competition-enhancing and antitrust regulations.

Agricultural trade policy. In light of the continuing importance of the United States as a supplier of food and agriculture products to the global market, domestic farm policies should promote productivity gains and sustainable natural resource management to assure continued viability of the food and agriculture sector. Policies should focus on preserving and enhancing family farming operations because small- and medium-sized farms are at least as efficient as larger commercial operations. They tend to provide sound management of soil, water, and wildlife; and decentralized land ownership produces more equitable economic opportunity for rural communities and greater social capital.[34]

With respect to agricultural trade policy, maintaining current inequities in the global trading system does not develop long-term, mutually beneficial relationships between developing and developed countries. Allowing higher value agricultural products from developing countries into U.S. markets without escalating tariffs and encouraging other industrialized countries to take similar steps could greatly benefit developing countries and enhance global public opinion regarding the trading system. In addition, the United States should end export subsidies (both explicit and hidden) and encourage other industrialized countries to eliminate policies that exacerbate global price fluctuations. The United States should also seek a global intellectual property rights framework that balances rights of seed companies and other plant breeders with farmers' rights to save and reuse seed.

NOTES

1. Accessed via http://www.worldbank.org/poverty/data/trends/income.htm on 15 October 1999.
2. A. Deaton, *The Analysis of Household Surveys: A Microeconomic Approach to Development Policy* (Baltimore and London: Johns Hopkins University Press for The World Bank, 1997).
3. Food and Agriculture Organization of the United Nations (FAO), *The State of Food Insecurity in the World*, 1999 (Rome: FAO, 1999).
4. Accessed via http://www.ers.usda.gov/briefing/foodsecurity on 6 June 2000.
5. United Nations Administrative Committee on Coordination/Subcommittee on Nutrition (ACC/SCN) and International Food Policy Research Institute (IFPRI), *Fourth Report on the World Nutrition Situation* (Geneva: ACC/SCN and Washington, D.C.: IFPRI, 2000); and World Health Organization, *Malnutrition Worldwide,* accessed via http://www.who.int/nut on 6 June 2000.
6. P. Pinstrup-Andersen, R. Pandya-Lorch, and M. W. Rosegrant, *World Food Prospects: Critical Issues for the Early Twenty-First Century*, 2020 Vision Food Policy Report (Washington, D.C.: IFPRI, 1999).
7. United Nations Children's Fund, *The State of the World's Children 1998* (New York: Oxford University Press, 1998).
8. ACC/SCN and IFPRI, note 5 above, pages 23–32; and ACC/SCN, *Third Report on the World Nutrition Situation* (Geneva: ACC/SCN, 1997).
9. Pinstrup-Andersen, Pandya-Lorch, and Rosegrant, note 6 above, pages 8–18.
10. Ibid., pages 23–4; E. Diaz-Bonilla and S. Robinson, eds., *Getting Ready for the Millennium Round Trade Negotiations*, 2020 Vision Focus 1 (Washington, D.C.: IFPRI, 1999); and P. Pinstrup-Andersen, R. Pandya-Lorch, and M. W. Rosegrant, *The World Food Situation: Recent Developments, Emerging Issues, and Long-Term Prospects*, 2020 Vision Food Policy Report (Washington, D.C.: IFPRI, 1997).
11. FAO, *The State of Food and Agriculture in Figures, 1999* (Rome: FAO, 1999).
12. P. Pinstrup-Andersen and M. J. Cohen, *Aid to Developing Country Agriculture: Investing in Poverty Reduction and New Export Opportunities*, 2020 Vision Brief no. 56 (Washington, D.C.: IFPRI, 1998).
13. FAO, *Investment in Agriculture: Evolution and Prospects,* World Food Sum-

mit technical background document no. 10 (Rome: FAO, 1996).

14. Accessed via http://www.oecd.org/dac on 6 June 2000.

15. Accessed via http://www.wfp.org on 6 June 2000.

16. Accessed via http://www.unhcr.ch and http://www.refugees.org on 19 July 2000.

17. E. Messer, M. J. Cohen, and J. D'Costa, "Food from Peace: Breaking the Links between Conflict and Hunger," *Food, Agriculture, and the Environment Discussion Paper* no. 24 (Washington, D.C.: IFPRI, 1998).

18. Pinstrup-Andersen, Pandya-Lorch, and Rosegrant, note 6 above, pages 24–6; "Global Study Reveals New Warning Signals: Degraded Agricultural Lands Threaten World's Food Production Capacity," *IFPRI News Release* (21 May 2000), accessed via http://www.cgiar.org/ifpri/pressrel/052500.htm on 6 June 2000.

19. E. C. Oerke, H. W. Dehne, F. Schonbeck, and A. Weber, *Crop Production and Crop Protection: Estimated Losses in Major Food and Cash Crops* (Amsterdam: Elsevier, 1994).

20. M. Yudelman, A. Ratta, and D. Nygaard, "Issues in Pest Management and Food Production: Looking to the Future," *Food, Agriculture, and the Environment Discussion Paper* no. 26 (Washington, D.C.: IFPRI, 1998); *1998–99 World Resources Guide to the Global Environment: Environmental Change and Human Health* (New York: Oxford University Press, 1998); and P. Pinstrup-Andersen and R. Pandya-Lorch, "Food for All in 2020: Can the World Be Fed without Damaging the Environment?" *Environmental Conservation* 23, no. 3 (1996): 226–34.

21. Pinstrup-Andersen, Pandya-Lorch, and Rosegrant, note 10 above, pages 23–5.

22. S. J. Scherr, *Soil Degradation: A Threat to Developing-Country Food Security by 2020?* 2020 Vision Brief no. 58 (Washington, D.C.: IFPRI, 1999).

23. A. F. McCalla and W. S. Ayres, *Rural Development: From Vision to Action* (Washington, D.C.: The World Bank, 1997).

24. C. L. Delgado et al., *Agricultural Growth Linkages in Sub-Saharan Africa*, Research Report no. 107 (Washington, D.C.: IFPRI, 1998).

25. FAO, note 13 above, page 33; and FAO, "Public Assistance and Agricultural Development in Africa," accessed via http://www.fao.org.docrep/meeting/x3977e.htm on 21 February 2000.

26. P. G. Pardey and J. M. Alston, *Revamping Agricultural R & D*, 2020 Vision Brief no. 24 (Washington, D.C.: IFPRI, 1996).

27. P. G. Pardey, J. M. Alston, J. E. Christian, and S. Fan. *Hidden Harvest: U.S. Benefits from International Research Aid,* Food Policy Report (Washington, D.C.: IFPRI, 1996). The 16 Future Harvest international agricultural research centers are supported by the Consultative Group on International Agricultural Research, an informal association of 58 governments, international organizations, and private foundations, that seeks to contribute to food security and poverty eradication in developing countries through research, partnership, capacity-building, and policy support.

28. For more on this topic, see G. Persley, ed., *Biotechnology for Developing Countries: Problems and Opportunities*, 2020 Vision Focus 2 (Washington, D.C.: IFPRI, 1999); P. Pinstrup-Andersen and M. J. Cohen, "Modern Biotechnology for Food and Agriculture: Risks and Opportunities for the Poor," in G. J. Persley and M. M. Lantin, eds., *Agricultural Biotechnology and the Poor* (Washington, D.C.: The World Bank, 2000); and P. Pinstrup-Andersen and M. J. Cohen, "Food Security in the 21st Century and the Role of Biotechnology," *Foresight* 1, no. 5 (1999): 399–412.

29. C. James, *Global Status of Commercialized Transgenic Crops: 1999*, ISAAA Briefs no. 12: Preview (Ithaca, N.Y.: International Service for the Acquisition of Agri-Biotech Applications, 1999).

30. Accessed via http://www.bls.gov, http://www.census.gov. and http://www.usda.gov/nass on 18 November 1999.

31. "Sachs on Development: Helping the World's Poorest," *The Economist*, 14 August 1999, 17–20.

32. I. Serageldin, "Biotechnology and Food Security in the 21st Century," *Science*, 16 July 1999, 387–9.

33. L. C. Smith and L. Haddad, "Overcoming Child Malnutrition in Developing Countries," *Food, Agriculture, and the Environment Discussion Paper* no. 30 (Washington, D.C.: IFPRI, 2000).

34. U.S. Department of Agriculture (USDA), *A Time to Act: A Report of the USDA Notional Commission on Small Farms* (Washington, D.C.: USDA, 1998).

Per Pinstrup-Andersen is director-general of the International Food Policy Research Institute in Washington. D.C. He is a member of several committees, including the National Research Council's Committee on Biotechnology, the Working Group on Biotechnology under the State Department's Advisory Committee on International Economic Policy, and the World Health Policy Forum and its General Council. He can be contacted at (202) 862-5633 or p.pinstrup-andersen@cgiar.org. This article is © The Aspen Institute and is printed and distributed with permission.

Marc Cohen contributed greatly to the preparation of this article.

Prepared for the Aspen Institute's 2000 Environmental Policy Forum. From *Environment,* July/August 2001, pp. 20-26. © 2001 by The Aspen Institute. Reprinted by permission.

Escaping Hunger,
Escaping Excess

The big myth of malnutrition is that it's a problem of poor countries. But in a world at once rich in food and filled with poverty, malnutrition now has many faces—all over the world.

by Gary Gardner and Brian Halweil

TODAY, ETHIOPIA AND ITS neighbors are once again in the grip of an unrelenting famine, which has left more than 16 million people on the brink of starvation. After a massive international mobilization to aid this region in the 1980s, the Horn of Africa has become synonymous with famine and malnutrition. But across the Atlantic Ocean, another country is currently facing an epidemic that has left not *tens* of millions, but more than *100* million people malnourished—a quarter of them morbidly so. This growing problem receives little attention as a public health disaster, despite warnings from health officials that malnourishment has reached epidemic levels and has left vast numbers of people sick, less productive, and far more likely to die prematurely.

In this country—the United States—55 percent of adults are overweight and 23 percent are obese. The medical expenses and lost wages caused by obesity cost the country an estimated $118 billion each year, the equivalent of 12 percent of the annual health budget. Being overweight and obese are major risk factors in coronary heart disease, cancer, stroke and diabetes. Together these diseases are the leading killers in the United States, accounting for half of all deaths.

Misconceptions of hunger and overeating abound worldwide. We tend to think of hunger as resulting from a desperate scarcity of food, and we imagine it occurring only in poor countries. However, in those nations in Africa and South Asia where hunger is most severe, there is often plenty of food to go around. And even food rich nations are home to many underfed people.

Meanwhile, as the concept of malnutrition stretches to encompass excess as well as deficiency, wealthy nations are seeing rates of malnourishment that rival those in desperately poor regions. And overeating is growing in poorer nations as well, even where hunger remains stubbornly high. In Colombia, for example, 41 percent of adults are overweight, a prevalence that rivals rates found in Europe. While hunger is a more acute problem and should be the highest nutritional concern, overeating is the fastest growing form of malnourishment in the world, according to the World Health Organization (WHO). For the first time in history, the number of overweight people rivals the number who are underweight, both estimated at 1.1 billion.

Because myth and misconception permeate the world's understanding of malnutrition, policy responses have been wildly off the mark in addressing the problem. Efforts to eliminate hunger often focus on technological quick fixes aimed at boosting crop yields and producing more food, for example, rather than addressing the socioeconomic causes of hunger, such as meager incomes, inequitable distribution of land, and the disenfranchisement of women. Efforts to reduce overeating single out affected individuals—through fad diets, diet drugs, or the like—while failing to promote prevention and education about healthy alternatives in a food environment full of heavily marketed, nutritionally suspect, "supersized" junk food. The result: half of humanity, in both rich and poor nations, is malnourished today, according to the WHO. And this is in spite of recent decades of global food surpluses.

Malnutrition has become a significant impediment to development in rich and poor countries alike. At the individual level, both hunger and obesity can reduce a person's physical fitness, increase susceptibility to illness, and shorten lifespan. In addition, children deprived of adequate nutrients during development can suffer from permanently reduced mental capacity. At the national level, poor eating hampers educational performance, curtails economic productivity, increases the burden of health care, and reduces general well-being. Confronting this epidemic of poor eating will have widespread benefits, but first the myths that obscure the causes of malnutrition must be dispelled.

The Scarcity Myth

IN THE EARLY 1980s, the world was flooded by news of hunger and death from the Horn of Africa. By 1985, nearly 300,000 people had died. But international observers paid little attention to the fact that in the midst of famine, these countries were exporting cotton, sugar cane, and other cash crops that had been grown on some of the country's best agricultural land. While only 30 percent of farmland in Ethiopia was affected by drought, ubiquitous images of emaciated people surrounded by parched land have served to reinforce the single largest myth about malnutrition: that hunger results from a national scarcity of food.

Indeed, for more than 40 years the world has produced regular and often bountiful food surpluses—large enough, in fact, to prompt major producing countries like the United States to pay farmers *not* to farm some of their land. Indeed, the Food and Agriculture Organization (FAO) estimates that 80 percent of hungry children in the developing world live in countries that produce food surpluses. And only about a quarter of the reduction in hunger between 1970 and 1995 could be attributed to increasing food availability per person, according to a study by the International Food Policy Research Institute (IFPRI).

This is not to say that scarcity might not one day become the principal source of hunger, as population growth and ongoing damage to farmland and water supplies shrink food availability per person in many countries. Countries like Nigeria and Pakistan, which are on track to double their populations in the next 50 years, have already seen stocks of surplus food erode steadily in the 1990s. And countries such as India, which overpump groundwater to prop up agricultural production, will be hard pressed to maintain self sufficiency once aquifers run dry or become uneconomical to pump. But for the billion or so people who are hungry *today*, the finger of blame points in other directions.

Hands down, the major cause of hunger is poverty—a lack of access to the goods and services essential for a healthy life. Where people are hungry, it's a good bet that they have little income, cannot gain title to land or qualify

Supersized

Food portions have steadily grown in recent decades. A standard serving of soda in the 1950s, for example, was a 6.5 ounce bottle. Today the industry standard is a 20-ounce bottle. "Supersizing" has evolved as a marketing strategy that costs food producers little, but appears to give significant added value to consumers. But this trend skews perceptions of normal servings: in the United States, one study found that participants consistently labeled as "medium" portions that were double or triple the size of recommended portions.

Food companies tend to push fatty or sweet foods for two reasons: they know we have an innate preference for them, and highly processed foods like white hamburger buns offer greater profits than more elemental products like fruits and vegetables. Adding more sugar, salt, fats, or oils (as typically concentrated in prepared mustard, ketchup, or pickles), can provide a tasty and profitable product that is often irresistible to consumers and companies alike.

The overeating phenomenon is quickly spreading around the world, in part due to heavy advertising. Food companies spend more than any other industry on advertising in the United States. Coca-Cola and McDonald's are among the 10 largest advertisers in the world. This strategy has paid off: McDonald's opens five new restaurants every day, four of them outside the United States.

Americans consume 70 kilograms of caloric sweeteners per year—75 percent more than in 1909. That is nearly 200 grams or 53 teaspoons a day, the equivalent of a 5 pound bag of sugar every week and a half. In Europe and North America, fat and sugar count for more than half of all caloric intake, squeezing complex carbohydrates like grains and vegetables down to about one-third of total calories.

for credit, have poor access to health care, or have little or no education. Worldwide, 150 million people were unemployed at the end of 1998, and as many as 900 million had jobs that paid less than a living wage. These billion-plus people largely overlap with the 1.1 billion people who are underweight, and for whom hunger is a chronic experience. And nearly 2 billion more teeter at the edge of hunger, surviving on just 2 dollars or less per day, a large share of which is spent on food.

Hunger, like its main root, poverty, disproportionately affects females. Girls in India, for example, are four times as likely to be acutely malnourished as boys. And while 25 percent of men in developing countries suffer from anemia, a condition of iron deficiency, the rate is 45 percent for women—and 60 percent for those who are pregnant. This gender bias stems from cultural prejudices in households and in societies at large. Most directly, lean rations at home are often dished out to father and sons before mother and daughters, even though females in developing countries typically work longer hours than males do. Gender bias is also manifest in education. Inequitable schooling opportunities for girls lead to economic insecurity: women represent two-thirds of the world's illiterate people and three-fifths of its poor. With fewer educational and economic opportunities than men, women tend to be hungrier and suffer from more nutrient deficiencies.

Any serious attack on hunger, therefore, will aim to reduce poverty, and will give special attention to women. The IFPRI study on curbing malnutrition found that improving women's education and status together accounted for more than half of the reduction in malnutrition between 1970 and 1995. Such nutritional leverage stems from a woman's pivotal role in the family. A woman "eats for two" when she is pregnant and when she is nursing; pull her out of poverty, and improvements in her nutrition are passed on to her infant. But there's more: studies show that provided with an income, a woman will spend nearly all of it on household needs, especially food. The same money in a man's pocket is likely to be spent in part—up to 25 percent—on non-family items, such as cigarettes or alcohol.

From this perspective, microcredit initiatives, such as those of the Bangladesh-originated Grameen Bank, offer a promising means of combatting hunger. These unconventional programs provide small loans of tens or hundreds of dollars to help very poor women generate income through basket-weaving, chicken-raising, or other small projects. As the loans lift women out of poverty, they also yield nutritional benefits: a 10 percent increase in a woman's Grameen borrowing, for example, has been shown to produce a 6 percent increase in the arm circumference of her children (a measure of nutritional well-being). It also increases by 20 percent the likelihood that her daughter will be enrolled in school, which lowers the girl's risk of suffering malnutrition as an adult.

International support for such programs could expand them dramatically. One option is the nonprofit Microcredit Summit's campaign to raise $22 billion to increase the number of microcredit beneficiaries from 8 million in the late 1990s to 100 million by 2005. Such investments are a high-leverage option for a nation's foreign aid commitment, given all of the benefits—improved nutrition, better health, and slower rates of population growth—that come from reducing poverty, especially among women.

At a broader social level, the journey out of poverty and hunger can be expedited through better access to land and agricultural credit. These measures are especially important for women, since they produce more than half of the world's food, and a large share of what is consumed in rural households in developing countries. In India, Nepal, and Thailand, less than 10 percent of women own land, and those who do often have small, marginal tracts. For landless women, credit is next to impossible to obtain: in five African countries—Kenya, Malawi, Sierra Leone, Zambia, and Zimbabwe—where women constitute a large share of farmers, they receive less than one percent of the loans provided in agriculture. This despite their exceptional creditworthiness: women typically pay their debts more faithfully than men do.

Women also need access to sound nutritional information as a way to avoid nutritional impoverishment and unnecessary food expenditures. Breastfeeding campaigns, for example, can highlight the many advantages of this free and wholesome method of infant feeding. Baby formulas are often prepared in unsanitary conditions or watered down to reduce costs. Campaigns to promote breastfeeding and restrict sales of formula have been estimated to reduce illness from diarrhea—a condition that robs infants of needed vitamins and minerals—by 8 to 20 percent. They have reduced deaths from diarrhea by 24 to 27 percent. Breastfeeding also acts as a natural contraceptive following pregnancy, spacing births at greater intervals and thereby easing the pressure to feed everyone in poor families.

Nutritional education efforts are also essential to fighting hunger, and the most successful programs involve entire communities by enlisting affected people and local leaders. The BIDANI program in the Philippines, for example, provides orientation and training for villagers to participate in nutritional "interventions," which have worked to elevate 82 percent of enrolled children to a higher nutritional status. A similar program in Gambia substantially cut the death rate among women and children by working with the highly respected women elders of the matriarchal Kabilo tribe to educate community members about child-feeding practices, hygiene, and maternal health care.

Important as these social initiatives are for improving nutrition, more direct action is often required to meet the needs of those who suffer from hunger today. Even here, however, creative approaches can empower women and aid entire communities. In one simple case in Benin, food aid is dispensed not directly to families, but to girls at school, who bring it home to their parents. The practice combats the cultural bias against girls found in many countries, which often results in their removal from school at a young age to help at home or to allow a brother to get an education. It achieves two critical nutritional goals: it gets food to families that need it, and it increases girls' future employment prospects, which in turn reduces the likelihood of future malnutrition.

The Prone-to-Obesity Myth

FOR THOSE WHO HAVE ACCESS to enough food, eating habits around the world are in the midst of the most significant change since the development of agriculture thousands of years ago. Since the turn of the century, traditional diets featuring whole grains, vegetables, and fruits have been supplanted by diets rich in meat, dairy products, and highly processed items that are loaded with fat and sugar. This shift, already entrenched in industrial countries and now accelerating in developing nations as incomes rise, has created an epidemic of overeating and sparked a largely misunderstood public health crisis worldwide. In the United States, the leader in this global surge toward larger waist sizes, more than half of all adults are now overweight—a condition that, like hunger, increases susceptibility to disease and disability, reduces worker productivity, and cuts lives short.

The proliferation of high-calorie, high-fat foods that are widely available, heavily promoted, low in cost and nutrition, and served in huge portions has created what Yale psychologist Kelly Brownell calls a "toxic food environment." Sweets and fats increasingly crowd out nutritionally complete foods that provide essential micronutrients. For instance, one-fifth of the "vegetables" eaten today in the United States are servings of french fries and potato chips. Our propensity to eat sweet and fatty foods may have served our ancestors well for weathering seasonal lean times, but amidst unbridled abundance for many, it has become a handicap. When these eating habits are combined with increasingly urbanized, automated, and more sedentary lifestyles, it becomes clear why gaining weight is often difficult to avoid.

Failure to recognize the existence of this negative food environment has created the widespread misconception that individuals are entirely to blame for overeating. The reality is most countries embrace policies and practices that promote mass overconsumption of unhealthy foods, but abandon citizens when it comes to dealing with the health implications. Because individuals are stigmatized as weak-willed or prone to obesity, prevailing efforts to curb overeating have focused on techno-fixes and diets, not prevention and nutrition education.

This end-of-the-pipe mentality manifests itself in a variety of ways: liposuction is now the leading form of cosmetic surgery in the United States with 400,000 operations performed each year; fad diet books top the bestseller lists; designer "foods" such as olestra promise worry-free consumption of nutritionally empty snacks; and laboratories scurry to find the human "fat gene" in an effort to engineer our way out of obesity. While the U.S. Agriculture Department spends $333 million each year to educate the public about nutrition, the U.S. diet and weight-loss industry records annual revenues of $33 *billion*. And the highly lucrative weight-loss business feeds off of a global food industry that now has significant influence over food choices around the world.

Indeed, consumers get the majority of their dietary cues about food from food companies, who spend more on advertising—$30 billion each year in the United States alone—than any other industry. The most heavily advertised foods, unfortunately, tend to be of dubious nutritional value. And food advertisers disproportionately target children, the least savvy consumers, in order to shape lifelong habits. In fact, in the United States, the average child watches 10,000 commercials each year, more than any other segment of the population. And more than 90 percent of these ads are for sugary cereals, candy, soda, or other junk food, according to surveys by the Center for Science in the Public Interest.

Targeting Women

Hungry children are often scarred for life, suffering impaired immune systems, neurological damage, and retarded physical growth. Infants that are underweight in utero will be five centimeters shorter and five kilograms lighter as adults.

Chronic hunger leaves children and adults more susceptible to infectious diseases. Among the five leading causes of child death in the developing world, 54 percent of cases have malnutrition as an underlying cause.

Conflict and military spending exacerbate hunger directly by disrupting economies and food production, and indirectly by diverting funds away from poverty alleviation to militaries.

Where hunger exists, women are invariably more malnourished than men. In India, for example, girls are four times as likely to be hungry or suffer from micronutrient deficiencies as boys are. Hungry women bear and raise hungry children. Because impoverished families are less able to care for their offspring, hunger is perpetuated across generations.

Women produce more than half of the world's food, and in rural areas they provide the lion's share of food consumed in their own homes. Yet, note who's in control here. Women often cannot obtain access to land, credit, or the social and political support that men can.

Numerous studies show that these ads work. They prompt children to more frequently request, purchase, and consume advertised foods, even when they become adults. And as kids fill up on items loaded with empty calories like soda or candy, more nutritious items are squeezed out of the diet. Marketing to children has intensified in recent years as food companies have begun to target the school environment. More than 5,000 U.S. schools—13 percent of the country's total—now have contracts with fast-food establishments to provide either food service, vending machines, or both. Since 1990, soda companies have offered millions of dollars to cash-strapped school districts in the United States for exclusive rights to sell their products in schools.

With industrial country markets increasingly saturated, many food corporations are now looking to developing countries for greater profits. Mexico recently surpassed the United States as the top per capita consumer of Coca-Cola, for example. And that company's 1998 annual report notes that Africa's rapid population growth and low per capita consumption of carbonated beverages make that continent "a land of opportunity for us." The number of U.S. fast-food restaurants operating around the world is also growing rapidly: four of the five McDonald's restaurants that open every day are located outside the United States.

Overeating is also becoming a problem even in countries where hunger and poverty persist. In China, for ex-

Nutrition Split

Every region in the world now has large numbers of hungry or overweight people — or both — as affluence spreads and poverty persists.

Some of the clearest evidence that hunger is caused by poverty and not regional food scarcity is the presence of hunger in the **United States**. In 1998, 10 percent of U.S. households, home to nearly one in five American children, were "food insecure" — hungry, on the edge of hunger, or worried about being hungry.

From 1980 to 2000, the share of children who are underweight in **Latin America and the Caribbean** has dropped from 14 percent to 6 percent. But it seems this region has simply traded one form of poor eating for another: in most Latin American nations, the overweight population now exceeds the underweight population.

(graphic continues on next page)

Millions of People

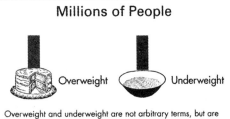

Overweight   Underweight

Overweight and underweight are not arbitrary terms, but are defined using body-mass index (BMI), a scale calibrated to reflect the health effects of weight gain. A healthy BMI ranges from 19 to 24; a BMI of 25 or above indicates "overweight" and brings increased risk of illnesses such as heart disease, stroke, diabetes, and cancer. A BMI above 30 signals "obesity" and even greater health risks. BMI is calculated as a person's weight in kilos divided by the square of height in meters.

ample, consumption of high-fat foods such as pork and soy oil (which is used for frying) both soared after the economic boom of the 1980s, while consumption of rice and starchy roots dropped—changes that were most pronounced among wealthier households. The parallel trend of urbanization in the developing world also means exposure to new foods and food advertising—particularly for highly processed and packaged items—and considerably more sedentary lifestyles. A recent study of 133 developing countries found that migration to the city—without any changes in income—can more than double per capita intake of sweeteners. Cash-squeezed households in Guayaquil, Ecuador, often spurn potatoes and fresh fruit juices in favor of fried plantains, potato chips, and soft drinks, replacing nutrient-dense foods with empty calories.

A world raised on Big Macs and soda isn't inevitable. But countering an increasingly ubiquitous toxic food environment will require dispelling the myths that sur-

Nutrition Split
(graphic continues from previous page)

European levels of overeating are not far behind those of North America. The share of the adult population in Russia, Germany, and the United Kingdom that is overweight is roughly half, while the share in other European nations tends to be slightly lower.

Much of the **Middle East** faces an overeating crisis of North American proportions. But in poorer, war-torn nations, like Iraq and the Sudan, hunger reaches the desperate levels found in southern Africa.

Along with Sub-Saharan Africa, **South Asia** is home to a massive concentration of hungry people. Some 44 percent of the region's children are underweight, while the shares in India, Bangladesh, and Afghanistan are well above this average. At the same time, among the urban upper-class of this region, obesity is a growing problem.

Like Latin America, **East and Southeast Asia** have seen significant decreases in the share of the population that is hungry. Yet hunger remains stubbornly fixed in some countries, and overeating is spreading rapidly. The share of adults who are overweight in China jumped by more than half—from 9 percent to 15 percent—between 1989 and 1992.

The share of the world's population that is underweight is in decline, except in **sub-Saharan Africa,** where 36 percent of children are underweight due to poverty and other social factors.

round overeating. Governments will have to recognize the existence of a health epidemic of overeating, and will have to work to counter the social pressures that promote poor eating habits. Empowering individuals through education about nutrition and healthy eating habits, particularly for children, is also essential.

If preventing overeating is the goal, rather than treating it after habits have been formed, then the school environment is an obvious place to start. In Singapore, for example, the nationwide Trim and Fit Scheme has reduced obesity among children by 33 to 50 percent, depending on the age group, by instituting changes in school catering and increasing nutrition and physical ed-

ucation for teachers and children. Similar programs in other countries have found comparable results, yet physical education programs in many nations are actually being scaled back.

Mass-media educational campaigns can also change long-standing nutritional habits in adults. Finland launched a campaign in the 1970s and 1980s to reduce the country's high incidence of coronary heart disease, which involved government-sponsored advertisements, national dietary guidelines, and regulations on food labeling. This broad, high-profile approach—it also advocated an end to smoking, and involved groups as diverse as farmers and the Finnish Heart Association—increased fruit and vegetable consumption per person two-fold and slashed mortality from coronary heart disease by 65 percent between 1969 and 1995. About half of the drop in mortality is credited to the lower levels of cholesterol induced by the nutrition education campaign.

A public health approach to overeating might also take some hints from successful campaigns against smoking, including warning labels and taxes to deter consumption. In Finland, the government now requires "heavily salted" to appear on foods high in sodium, while allowing low-sodium foods to bear the label "reduced salt content." A complement to the "low-fat" labels that grace so many new food products would be a more ominous "high-fat" or "high-sugar" label.

Consumption of nutrient-poor foods can be further reduced by fiscal tools. Yale's Kelly Brownell advocates adoption of a tax on food based on the nutrient value per calorie. Fatty and sugary foods low in nutrients and loaded with calories would be taxed the most, while fruits and vegetables might escape taxation entirely. The idea is to discourage consumption of unhealthy foods—and to raise revenue to promote healthier alternatives, nutrition education, or exercise programs, in essence to make it easier and cheaper to eat well. Large-scale cafeteria and vending machine studies show how powerful an influence price has on buying choices—reducing the price and increasing the selection of fruit, salad, and other healthy choices can often double or triple purchase of these items, even as total food purchases remain the same.

Such a tax is also justified as the cost of overeating to society grows. Graham Colditz at Harvard estimates the direct costs (hospital stays, medicine, treatment, and visits to the doctor) and indirect costs (reduced productivity, missed workdays, disability pensions) of obesity in the United States to be $118 billion annually. This sum, equal to nearly 12 percent of the U.S. annual health budget, is more than double the $47 billion in costs attributable to cigarette smoking—a better known and heavily taxed drag on public health. Fiscal measures to reduce overeating may be most attractive to developing countries, which must tackle growing caseloads of costly chronic diseases even as they struggle to eradicate infectious illness.

Putting the Pieces Together

THE EFFECTS OF POOR NUTRITION run deep into every aspect of a community, curtailing performance at school and work, increasing the cost of health care, and reducing health and well-being. By the same token, improving nutrition promises to have equally far-reaching, positive impacts on regions that choose to address the problem. Better eating can set into motion a host of other benefits, many in areas seemingly unrelated to food.

For this to occur, however, efforts to improve nutrition must be integrated into all aspects of a country's development decisions—from health care priorities to transportation funding to curricula planning for schools. A cleaner water supply, for example, would reduce the incidence of intestinal parasites that hamper the body's capacity to absorb micronutrients. Thus a ministry of public works dedicated to increasing access to clean water is a logical partner in a campaign to reduce micronutrient deficiencies. Similarly, transportation officials who promote bicycle commuting, ministers of culture who discourage TV watching, and an agriculture ministry that promotes nutritional education are all promoting lifestyles that, in conjunction with better eating, can reduce incidences of obesity.

There are numerous less obvious means, as well, by which nutritional improvement can be woven into daily life. To begin with, smart nutrition policies can be added to already-existing social programs. Health, education, and agricultural extension programs already reach deep into nutritionally vulnerable populations through existing networks of clinics, schools, and rural development offices. Nutrition is a natural outgrowth of their current responsibilities. Clinic staff, for example, could promote breast-feeding, and extension agents could encourage home gardening. Such partnering is cost effective, not only because it uses existing infrastructure, but also because it often reduces the need for the original service. Women educated about breastfeeding on a prenatal visit, for example, are less likely to return months later with an infant suffering from diarrhea.

Programs intended to eradicate poverty, from microlending to employment creation, are most likely to raise nutritional levels when accompanied by education about health and nutrition. A "Credit with Education" program initiated in Ghana by the international group Freedom from Hunger coupled lending with education about breastfeeding, child feeding, diarrhea prevention, immunization, and family planning. A three-year follow-up study documented improved health and nutrition practices, fewer and shorter-lived episodes of food shortages, and dramatic improvements in childrens' nutrition among the participants compared with the control groups. Barbara MkNelly, the program's coordinator, warns that "simply improving a family's ability to buy food is no guarantee that poor baby-feeding practices, dietary choices or living conditions will not undercut nutritional gains."

The city of Curitiba, Brazil has even found links between nutrition and the city's waste flows. Concerned about the city's growing waste burden, and about malnutrition among the poorest sectors of the population, officials established a recycling program for organic waste that benefits farmers, the urban poor, and the city in general. City residents separate their organic waste from the rest of their garbage, bag it, then exchange it for fresh fruits and vegetables from local farmers at a city center. The city reduces its waste flow, farmers reduce their dependence on chemical fertilizer, and the urban poor get a steady supply of nutritious foods.

In any society, but especially where food cues come primarily from advertising, education is critical to making progress toward good nutrition. In the United States, the Berkeley Food System Project, for example, not only teaches kids about healthy eating, but promotes the use of vegetable gardens in school to help children learn about food at the source. The gardens also supply some of the food for school cafeterias, which were required in 1999 to begin serving all-organic lunches. The project encourages schools to incorporate this comprehensive view of food into their classwork. Janet Brown, who spearheaded the project for the Center for Ecoliteracy, explains that kids weaned on packaged and processed foods often shy away from fruits and veggies, because they have not been properly introduced. "But when a child pops a cherry tomato that she helped to grow into her mouth, then introducing a salad bar in the cafeteria is likely to be more successful."

Nutritional literacy is not just for kids, however. Doctors, nurses, and other health care professionals are well positioned to educate patients about the links between diet and health, and can be instrumental in improving eating habits. But modern medical systems often de-emphasize the role of nutrition: in the United States, only 23 percent of medical schools required that students take a separate course in nutrition in 1994. Doctors poorly trained in nutrition are less likely to take a preventive approach to health care, such as encouraging greater consumption of fruits and vegetables or increased physical activity, and are more likely to deal only with the consequences of poor eating—prescribing a cholesterol-lowering drug, for example, or scheduling bypass surgery. A recent U.S. survey by the Centers for Disease Control found "less than half of obese adults report being advised to lose weight by health care professionals."

Beyond educating medical professionals, health care as a whole could integrate nutrition by recognizing obesity as a disease and covering weight-loss programs and other nutritional interventions. Covering these expenses would not only reduce illness and patient suffering, but is likely to cut health care costs. An encouraging first step in this direction is Mutual of Omaha's decision to cover intensive dietary and lifestyle modification program of patients with heart disease, an initiative they hope will eliminate costly prescriptions and prevent surgeries months or years down the road. A local next step for the industry might be to cover regular nutrition checkups, akin to dental check-ups, as part of a basic insurance coverage.

Where communities have lost access to healthy food options, improving diets may require involving players throughout the food chain. Support of urban agriculture and urban farmers' markets has proven effective in getting good food to low-income urbanites. Urban gardens in Cuba, which meet 30 percent of the vegetable demand in some cities, have prospered under government nurturing. In the nutritionally impoverished inner cities of wealthier nations, farmers' markets are often the only source of fresh produce, as green grocers and supermar-

Where's the Nutrition?

Food advertisers disproportionately target children, the least savvy consumers. In the United States, the average child is bombarded with 10,000 commercials each year—90 percent of them for sugary cereals, candy, or other junk foods.

Junk foods often displace more nutritious foods, providing only "empty calories"—energy with little nutritional value. In the United Kingdom, per capita consumption of snack foods is up by nearly a quarter in the past five years—snack foods are now a $3.6 billion industry.

Eating in industrial countries centers less than ever before on home and family. In 1998, just 38 percent of meals in U.S. homes were homemade, and one out of every three meals were eaten outside of the home.

Nutritionally poor foods are invading U.S. schools. Fast food companies have contracts, often worth millions of dollars, to provide food service or vending machines, at more than 5,000 U.S. schools. One deal prompted a Colorado school district to push Coca-Cola consumption, even in classrooms, when sales fell below contractual obligations.

kets have left for the more affluent suburbs, and as fast food joints and convenience stores have replaced them. The Toronto Food Policy Council has used both farmers' markets and produce delivery schemes to connect local farmers and low-income urban residents, many of whom are single mothers. Some 70 percent of those buying food now eat more vegetables than they did when the program began in the early 1990s; 21 percent eat a greater variety; and 16 percent now try new foods. More people also know about the recommended five or more servings a day of fruit and veggies.

Eliminating poor eating is the business of fiscal authorities as well. The food tax advocated by Kelly Brownell could raise funds for nutritional interventions. Michael Jacobsen, director of the Center for Science in the Public Interest, notes that even small taxes could generate sufficient revenues to fund "television advertisements, physical education teachers, bicycle paths, swimming pools, and other obesity prevention measures." In the United States, a 2/3-cent tax per can of soda, a 5 percent tax on new televisions and video equipment, a $65 tax on each new motor vehicle, or an extra penny tax per gallon of gasoline would each raise roughly $1 billion each year.

Even without such a tax, authorities in some countries have begun to encourage lifestyle changes that are important complements to good nutrition. Australia's Department of Transport and Regional Services, Department of Health and Aged Care, and Department of Environment and Heritage teamed up in 1999 to promote the country's National Bicycling Strategy, which seeks to raise the level of cycling in the country. The involvement of this diverse set of government agencies demonstrates the broad impact that a commitment to good nutrition can have. More cycling means more exercise, an indispensable tool in the fight against overweight. But it can also mean cleaner air, less congested cities, and cheaper transportation infrastructure.

A final part of reshaping the food environment is re-cultivating an appreciation of food as a cultural and nutritional treasure. The consumer culture, applied to eating, emphasizes brand allegiance and megameals, often at the expense of nutrition and health. Groups like the Slow Food Movement, based in Italy, and the Oldways Preservation and Exchange Trust in the United States, offer a postmodern critique of today's culinary norm by promoting a return to the art of cooking traditional foods and of socializing around food. Their work, which targets chefs as well as consumers, is the kind of cultural intervention that could help more people shift to a healthy diet, similar to the change in consciousness that encouraged a shift away from smoking in the United States. Government encouragement of these groups, perhaps through assistance with marketing and promotional activities, would insure that this important work benefits everyone, not just the affluent.

The experience of the Slow Food Movement and Oldways shows that as people care more about their food choices, their concerns are likely to evolve well beyond nutritional value. Health-conscious consumers often gravitate toward organic produce, in an effort to avoid agrochemical residues and to stop promoting farm practices that deplete the soil or pollute waterways. Many also reduce their consumption of animal products, which can reduce their intake of fat and cholesterol, but also eases the pressure on land and water resources. And these consumers are likely to seek out local food sources, which offer superior freshness and quality, as well as the opportunity to know the farmer and his methods.

The far-reaching effects of nutrition make it a central factor in personal and national development. Poor eating is as much a drag on national economic activity as it is on

Hidden Hunger

Hunger has been alleviated somewhat in the past 20 years, except in Africa.

Micronutrient deficiencies plague between 2 and 3.5 billion people around the world, including a considerable number of both the 1.1 billion who are hungry and the 1.1 billion who are overweight. Micronutrients—vitamins and minerals such as iron, calcium, and vitamins A through E—are crucial elements of a healthy diet.

Food aid is not the long-term answer for most of the world's hungry. Nearly 80 percent of all malnourished children in the developing world live in countries that have food surpluses. Today, hunger is the product of human decisions—people are denied access to food as a result of poverty and other social inequities, not as a result of net scarcity.

Deficiencies in nutrients such as iodine can stunt physical and mental growth. More than 740 million people—13 percent of the world—suffer from iodine deficiency, which is the most common preventable cause of mental retardation. Vitamin A deficiency is the world's leading cause of blindness. Iron deficiency, prevalent in 56 percent of women in developing countries who are pregnant, causes anemia, which can stunt the development of the fetus.

personal health. The reverse is also true: development choices, such as whether girls have as many years of schooling as boys, or whether food corporations are free to advertise without limit to young consumers, heavily influence what and how we eat.

Gary Gardner is a senior researcher and Brian Halweil is a staff researcher at the Worldwatch Institute. They are co-authors of Worldwatch Paper 150, *Overfed and Underfed: The Global Epidemic of Malnutrition* (2000).

Growing more Food with less Water

by Sandra Postel

Six thousand years ago farmers in Mesopotamia dug a ditch to divert water from the Euphrates River. With that successful effort to satisfy their thirsty crops, they went on to form the world's first irrigation-based civilization. This story of the ancient Sumerians is well known. What is not so well known is that Sumeria was one of the earliest civilizations to crumble in part because of the consequences of irrigation.

Sumerian farmers harvested plentiful wheat and barley crops for some 2,000 years thanks to the extra water brought in from the river, but the soil eventually succumbed to salinization—the toxic buildup of salts and other impurities left behind when water evaporates. Many historians argue that the poisoned soil, which could not support sufficient food production, figured prominently in the society's decline.

Far more people depend on irrigation in the modern world than did in ancient Sumeria. About 40 percent of the world's food now grows in irrigated soils, which make up 18 percent of global cropland [*see* "Top 10 Irrigators Worldwide"]. Farmers who irrigate can typically reap two or three harvests every year *and* get higher crop yields. As a result, the spread of irrigation has been a key factor behind the near tripling of global grain production since 1950. Done correctly, irrigation will continue to play a leading role in feeding the world, but as history shows, dependence on irrigated agriculture also entails significant risks.

Today irrigation accounts for two thirds of water use worldwide and as much as 90 percent in many developing countries. Meeting the crop demands projected for 2025, when the planet's population is expected to reach eight billion, could require an additional 192 cubic miles of water—a volume nearly equivalent to the annual flow of the Nile 10 times over. No one yet knows how to supply that much additional water in a way that protects supplies for future use.

Severe water scarcity presents the single biggest threat to future food production. Even now many freshwater sources—underground aquifers and rivers—are stressed beyond their limits.

As much as 8 percent of food crops grows on farms that use groundwater faster than the aquifers are replenished, and many large rivers are so heavily diverted that they don't reach the sea for much of the year. As the number of urban dwellers climbs to five billion by 2025, farmers will have to compete even more aggressively with cities and industry for shrinking resources.

If the world hopes to feed its burgeoning population, irrigation must become less wasteful and more widespread

Despite these challenges, agricultural specialists are counting on irrigated land to produce most of the additional food that will be needed worldwide. Better management of soil and water, along with creative cropping patterns, can boost production from cropland that is watered only by rainfall, but the heaviest burden will fall on irrigated land. To fulfill its potential, irrigated agriculture requires a thorough redesign organized around two primary goals: cut water demands of mainstream agriculture and bring low-cost irrigation to poor farmers.

Fortunately, a great deal of room exists for improving the productivity of water used in agriculture. A first line of attack is to increase irrigation efficiency. At present, most farmers irrigate their crops by flooding their fields or channeling the water down parallel furrows, relying on gravity to move the water across the land. The plants absorb only a small fraction of the water; the rest drains into rivers or aquifers, or evaporates. In many locations this practice not only wastes and pollutes water but also degrades the land through erosion, waterlogging and salinization. More efficient and environmentally sound technologies exist that could reduce water demand on farms by up to 50 percent.

Drip systems rank high among irrigation technologies with significant untapped potential. Unlike flooding techniques, drip systems enable farmers to deliver water directly to the plants' roots drop by drop, nearly eliminating waste. The water travels at low pressure through a network of perforated plastic tubing installed on or below the surface of the soil, and it emerges through small holes at a slow but steady pace. Because the plants enjoy an ideal moisture environment, drip irrigation usually offers the added bonus of higher crop yields. Studies in India, Israel, Jordan, Spain and the U.S. have shown time and again that drip irrigation reduces water use by 30 to 70 percent and increases crop yield by 20 to 90 percent compared with flooding methods.

Technologies exist that could reduce water demand on farms by up to 50 percent

Sprinklers can perform almost as well as drip methods when they are designed properly. Traditional high-pressure irrigation sprinklers spray water high into the air to cover as large a land area as possible. The problem is that the more time the water spends in the air, the more of it evaporates and blows off course before reaching the plants. In contrast, new low-energy sprinklers deliver water in small doses through nozzles positioned just above the ground. Numerous farmers in Texas who have installed such sprinklers have found that their plants absorb 90 to 95 percent of the water that leaves the sprinkler nozzle.

Despite these impressive payoffs, sprinklers service only 10 to 15 percent of the world's irrigated fields, and drip systems account for just over 1 percent. The higher costs of these technologies (relative to simple flooding methods) have been a barrier to their spread, but so has the prevalence of national water policies that discourage rather than foster efficient water use. Many governments have set very low prices for publicly supplied irrigation, leaving farmers with little motivation to invest in ways to conserve water or to improve efficiency. Most authorities have also failed to regulate groundwater pumping, even in regions where aquifers are overtapped. Farmers might be inclined to conserve their own water supplies if they could profit from selling the surplus, but a number of countries prohibit or discourage this practice.

Efforts aside from irrigation technologies can also help reduce agricultural demand for water. Much potential lies in scheduling the timing of irrigation to more precisely match plants' water needs. Measurements of climate factors such as temperature and precipitation can be fed into a computer that calculates how much water a typical plant is consuming. Farmers can use this figure to determine, quite accurately, when and how much to irrigate their particular crops throughout the growing season. A 1995 survey conducted by the University of California at Berkeley found that, on average, farmers in California who used this tool reduced water use by 13 percent and achieved an 8 percent increase in yield—a big gain in water productivity.

An obvious way to get more benefit out of water is to use it more than once. Some communities use recycled wastewater. Treated wastewater accounts for 30 percent of Israel's agricultural water supply, for instance, and this share is expected to climb to 80 percent by 2025. Developing new crop varieties offers potential as well. In the quest for higher yields, scientists have already exploited many of the most fruitful agronomic options for growing more food with the same amount of water. The hybrid wheat and rice varieties that spawned the green revolution, for example, were bred to allocate more of the plants' energy—and thus their water uptake—into edible grain. The widespread adoption of high-yielding and early-maturing rice varieties has led to a roughly threefold increase in the amount of rice harvested per unit of water consumed—a tremendous achievement. No strategy in sight—neither conventional breeding techniques nor genetic engineering—could repeat those gains on such a grand scale, but modest improvements are likely.

Yet another way to do more with less water is to reconfigure our diets. The typical North American diet, with its large share of animal products, requires twice as much water to produce as the less meat-intensive diets common in many Asian and some European countries. Eating lower on the food chain could allow the same volume of water to feed two Americans instead of one, with no loss in overall nutrition.

Reducing the water demands of mainstream agriculture is critical, but irrigation will never reach its potential to alleviate rural hunger and poverty without additional efforts. Among the world's approximately 800 million undernourished people are millions of poor farm families who could benefit dramatically from access to irrigation water or to technologies that enable them to use local water more productively.

Top 10 Irrigators Worldwide

SOURCE: UN FAO AGROSTAT database, 1998

Most of these people live in Asia and Africa, where long dry seasons make crop production difficult or impossible without irrigation. For them, conventional irrigation technologies are too expensive for their small plots, which typically encompass fewer than five acres. Even the least expensive motorized pumps that are made for tapping groundwater cost about $350, far out of reach for farmers earning barely that much in a year. Where affordable irrigation technologies have been made available, however, they have proved remarkably successful.

I traveled to Bangladesh in 1998 to see one of these successes firsthand. Torrential rains drench Bangladesh during the monsoon months, but the country receives very little precipitation the rest of the year. Many fields lie fallow during the dry season, even though groundwater lies less than 20 feet below the surface. Over the past 17 years a foot-operated device called a treadle pump has transformed much of this land into productive, year-round farms.

To an affluent Westerner, this pump resembles a StairMaster exercise machine and is operated in much the same way. The user pedals up and down on two long bamboo poles, or treadles, which in turn activate two steel cylinders. Suction pulls groundwater into the cylinders and then dispenses it into a channel in the field. Families I spoke with said they often treadled four to six hours a day to irrigate their rice paddies and vegetable plots. But the hard work paid off: not only were they no longer hungry during the dry season, but they had surplus vegetables to take to market.

Costing less than $35, the treadle pump has increased the average net income for these farmers—which is often as little as a dollar a day—by $100 a year. To date, Bangladeshi farmers have purchased some 1.2 million treadle pumps, raising the productivity of more than 600,000 acres of farmland. Manufactured and marketed locally, the pumps are injecting at least an additional $350 million a year into the Bangladeshi economy.

In other impoverished and water-scarce regions, poor farmers are reaping the benefits of newly designed low-cost drip and sprinkler systems. Beginning with a $5 bucket kit for home gardens, a spectrum of drip systems keyed to different income levels and farm sizes is now enabling farmers with limited access to water to irrigate their land efficiently. In 1998 I spoke with farmers in the lower Himalayas of northern India, where crops are grown on terraces and irrigated with a scarce communal water supply. They expected to double their planted area with the increased efficiency brought about by affordable drip systems.

Bringing these low-cost irrigation technologies into more widespread use requires the creation of local, private-sector supply chains—including manufacturers, retailers and installers—as well as special innovations in marketing. The treadle pump has succeeded in Bangladesh in part because local businesses manufactured and sold the product and marketing specialists reached out to poor farmers with creative techniques, including an open-air movie and village demonstrations. The challenge is great, but so is the potential payoff. Paul Polak, a pioneer in the field of low-cost irrigation and president of International Development Enterprises in Lakewood, Colo., believes a realistic goal for the next 15 years is to reduce the hunger and poverty of 150 million of the world's poorest rural people through the spread of affordable small-farm irrigation techniques. Such an accomplishment would boost net income among the rural poor by an estimated $3 billion a year.

Over the next quarter of a century the number of people living in water-stressed countries will climb from 500 million to three billion. New technologies can help farmers around the world supply food for the growing population while simultaneously protecting rivers, lakes and aquifers. But broader societal changes—including slower population growth and reduced consumption—will also be necessary. Beginning with Sumeria, history warns against complacency when it comes to our agricultural foundation. With so many threats to the sustainability and productivity of our modern irrigation base now evident, it is a lesson worth heeding.

Further Information

SALT AND SILT IN ANCIENT MESOPOTAMIAN AGRICULTURE. Thorkild Jacobsen and Robert M. Adams in *Science*, Vol. 128, pages 1251–1258; November 21, 1958.

PILLAR OF SAND: CAN THE IRRIGATION MIRACLE LAST? Sandra Postel. W. W. Norton, 1999.

GROUNDWATER IN RURAL DEVELOPMENT. Stephen Foster et al. Technical paper No. 463. World Bank, Washington, D.C., 2000.

Irrigation and land-use databases are maintained by the United Nations Food and Agriculture Organization at http://apps.fao.org

SANDRA POSTEL directs the Global Water Policy Project in Amherst, Mass., and is a visiting senior lecturer in environmental studies at Mount Holyoke College. She is also a senior fellow of the Worldwatch Institute, where she served as vice president for research from 1988 to 1994.

UNIT 3

Energy: Present and Future Problems

Unit Selections

11. **Energy: A Brighter Future?** The Economist
12. **Oil, Profit$, and the Question of Alternative Energy**, Richard Rosentreter
13. **Here Comes the Sun: Whatever Happened to Solar Energy?** Eric Weltman
14. **Going to Work for Wind Power**, Michael Renner
15. **Power Struggle: California's Engineered Energy Crisis and the Potential of Public Power**, Harvey Wasserman

Key Points to Consider

• What are some of the differences between energy conservation produced by government regulation and that arising from market forces? Are there elements in the energy market that make it a particularly useful mechanism for directing energy conservation?

• What is the relationship between the economic and political power of large energy corporations and the automotive industry and the weakening trend toward research and development of alternative energy sources? Are there situations in which it can be to the advantage of multinational corporations to encourage the use of solar power and other alternative energies?

• What are some of the major benefits of such alternate energy sources as solar power and wind power? Do these energy alternatives really have a chance at competing with fossil fuels for a share of the global energy market? Defend your answer.

• In the context of the California energy crisis of 2000, assess the debate between those who claim energy shortages produced high energy costs and those who claim that the energy shortages were created to produce artificially high costs. Does the nature of the debate suggest something about the future ownership and control of energy resources and delivery systems? Explain.

 **Links: www.dushkin.com/online/**
These sites are annotated in the World Wide Web pages.

Alternative Energy Institute, Inc.
http://www.altenergy.org

Communications for a Sustainable Future
http://csf.colorado.edu

Energy and the Environment: Resources for a Networked World
http://zebu.uoregon.edu/energy.html

Institute for Global Communication/EcoNet
http://www.igc.org/igc/gateway/

U.S. Department of Energy
http://www.energy.gov

There has been a tendency, particularly in the developed nations of the world, to view the present high standard of living as exclusively the benefit of a high-technology society. In the "techno-optimism" of the post–World War II years, prominent scientists described the technical-industrial civilization of the future as being limited only by a lack of enough trained engineers and scientists to build and maintain it. This euphoria reached its climax in July 1969 when American astronauts walked upon the surface of the Moon, an accomplishment brought about solely by American technology–or so it was supposed. It cannot be denied that technology has been important in raising standards of living and enabling Moon landings, but how much of the growth in living standards and how many outstanding and dramatic feats of space exploration have been the result of technology alone? The answer is few–for in many of humankind's recent successes, the contributions of technology to growth have been just as important as the availability of incredibly cheap energy resources, particularly petroleum, natural gas, and coal.

As the world's supply of inexpensive fossil fuels dwindles and becomes more important as a factor in international conflict, it becomes increasingly clear that the energy dilemma is the most serious economic and environmental threat facing the Western world and its high standard of living. With the exception of the specter of global climate change, the scarcity and cost of conventional (fossil fuel) energy is probably the most serious threat to economic growth and stability in the rest of the world as well. The economic dimensions of the energy problem are rooted in the instabilities of monetary systems produced by and dependent upon inexpensive energy. The environmental dimensions of the problem are even more complex, ranging from the hazards posed by the development of such alternative sources as nuclear power to the inability of developing world farmers to purchase necessary fertilizer produced from petroleum, which has suddenly become very costly, and to the enhanced greenhouse effect created by fossil fuel consumption. The only answers to the problems of dwindling and geographically vulnerable, inexpensive energy supplies are conservation and sustainable energy technology. Both require a massive readjustment of thinking, away from the exuberant notion that technology can solve any problem. The difficulty with conservation, of course, is a philosophical one that grows out of the still-prevailing optimism about high technology. Conservation is not as exciting as putting a man on the Moon. Its tactical applications—caulking windows and insulating attics—are dog-paddle technologies to people accustomed to the crawl stroke. Does a solution to this problem entail the technological fixes of which many are so enamored? Probably not, as it appears that the accelerating energy demands of the world's developing nations will most likely be first met by increased reliance on traditional (and still relatively cheap) fossil fuels. Although there is a need to reduce this reliance, there are few ready alternatives available to the poorer, developing countries. It would appear that conservation is the only option.

But market forces operate in intriguing patterns and, perhaps, the picture for the immediate energy future is not as bleak as we might have imagined just a few years ago. In the first article in this section, the editors of *The Economist*, in a major survey of energy, note that there may be "A Brighter Future?" for energy conservation and, hence, for energy supply. While conservation mechanisms have been driven by government intervention (tax breaks, etc.) since the 1970s, it has become increasingly apparent that there are powerful market forces at work in energy conservation as well. Indeed, one of the authorities cited in the article suggests that the best thing that governments can do now in terms of conservation is "get out of the way" of those market forces. But conservation is probably not enough to solve the environmental problems related to fossil fuel use. In "Oil, Profit$, and the Question of Alternative Energy," columnist Richard Rosentreter notes that, even with the recent increases in oil and gas prices, the available alternative energy sources such as solar energy have not become a focal point for development. Rosentreter suggests that since the money and power of "Big Oil" is devoted to fossil fuel exploration and development, little is left over for alternative energy technologies, which are viewed as less profitable. The question of solar energy is also addressed in the third article in the unit. Environmental writer Eric Weltman, in "Here Comes the Sun: Whatever Happened to Solar Energy?" asks much the same question as did Rosentreter. Rather than pointing to economic answers, however, Weltman concludes that the explanation for the decline in interest and development in solar energy is political. The real issue, he notes, is the unfavorable political environment for renewables.

The concluding two articles in the section discuss situations in which political relevance and economic incentives for the development of alternative energy are present. In "Going to Work for Wind Power," Michael Renner, a senior researcher at the Worldwatch Institute, describes how newer technologies in the manufacture and use of wind-driven turbine electrical generators have begun to solve problems of low efficiency in wind-generated electricity and, as an added and perhaps unexpected bonus, have begun to employ significant numbers of workers as well. Renner points out that the new generation of windmills is "not your grandfather's windmill," and that their increasing use will free up money formerly invested in fossil fuels that can now be applied to alternative energy developments. In the final article in this section on energy, author Harvey Wasserman (The Last Energy War) takes on the difficult task of analyzing the recent energy crisis in California and what it means for the future of electricity as a public utility. While the conventional explanation for the supposed shortages that created "rolling blackouts" during California's summer of 2000 was that demand for electricity exceeded supply, Wasserman contends that the actual cause was a withholding of supply to create shortages that could, in turn, be used as an excuse for price/rate hikes. While industry experts will dispute Wasserman's contention, the debate does raise an intriguing question: Should energy be viewed as a public or private commodity? Answers to energy questions and issues are as diverse as the world's geography. But all the answers require a reorientation of thought and the action of committed groups of people who have the capacity to change the dominant direction of a culture.

ENERGY:
A brighter future?

The world of energy is being turned upside down. The best thing governments can do is to get out of the way, says **Vijay Vaitheeswaran**

FOR most of the past two decades, the world has enjoyed exceptionally low and stable energy prices, but for the past couple of years or so, world oil markets have been on an unnerving roller-coaster ride: prices collapsed to around $10 a barrel two years ago, and soared to a ten-year high of over $35 last year. It was those peaks that set off a political crisis over petrol prices and shortages in America's mid-western states last summer, and that provoked the fuel riots which paralysed several European countries in September. Now oil prices are lower, but remain volatile.

Controversy over the environmental impact of fossil-fuel use has added an extra layer of complication. Last November a ministerial summit on the Kyoto Protocol, a UN treaty among industrial countries to curb global warming, broke up in rancorous disarray. Just a few weeks earlier, the California Air Resources Board had delighted greens and outraged car manufacturers by unanimously upholding its controversial "zero-emissions" mandate, which requires 10% of the new cars sold in the state by big manufacturers to meet the state's definition of zero emissions by 2004. Greenery surfaced as a big factor in the European fuel protests too, this time at the opposite end: rather than blaming the price-fixers at OPEC, most rioters—especially in Britain—attacked their own governments for levying hefty fuel taxes in the name of protecting the environment.

Most recently, attention has focused on the turmoil in the gas and power markets. Thanks to under-investment in gas production in recent years, low stocks and soaring demand, gas prices have skyrocketed in the past year. A number of energy gurus think the natural-gas market may be facing a decade-long problem.

Such experts usually point to California, where utilities have been brought to the verge of bankruptcy by a botched deregulation of the power industry (of which more later). This has left the firms exposed to spot prices for electricity fired by natural gas which have been as much as ten times as high as a year ago. The resulting debts, and the utilities' attempts to recover them from unwilling ratepayers, have caused a political crisis in the country's biggest and richest state and raised fears of a possible recession there.

But despite the recent volatility in energy markets, comparisons with the oil shocks of recent decades are vastly overblown. For one thing, the causes of recent energy crises are quite different from those that produced the oil shocks of the 1970s and the lesser upsets during the Gulf war a decade ago. Today's woes come at a time of peace in energy-producing regions. OPEC has been working with oil-consuming nations to try to stabilise energy prices. The spikes in natural-gas prices are causing short-term pain, but the price signals they send are already encouraging the development of more gas fields.

No need to panic

Also, in a reversal of the conventional wisdom of two decades ago, most experts now believe that oil and especially natural gas will remain plentiful for decades hence, and that the means of converting those fuels into useful energy, such as internal combustion engines and combined-cycle gas turbines, will grow ever more efficient. What is more, the world has become much less vulnerable to oil shocks: thanks to conservation, fuel switching and improvements in efficiency, oil's share of industrial countries' imports, and their economies' reliance on it, has shrunk significantly.

Three powerful factors are now combining to shape the future of the energy industry: market forces, greenery and technological innovation. None of them is new, but together they are exerting strong pressure for change. Yet the industry's incumbents tend to resist change because they have much to lose from it; and given the sector's enormous and long-lived stock of fixed assets, a turnaround is bound to take time. And, confusingly, some of the forces for change pull in opposite directions: rising environmental standards may favour renewable energy, for example, but market reforms may choke off subsidies for it at the same time.

The [*Economist's*] survey will argue that energy is indeed on the cusp of dramatic change. The sections [that are in the*Economist's* Survey: Energy, *The Economist*, February 10, 2001], show how the cross-currents at work today are reshaping the en-

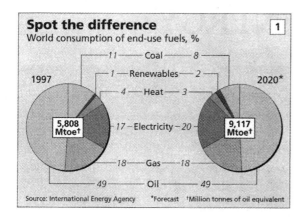

Spot the difference
World consumption of end-use fuels, %

1997   5,808 Mtoe†
2020*   9,117 Mtoe†

	1997	2020*
Coal	11	8
Renewables	1	2
Heat	4	3
Electricity	17	20
Gas	18	18
Oil	49	49

Source: International Energy Agency   *Forecast   †Million tonnes of oil equivalent

ergy world, from the liberalisation of power markets to the greening of the world's oil giants and the advance of disruptive innovations such as fuel cells and distributed power. These changes will force energy companies to think hard about what they are truly good at, where tomorrow's competitive threats may come from, and what their customers really want. They will also help bring modern energy to poor countries that so far have been left out. Ultimately, they may even propel the world towards a cleaner and more sustainable technology: hydrogen energy.

Hand over to the market

However, whether the world realises the full potential of these prospects depends crucially on one factor: government. The invisible hand may be ascendant, but that does not mean the visible one has become irrelevant. On the contrary, during the transition to liberalised energy markets the role of regulators and officials is vitally important. And as California's sad example shows, governments can make a big difference by getting it wrong.

All three main forces of change closely involve government. In pushing for deregulation, regulators must be willing to trust market forces: they must make the rules of the game clear and refrain from arbitrary interference during the transition, yet remain on the look-out for market abuses. And ultimately they must yield most of their powers to the market.

In advocating greenery to meet their citizens' rising expectations, governments must be careful to avoid the distorting effects of such measures as excessive petrol taxes and flip-flopping environmental standards. There is a good case for some government support for renewable energy and other alternatives to fossil fuels as an insurance policy against the possibility of distant hazards such as global warming and oil depletion. However, the final test for all such technologies must remain the marketplace.

When it comes to clean technology, the most effective boost that bureaucrats can give to a sustainable energy future is to avoid picking winners. Instead, they would do better to provide a level playing field by scrapping the huge and usually hidden subsidies for fossil fuels, and by introducing measures such as carbon taxes so that the price of fossil fuels reflects the costs they impose on the environment and human health. Governments should also ensure that incumbents do not obstruct the entry of nimble newcomers, and keep open a range of options for producing energy, including running existing nuclear plants to the end of their useful life. They should provide strong incentives for firms to invest in today's creaking electricity grids, but also remove barriers to the spread of distributed generation.

Lastly, the governments of the rich world should do much more to help the poorer part meet its energy needs by leapfrogging to clean technologies. Most of the growth in both energy demand and in emissions will soon come from developing countries. If they invest in yesterday's dirty and inefficient technologies, they will be locked into them for decades to come—and the whole world will suffer the consequences. Lack of energy, especially modern fuels, in the developing world is likely to depress the productivity of billions of its workers, and so hold back future global economic growth.

Taken together, these prescriptions suggest that successful reform will be a tricky balancing act. But without it, the future would look much dimmer.

From *The Economist*, February 10, 2001, Survey 5-6. © 2001 by The Economist, Ltd. Distributed by the New York Times Special Features. Reprinted by permission.

Oil, Profit$, and the Question of Alternative Energy

As oil and gas prices continue to rise, the sun has apparently set on the development of solar power and other forms of alternative energy, despite official claims that the United States is committed to making them a success. The explosion in oil and gas prices has been attributed to numerous causes, but little attention has been given to the lackadaisical effort to develop alternative fuel sources and the continuous quest by the oil industry to discover more oil. Big oil has both money and power, and it shouldn't be any surprise how much can be accomplished, or prevented, with such a potent combination.

by Richard Rosentreter

The application of solar power is not a new idea. The ancient Greeks and Romans developed mirrors that would direct the sun's rays and cause a target to burst into flames within seconds. Nearly two centuries ago in 1839, Edmund Becqurel, a French experimental physicist, discovered that sunlight could produce electricity—almost fifty years before the first successful internal combustion engine was built.

During the late 1800s, harnessing the sun's rays to produce hot water was a booming business in the United States. Although the Industrial Revolution was in high gear and remarkable discoveries and inventions abounded, it took over 100 years for the first photovoltaic cell (a cell capable of producing wattage when exposed to radiant energy) to be developed by Bell Laboratories in 1954. Considering that photovoltaic cells have been the exclusive power source for satellites since the 1960s, and how rapidly television evolved during an era known as the Atomic Age, it is a wonder that solar technology hasn't advanced further.

The utilization of solar energy was briefly resurrected during the 1970s when the United States appeared to be committed to pursuing a technology that had the potential to reduce our dependency on fossil fuels. In April 1977, in the midst of an "energy crisis," President Jimmy Carter began a bold initiative to develop solar energy and other alternative fuels when he unveiled his National Energy Plan, which included setting an example by placing solar panels on the White House. Carter announced a "national goal of achieving 20 percent of the nation's energy from the sun and other renewable resources by the year 2000," and he introduced legislation that would provide homeowners with tax breaks for investing in this promising technology. In 1979, Congress followed Carter's lead and approved a $20 billion development fund for synthetic and alternative fuels. It appeared that the alternative energy industry was finally getting the financial backing it needed to have a profound impact on the nation's energy needs.

During the period in which financial support for solar energy was growing and a "windfall tax" on the profits of the oil industry was imposed, the proponents of big oil were gathering their own resources on Capitol Hill. Political action committees (PACs) that were affiliated with oil and gas interests began to sprout and, from 1977 to 1979, they contributed over $2.6 million to House and Senate candidates. A report by Alan Berlow and Laura Weiss in *Congressional Quarterly* concluded that most of the money went to candidates "with strong pro-industry voting." Support for alternative energy took a downward spiral when Ronald Reagan (a former spokesperson for General Electric) was elected U.S. president and became a staunch ally of corporate America.

By the late 1970s, oil companies had bought out many of the patents for photovoltaic cells, and corporate giants like Atlantic Richfield, Amoco, Exxon, and Mobil took control of solar power companies. This trend would lead Alfred Dougherty, former director of the Federal Trade Commission's bureau of competition to warn, "If the oil companies control substantial amounts of substitute fuels… They may slow the pace of production of alternative fuels in order to protect the value of their oil and gas reserves." Edwin Rothschild, a spokesperson for the Citizen Energy Labor Coalition, was concerned that the big oil companies "see solar power as a competing source of energy, and they want to control it and slow it down." However, ownership of solar technology by big oil was only the first step in the methodical dismantling of the alternative energy renaissance.

The Reagan administration would continue the squeeze on alternative energy as tax credits for residential investment in solar and wind power became "obsolete," as it was deemed to be "the responsibility of the private sector to develop and introduce new solar technologies." The $684 million requested by Carter for alternative energy in fiscal 1982 was slashed to $83 million in Reagan's 1983 proposal. What was transpiring in the realm of solar technology development didn't go unnoticed by the science community. In the November 1981 *New Scientist* article "Big Oil Reaches for the Sun," Ralph Flood reports that the "energy policy under the Reagan administration seems designed to accelerate the oil companies' control." The solar panels on the White House were discarded.

The lack of support for the development of alternative energy continued despite the findings of a study released in 1980 by the National Academy of Sciences at the request of the Energy Research and Development Administration. The study concluded that "a costly push by the government would lead to rapid growth in solar-related energy sources" and, if this occurred, "renewable energy sources could account for a quarter or more of the nation's energy needs within thirty years." Instead, the Reagan administration preferred to rapidly expand military aid and sent billions of dollars to foreign countries. It also greatly increased the federal budget for the research and production of atomic weapons.

Despite its promise, solar power faced several hardships and questions. Many fly-by-night companies popped up hoping to jump on the bandwagon and make a quick profit, which led to shoddy construction and customer dissatisfaction. The solar industry during this early stage lacked structure, and there was enormous confusion as industry representatives scrambled to establish guidelines for advertising claims, technical specifications, and warranty standards, which contributed to a higher cost.

Regardless of the problems the solar industry faced, greater financial support and stricter guidelines by the

government could have offset many of the obstacles. There were some who argued that solar energy could only be harnessed during bright sunshine, which added to the skepticism surrounding it. However, in 1982 Charles Wurmfeld, former president of the New York State Professional Engineers Society, invented and patented an engine that drew energy from the ordinary atmosphere, not requiring bright sunshine. He believed his invention would save one-third of the oil burned in this country.

Two decades later, we are once again in the midst of an energy-related controversy. The dramatic rise in oil prices has been blamed on the Organization of Petroleum Exporting Countries (OPEC), pipeline problems, tougher standards imposed by the Environmental Protection Agency (EPA), and even consumer choice in vehicles. However, just as they did in 1979, oil companies are reporting large profit increases during this recent "energy crisis." Chevron Corporation announced that it earned a record $1.1 billion as surging oil and gas prices boosted first-quarter profits in 2000 nearly fourfold. BP Amoco said its profits more than doubled because of rising oil prices and cost-cutting measures. Exxon Mobil, Texaco, and Conoco also reported between two and four times more profits than in 1999.

During a House hearing in June, oil company profits were updated: Texaco profits were up 473 percent; Conoco, 371 percent; BP Amoco, 290 percent; and Chevron, 271 percent. The list goes on, and it is predicted that the oil business will continue to be lucrative. According to Sam Fletcher in the February 21 *Oil and Gas Journal*, the rise in oil prices is "fueling a recovery among producers… that is expected to propel the oil and gas industry through 2000 and beyond."

Analysts across the country agree with this prediction. According to Deutsche Banc Alex Brown, a major investment banking firm in Baltimore, Maryland, there is a "bullish outlook for the commodity." David Garcia, a financial analyst, says, "These are the best industry conditions for at least fifteen years." And Fadel Gheit of Fahnestock and Company says, "The industry has never looked better."

Motorists paying high prices at the pump would beg to differ. According to Steve Leisman in his April 26 *Wall Street Journal* article, "High Prices Give Oil Companies a First-Quarter Boost," analysts describe the stocks of several big oil companies as "increasingly attractive." This attractiveness relies on profitability, which relies on finding more oil reserves. According to a June 1 article in the *Houston Chronicle*, shareholders of Exxon Mobil recently rejected overwhelmingly proposals to invest in renewable energy and study potential harm from oil drilling in Alaska.

Running on Empty

Phillip M. Morse

With the flap this summer over gas prices in the United States, I feel impelled to ask why anyone is actually surprised. Petroleum is a finite resource and hence, before it runs out, it's only natural that the price should go up. Current estimates, based upon today's rate of consumption, indicate that cheap oil will be gone within fifty years. Therefore, as that time approaches, local conditions and price gouging can be expected to cause painful spikes in what U.S. consumers pay at the pump.

We must adjust to reality by reducing demand and converting to alternative sources. Higher oil prices today are a wakeup call to the fact that infinite growth is impossible and population growth is approaching the limits of what the environment can acceptably support.

The standard of living of individual societies results from conversion of available natural resources into useful products through technology. These natural resources have limits that can be temporarily exceeded by importing them from regions with a surplus. But when such reserves have also been consumed, only the development of new technology to utilize renewable resources will preserve or improve a traditional standard of living.

Unfortunately for the people with underdeveloped economies, resources that are necessary to raise their living standard are being consumed by the developed economies. For the people living in the underdeveloped economies to raise their standard of living now, new technological advances are required on a magnitude comparable to those that made fools of earlier predictors.

World population was about one billion in 1800 when Thomas Malthus predicted that starvation would halt population growth. He was wrong because he didn't anticipate the technological advancements of the Industrial Revolution, which raised Western living standards and increased life expectancy. World population was about four billion in 1960 when Paul Ehrlich wrote his book *The Population Bomb*, in which he also predicted that starvation would stop population growth. He, too, was wrong because he didn't anticipate the technology that became known as the green revolution.

But both Malthus and Ehrlich were right in terms of the underdeveloped economies, which are still very nearly in the state they envisioned. Technology gave Western societies the two revolutions that developed their economies, raised their middle-class standards of living, and reduced their death rates, resulting in rapid population growth. World population has now passed six billion and is expected to peak at ten billion within the next 100 years. But there is a tradeoff between population size and the living standard. The higher the standard of living, the less people the environment can sustain with its finite resources. Some estimate that two billion is the maximum world population sustainable with an acceptable standard of living. At two billion, poverty and starvation might be ended throughout the world. But at over ten billion, virtually all will be at a starvation level of existence.

In the United States, we behave as though there are no limits. An automobile driver who runs out of gas in the middle of nowhere is thought to be irresponsible, but such common sense isn't applied to our civilization. Politicians like to give the impression that they can influence the price of fuel by government action. But in the long run it is a supply-and-demand problem, and when the demand exceeds supply the price defies controls. Rationing will occur either through higher prices or long lines at gas stations.

At present, alternatives to oil aren't cost-effective, but rising oil prices will eventually resolve that problem. We are entering a time of transition to the post-oil era. There is a whole new growth industry waiting in the wings based upon energy alternatives stimulated by higher oil prices. For example, hydrogen fuel cells and photovoltaic cells have yet to be applied to their full potential.

These alternative energy sources and more must be applied now by new technology to transportation, heating, food production, plastic products, and related industries before the price of petroleum becomes prohibitive.

Change is coming whether we like it or not, forced on us by the natural increase in the price of petroleum. But if the producing countries are persuaded to temporarily increase their production to keep prices low, they will run out sooner rather than later. If the price increase is gradual, the transition could be smooth; but if it is manipulated, when change does come it could be catastrophic for Western societies. And while it might have less effect on the underdeveloped economies, it will make it harder for them to catch up with the developed world.

Such fundamental changes should take many years, but we have less than fifty to complete the transition. The price of petroleum will probably go ballistic over the next ten years—and this seems to have already begun. We must learn how to function on the energy we receive daily from the sun. We must allow and encourage our population to shrink back to a sustainable size while we preserve a tolerable standard of living and the freedom of individual choice.

As nations become increasingly dependent upon oil controlled by unsympathetic suppliers, national security crises are imminent—perhaps even wars will be fought to control the remaining supply. The Gulf War may be a preview of things to come. Saddam Hussein's goal was to control the world's largest supply of oil. He was stopped in time. But the next Gulf War may not be resolved so easily.

At some point an epiphany will occur, and the sooner it does the better. But as Mark Twain said, "People refuse to see the handwriting on the wall until their backs are against it."

Philip M. Morse can be reached by e-mail at pmmorse@worldpath.net

While the development of alternative energy sources continues to lag, supporters of the oil industry continue to promote the use of fossil fuels. During a recent House hearing on high gas prices, representatives from the oil industry argued that a possible solution would be to begin drilling in environmentally sensitive regions, such as the Arctic and Rocky Mountains. According to Red Cavaney, president and chief executive officer of the American Petroleum Institute (API), the nation's energy woes could be resolved if the oil companies were allowed to drill in areas that have been safeguarded by environmental protection legislation.

Already there is congressional support to begin such action. Senate Energy and Natural Resources Chair Frank Murkowski (Republican–Alaska) proposed legislation that would allow oil companies to drill in the Arctic National Wildlife Preserve. A Senate budget measure has already projected $1.2 billion in royalties from the Alaska refuge in 2005, and it recently voted in favor of opening the region to commercial drilling.

Currently, American oil corporations are sinking millions of dollars into exploration and gaining access to large oil deposits in the Caspian Sea region. According to Jeanne Whalen in the March 4, 2000, *Wall Street Journal*, the Caspian holds as much as 2.2 billion barrels, and Michael Davis reports in the April 22 *Houston Chronicle* that Conoco and Exxon Mobil have received the green light to proceed with an estimated $5 billion oil development project in that region and will pay $75 million for the right to develop in the Azerbaijan fields of the Caspian. Secretary of Energy Bill Richardson, who recently addressed the House of Representatives, summed up the rationale for the movement toward searching for oil when he said that the "world's thirst for oil is steadily rising" and "demand will continue to grow."

Imagine if the money spent by the U.S. government on foreign aid and by oil corporations on fossil fuel exploration were invested in the development of alternative fuels. Apparently loyalty to fossil fuels is too deep. According to a recent analysis by the Congressional Research Service, reported in the March/April issue of *Mother Jones*, seventy-seven cents of every energy research dollar from 1973 to 1997 went to nuclear and fossil fuels; just fourteen cents went to alternative energy.

The adverse effects that burning fossil fuels have had on our environment and their contribution to global warming are well documented. Yet, according to Exxon Mobil's 1999 annual report, the company acknowledged the public's concern over "climate changes" due to the use of fossil fuels but said that the projected serious effects "rely on speculative assumptions and results from unproven models." Exxon Mobil doesn't believe that "the current scientific understanding justifies mandatory restrictions on the use of fossil fuels" as it is certain "that significant economic harm would result from restricting fuel availability to consumers." The fact is that the company would lose profits—just as any other oil company

would if alternative energy sources were developed. Oil companies are searching for more oil reserves, not alternative sources for fossil fuels. More oil means more profit.

While there is no concrete evidence that proves oil companies purposely sabotaged the progress of alternative energy, there are clues that point to political favoritism on their behalf. The U.S. Senate recently passed an amendment by Senator Kay Bailey Hutchinson (Republican–Texas) that saves oil companies from paying $66 million a year in royalties to the government. Opposing the amendment was Senator Russ Feingold (Democrat–Wisconsin), who said, "I am very concerned that Congress is abdicating its responsibility." Feingold cited soft money political contributions from oil giants—including Exxon, Chevron, BP Oil, and Amoco—that totaled over $2 million during a two-year span. In all likelihood, the oil industry will use the latest "oil crisis" as leverage to promote continued legislation in its favor.

As for future legislation, the current poll leader for the presidency, Republican George W. Bush—a former oil company executive who has amassed "substantial financial contributions" from the industry—has also been criticized for being a "tool of oil interests." It is highly unlikely that alternative fuels have a chance for further development if Bush—who, according to Joby Warrick in the April 4 *Washington Post*, "supports oil exploration in the Arctic National Wildlife Preserve while opposing the United Nation's 1997 Kyoto protocol that requires industrialized countries to cut emissions of greenhouse gases blamed for global warming"—lands in the White House in 2001.

Nearly two decades ago, a group of students from Crowder College in Neosho, Montana, built a car at the cost of about $5,000 that traveled across the continental United States powered only by the sun. Just last year, students from the University of Oklahoma built a solar-powered vehicle that won a biennial intercollegiate competition which provides them an opportunity to "design, build, and race solar-powered cars." Modern science technology has given humanity the ability to access hundreds of channels on cable television, develop computers to communicate on a global scale, clone animals, and produce state-of-the-art weapons of mass destruction. Imagine if the inventors and scientists of the world focused their minds and energy on developing alternative energy sources for the public good.

Time magazine recently published a special issue entitled "The Future of Technology." In a section depicting the future of the automobile, surprisingly, there was no mention of cars powered by solar technology or any other alternative fuels. Could it be that the fate of alternative energy has been sealed by an industry that would crumble if it were to face competition from other sources?

At the end of the movie *Back to the Future*, a brilliant scientist called "Doc" refueled a DeLorean with a handful of household trash. It is a fantastic concept that falls into the same category as solar energy because it is nonprofitable. According to the July 7 *Fortune 500*, Exxon Mobil, Ford Motors, and General Motors are some of the top profit-making corporations in the United States, and they wield a great amount of economic and political clout. Although there are alternative energy sources and related technologies available for development to meet our growing energy needs, there is currently not enough profit in them to be an attractive alternative for corporations. Perhaps renewable energy, too, is destined to become fossilized.

Richard Rosentreter holds an associate's degree in journalism from Suffolk County Community College in New York, where he has been cited for his investigative journalism, and a bachelor's degree in English from Southampton College at Long Island University, New York, where he regularly writes a column on social and political issues. He can be reached by e-mail at richinsight@hotmail.com.

Here Comes the Sun

Whatever happened to solar energy?

By Eric Weltman

Looking back, the worst of the '70s—polyester, Nixon, disco—is remembered, even celebrated. But it's forgotten that in the same decade, amidst the oil shocks and nuclear debacles, the future seemed to belong to solar energy. Indeed, with gas-guzzling SUVs clogging our roads, it's difficult to remember the urgency of that time.

There even once was a day called Sun Day. The idea, recalls organizer Denis Hayes, was "to convey to the American public that there are options, that it is possible to run a modern industrial state on sunshine." On May 3, 1978, Sun Day began with a sunrise ceremony at the United Nations led by Ambassador Andrew Young and continued with hundreds of events across the country. President Carter used the occasion to announce an additional $100 million in federal solar spending and the installation of a solar water heater on the White House roof. The White House Council on Environmental Quality ambitiously declared, "A national goal of providing significantly more than half of our energy by solar sources by the year 2020 should be achievable."

But then the '80s happened. With the election of Ronald Reagan, solar energy entered a dark age of malign neglect. Reagan eliminated tax credits for solar energy and removed the solar panels from the White House roof. Federal research-and-development funding for solar power fell from $557 million in 1980 to $81 million in 1990. At the same time, oil prices plummeted, diminishing demand for alternatives and taking energy off the agenda of the nation and much of the environmental movement. "If oil had remained expensive," Hayes says, "everything would have fallen into place."

Consequently, things now look a lot different than the sunny optimists of the '70s predicted. Consumers pay more for a gallon of bottled water than they do for a gallon of gas, while, at $20 a ton, coal is cheaper than topsoil. The universe of renewable energy sources—including solar, wind and geothermal power (but not hydroelectric)—provides only 2.1 percent of the nation's electricity. The future isn't much brighter: Absent any new policies, according to federal projections, by 2020 renewable energy is expected to provide just 3 percent of the nation's electricity.

Of course, the problems of fossil fuels—toxic spills, mining waste, acid rain, smog, etc.—haven't gone away. Meanwhile, a new problem has emerged: global climate change, with its multiple threats of rising sea levels, disrupting agriculture, increasing weather-related disasters and spreading infectious diseases. The scientific consensus is that climate change is happening, and its chief source is carbon dioxide released by the combustion of fossil fuels. The United States accounts for about a quarter of the world's energy consumption, so it's no surprise that this country also is responsible for 24 percent of carbon dioxide emissions—the largest source of which is power plants.

In the past, the world would have turned to the United States for renewable energy solutions. After all, the United States invented photovoltaic (PV) panels, devices that turn sunlight into electricity, and, in the '80s, California produced more than 90 percent of the world's wind energy. But the torch has been seized by Europe and Japan, which support renewable energy with a range of tax benefits, mandates and pricing programs. In fact, the European Union prohibits subsidies for fuels other than renewables. Meanwhile, according to the Worldwatch Institute, wind power is the most rapidly growing source of energy in the world, increasing 20 percent per year since 1990. The Danes have captured half of the market for wind technologies.

But the potential for renewables remains great in the United States. The Solar Energy Industries Association (SEIA) claims that PV panels covering 0.3 percent of the country, a quarter of the land occupied by railroads, could provide all of the nation's electricity. Likewise, the 11 states stretching between North Dakota and Texas have been dubbed the "Saudi Arabia of wind energy," with enough gusts to supply more than the nation's electricity consumption.

The American public continues to support renewable energy in survey after survey. Yet renewables face the continued obstacle of the political power of the utility, nuclear and fossil fuels industries. That clout has translated, among other things, into billions of dollars in subsidies in the form of federal research and development, tax benefits and ratepayer bailouts. However, there are several reasons that renewables may finally earn their day in the sun. One is the rising concern over the global climate change, which has spurred the interest of environmental organizations and the foundations that fund them. Another is the restructuring of the electric utility industry.

A growing number of states are dismantling the monopolies that have controlled the entire process of electricity generation and distribution, and Congress is contemplating national restructuring legislation. Large industrial customers tout the lower energy prices they say will result from opening up the market to competition. Environmentalists see opportunities, perils and problems in the rush to deregulate.

Deregulation has allowed some consumers to choose "green energy." Since electricity from all sources is mixed up in the utility grid, no one can guarantee that a particular home will receive "green electrons." However, consumers can sign up to pay their bills to companies that generate their electricity from renewable sources. In California, the municipal governments of San Diego, Santa Monica and San Jose, as well as the Los Angeles Dodgers, have opted for green power.

The vision of true energy independence—households generating their own energy—has been advanced by developments in PV "solar roofing shingles." A top item on the solar industry's wish list is federal legislation requiring "net metering," which would allow solar-powered homes to cut their bills by sending extra electricity to utilities and running their meters backward. Meanwhile, David Morris of the Institute for Local Self-Reliance advocates more research and development in batteries that could store excess "home-grown" energy, which could "potentially make obsolete a trillion dollars in transmission and distribution lines."

But environmentalists are quick to point out the limits of green consumerism. First, utilities are insisting that ratepayers bail them out for billions of dollars invested in nuclear boondoggles that would otherwise die a quick death in a competitive market. Furthermore, since California's markets opened up in 1998, only 1 percent of the state's consumers have chosen green energy—and that's with a subsidy scheduled to end in 2001. The most optimistic marketers expect that 20 percent of residential customers and 10 percent of commercial customers will choose green power. As Rob Sargent of the Massachusetts Public Interest Research Group (PIRG) notes, "The utilities would like nothing better than to use consumer choice as an argument against policies requiring renewable energy."

The utilities argue that in a competitive marketplace there should be no restrictions on giving customers what they want: cheap power. But environmentalists counter that the price of this power doesn't include the environmental costs born by society, including dirty air, dangerous wastes and climate change. Their concern that a focus on narrow, short-term costs could have dire social consequences has already been realized: utilities have slashed investments in energy efficiency by half since the mid-'90s.

Environmentalists have fought some tenacious state-by-state battles to incorporate green energy policies into utility restructuring, with mixed results. Fourteen states have established "public benefits trusts," which tax electricity use to fund renewable energy, energy efficiency and low-income energy programs. Eleven states require that a certain percentage of their electricity be generated by renewables, but these "Renewable Portfolio Standards" (RPS) are largely unambitious. For example, Arizona requires that solar energy supply just 1 percent of the state's power by 2002.

In Congress, environmentalists are supporting a bill introduced by Vermont Republican Sen. Jim Jeffords to add a shade of green to federal restructuring legislation. The bill would establish an RPS of 20 percent renewable energy use (excluding hydropower) by 2020, create a public benefits trust, require utilities to tell consumers how much pollution they produce and place a cap on emissions of carbon dioxide and other pollutants. It's a much more ambitious bill than the Clinton administration's, which would establish an RPS of only 7.5 percent by 2010. But the Jeffords bill doesn't have universal support among environmentalists, some of whom criticize its failure to prohibit nuclear bailouts and its allowance of emissions trading.

The biggest problem, though, is that it doesn't stand a chance of passing. The bill doesn't even have the full support of the 151-member House Renewable Energy Caucus or the newly formed 24-member Senate caucus. "If you took a vote today," says Ken Bossong of the Sun Day Campaign, "it would go down in flames."

The bill's poor prospects, Bossong says, stem from the lack of grassroots momentum behind it. Scott Denman of the Safe Energy Communication Council adds, "The movement needs to develop a political base and be much more politically aggressive."

Energy advocacy lost its populist edge in the '80s, Sargent says, when environmentalists and utilities began collaborating, chiefly to promote investments in energy efficiency. Getting a "seat at the table" was a positive thing, he says, but it encouraged environmentalists to forget that "our power is derived from the size of our constituency, not from our access."

Environmentalists trying to overcome this mistake face several key challenges. Today's movement lacks the built-in activist base of opponents to nuclear power that existed in the '70s, when more than 50 nuclear power plants were under construction. Now energy is so far off society's radar screen, most people don't even know where their power comes from. "Most people we talk to think their electricity comes from hydropower," says Andrea Kavanagh of the National Environmental Trust (NET).

The most ambitious effort to re-energize the movement is Earth Day 2000, chaired by Hayes, an original Earth Day and Sun Day organizer. The focus of Earth Day 2000 is energy and Hayes hopes that the month-long series of events will provide a spark missing from the issue. Lacking the immediate context of an energy crisis or the Three-Mile Island disaster, he says, "we kind of have to create the timeliness of the issue ourselves."

His Earth Day Network claims to have nearly 3,000 groups in 163 countries involved thus far and boasts a flagship event on the mall in Washington on April 22 featuring actor Leonardo DiCaprio. The Earth Day agenda—endorsed by about 500 organizations, including the Natural Resources Defense Council (NRDC) and U.S. PIRG—calls for quadrupling federal investments in renewable energy and efficiency in five years and halting subsidies for fossil fuels and nuclear power, with a goal of producing at least one-third of the nation's energy from renewables by 2020. In the "changed environment" after Earth Day 2000, Hayes hopes, there will [be] an opportunity for the environmental movement to achieve such goals.

Not everyone shares his optimism, however. Kalee Kreider of NET acknowledges the tremendous boost recycling received from Earth Day 1980's focus on solid waste, but says she's "not going to plan on a similar bounce for energy." Citing the lack of infrastructure to build on any momentum, Bossong predicts, "A lot of money will be spent, a modest amount of media coverage will be generated and probably nothing will happen."

That said, Bossong himself maintains a database of about 1,000 organizations across the country that have some level of involvement in clean energy issues, including several national outfits with extensive field operations. PIRG, with 27 state-based organizations and six U.S. PIRG field offices, has campaigns to clean up dirty power plants and promote clean energy. The Sierra Club, with 65 chapters, is focusing its energy program on transportation. Ozone Action has paired up with the International Council on Local Environmental Initiatives to help municipal officials take action against global warming.

The new kid on the block is NET, a group started by the Pew Charitable Trusts in 1994, which organized a "Pollution Solutions" bus tour of 36 cities this fall to demonstrate how people can consume less energy. "You have to build people toward political action," Kreider says. "Most people, as a first dipping of their toes in the energy issue, are not prepared to slam their senator or take on a multinational corporation."

In addition to national organizations, Bossong's database includes approximately 800 state and regional organizations, from the Northeast Sustainable Energy Association to the Northwest Energy Coalition. Local activists—whether they're campaigning to clean up dirty power plants in Massachusetts or to stop nuclear waste storage in Minnesota—are ready to be plugged into national clean energy initiatives. "These battles are creating a constituency for clean energy in a way that I've not seen in a long time," Sargent says.

Of course, some groups differ on how the movement should proceed. The NRDC's Ralph Cavanaugh cautions that the fossil fuel industry is "not monolithic," pointing to the key support of some utilities in winning a recent extension of wind energy tax credits and British Petroleum's ownership of one of the nation's largest PV manufacturers. While acknowledging that the coal mining industry has been "unsupportive," Cavanaugh notes that it "is a declining force both economically and politically." He says, "I don't find a lot of organized opposition to renewable energy in general."

But don't tell that to U.S. PIRG, Friends of the Earth or Taxpayers for Common Sense, who have been battling to reduce federal subsidies for fossil fuels. The fiscal year 2000 budget contains $1.5 million more for coal research and development than last year, a total of $124 million. Likewise, Rebecca Stanfield of U.S. PIRG maintains that the utility industry's opposition to a national renewables mandate has been "relentless." Sargent adds, "There are some in the environmental community who place too much faith in the goodwill and enlightenment of corporate leaders and are unwilling to point the finger at our enemies."

Another question is what place renewable energy has in the environmental movement's clean air agenda. Many environ-mentalists, some reluctantly, acknowledge that relying more on natural gas, which contains less carbon and other pollutants than coal and oil, is essential to combat global climate change. But a NET fact sheet—to the chagrin of other advocates—goes so far as to declare natural gas a "solution for today," while renewable energy is a "solution for tomorrow."

There are also different views on how fast the renewable energy industry can increase production to meet clean air needs. "You can only ramp up new technologies and industries so fast without creating bottlenecks, increasing costs and creating a backlash," argues Alan Nogee of the Union of Concerned Scientists, who supports the RPS standard in the Jeffords bill. Ironically, he points to nuclear power "as a really good example of an industry that created a lot of its own problems by growing too fast." Hayes disagrees: "Technically, we can do pretty much what we want to pay to do."

The real issue is political will: The political environment for renewables is not good. Clinton's hallmark has been programs that are big on pronouncements and goals but lack the cash to translate them into action. Members of Congress can greenwash themselves by joining their green energy caucus and then, as Denman says, "stab sustainable energy in the back." The fiscal year 2000 budget of $247 million for renewables research and development is a decrease of nearly $20 million from last year, which Scott Sklar of SEIA blames on pre-election shenanigans by congressional Republicans trying to embarrass Vice President Al Gore. However, the environmentalists' own budget recommendations also were down from the previous year. "We slightly tempered our request to make sure we were politically relevant," Sklar says. "It didn't work."

The problem with what is politically relevant is that it may not be enough to save the planet. The Jeffords bill, for example, would freeze utility carbon dioxide emissions at 2000 levels by 2020, while scientists say that emissions reductions of more than 60 percent are necessary to stabilize carbon levels in the atmosphere—and the sooner those cuts are made, the better. Ross Gelbspan, author of *The Heat Is On*, a best-selling book on global warming, charges that environmental organizations involved in the climate change negotiations "are more concerned with their access to government officials than solving the problems with global warming."

To counter this troubling disconnect, Gelbspan and a group of energy experts have proposed their own "World Energy Modernization Plan." The plan calls for the creation of a 0.25 percent tax on international currency transactions, yielding $150 to $200 billion for a fund to promote the global adoption of renewable and energy efficient technologies. "Even if people reject the details of the plan," Gelbspan says, "our hope is that it communicates the scope and scale of what's needed to deal with this crisis. The science on what needs to be done is unambiguous."

Eric Weltman is a writer and activist in Cambridge, Massachusetts.

Going to Work for Wind Power

The renewable energy of the future is already beginning to generate new jobs to replace the ones that are disappearing in the older energy sectors.

by Michael Renner

This is not your grandfather's windmill

Think of the Netherlands, and what may come to mind is a quaint countryside of historic canal houses, fields of tulips, and—of course—those ubiquitous windmills. Though the Netherlands today is a highly urban and technologically sophisticated nation, that image of the "old" country still plays a large role in the country's economy—as a lure to millions of tourists. It's fascinating to consider that these windmills were, for centuries, the main sources of mechanical energy before the dawn of the fossil fuel age—that such silent, pleasant-looking contraptions could have provided the power needed to pump water, grind grain, saw timber, and do a wide range of other tasks now done by loud, polluting machines. To the tourists, the relation between these quaint windmills and the modern diesel turbines or giant coal-burning power plants that have replaced them may seem as distant as that of schooners to speedboats.

Enter the new high-tech wind generators of today, which began appearing two decades ago and have proliferated in the Netherlands and in some 40 other countries so far. Unlike their predecessors, the modern wind turbines do not directly operate pumps, sluice-gates, or grindstones, but generate the basic commodity—electricity—needed to run any modern industrial economy. These new wind turbines are as different from the old windmills in their use of wind as a telephone wire is different from a 19th-century church bell in its use of copper.

While providing a means of reducing global-warming gases and other air pollution in a way that is now becoming competitive with coal and oil in sheer cost per kilowatt-hour, the new wind-power also offers an advantage that has been largely ignored during the last few years of booming stock markets—but that will prove enormously important as the 21st century un-

folds: it is not only a clean, competitive energy source but is a rich source of new employment. Whereas some defenders of the entrenched oil and coal interests predict that major efforts to stabilize climate change will spell economic doom, the evident capacity of wind power to deliver cost-effective power *and* new employment makes a compelling case that good environmental policy can also be good economic policy.

As far back as 200 B.C., windmills were used to pump water in China and to grind grain in Persia and the Middle East. In medieval Europe, merchants and crusaders returning from the Holy Land introduced this technology to their homelands, and windmills were erected in numerous places on the continent. By the early 15th century, in England alone, the use of animal power to grind grain—cattle pulling large stones in circles—had been supplanted by some 10,000 windmills. But it was in the Netherlands that windmill design evolved most over the ensuing centuries, producing incremental improvements in aerodynamic lift, rotor efficiency, and rotor speed. The Dutch relied on wind power to help drain the numerous lakes and marshes that made the Rhine river delta barely habitable and to hold their own against frequent and devastating floods. From the Netherlands, England, and elsewhere in Europe, wind technology reached the New World with the waves of settlers crossing the Atlantic. In the late 19th century, windmills were used on a massive scale to pump water for farms and ranches in the American West. Between 1850 and 1970, over 6 million mostly small units were installed in the United States.

Predictably, when it became apparent that electricity would be the elixir of the new industrial economy, efforts were made to put wind energy to use in generating it. Wind-electric machines first appeared in Denmark and the United States around 1890. The development of a utility-scale system was first undertaken in Russia in 1931 with the 100 kilowatt Balaclava wind generator on the shore of the Caspian Sea. Operating for about two years, it generated a cumulative 200,000 kWh of electricity. During the next few decades, experimental wind-power machines were built in the United States, Denmark, France, Netherlands, Germany, and Great Britain.

Despite these efforts to "modernize" wind energy use, windmills were eventually retired from active service and preserved only as tourist sites. A principal reason for their demise was the invention of the steam engine, which had to be powered by heat—and which thus created a huge new market for coal. The steam engine was soon joined by a plethora of other coal- and oil-driven machines. Wind-powered machines went into a gradual decline, first in Europe and then in North America. In 1895, there were still some 30,000 windmills operating in Germany, providing the equivalent of 87 megawatts of power, but this amounted to only 1.8 percent of the country's total power requirements—compared with 78 percent provided by steam engines.

Moving into the 20th century, the world's industrial economies developed appetites for growing amounts of coal, oil, natural gas and, later, nuclear power. By then, it was clear that fossil fuels were simply too convenient to compete with; whereas wind could only be used on site—and only when the wind was blowing—coal or oil could be transported anywhere and used anytime. It took another half-century for the environmental costs of coal and oil to become a serious issue, but by then there was a new competitor on the horizon—nuclear power, which was initially expected to prove "too cheap to meter." Substantial subsidies cemented these energy sources' advantage.

It was only with the advent of the modern environmental movement that some economists began to reassess the economics of the prevailing energy system, and to recognize that the sizable environmental and health costs—the burdens of air pollution, acid rain, climate change, toxic mining and radioactive wastes, "black lung," and respiratory diseases—were not being accounted for by conventional measures of cost per kilowatt-hour. Instead, they were "externalized"—not accounted for on any balance sheet. But at the same time that environmentalists were making this argument, defenders of the status quo were making a counter-argument: that industrial reforms made for environmental reasons would have prohibitively damaging impacts on the economy because they would take away jobs. Restricting clearcutting of forests, for example, would take jobs away from loggers; restricting fishing of depleted species would take jobs away from fishermen; and so on. In the energy sector, it was said, cutting back on coal and oil would take jobs away from miners and refiners.

Since that argument was first promulgated, however, an ironic shift has occurred. In the coal and oil businesses, massive job losses—counted in the hundreds of thousands—have occurred in the past decade without their having been driven by environmental regulation and despite the continuing preferential subsidies they have received. Meanwhile, wind power is beginning to benefit from technological advances that will diminish its historic disadvantages of not being subject to transport and storage. Wind is now poised to compete economically with coal and oil on even terms in many places—and to do so not only with the advantage of being environmentally benign, but with the important added advantage of providing more jobs per unit of cost than the fossil-fuel industries it now challenges.

Wanted, To Run With the Wind

It was only in the wake of the oil crises of the 1970s that interest in wind turbines revived after more than half a century of dormancy, setting the stage for the emergence of a whole new, futuristic tech wind energy sector. It took a decade or so to take hold, but since the beginning of the 1990s, the new sector has been growing at a breathtaking rate. Worldwide installed generating capacity grew from about 2,000 megawatts in 1990 to 15,000 megawatts by mid-2000, an average growth rate of 24 percent per year. That's still tiny in absolute terms, but it's comparable to the position automobiles were in a century ago. And the prospects for continued expansion are good. Electricity from the wind is now rapidly closing the price gap with conventional power plants. In October 1999, the European Wind Energy Association, the Forum for Energy and Development, and Greenpeace International jointly released a study, *Windforce 10*, that contends that wind energy could meet 10 percent of the world's electricity demand by the year 2020. Under their scenario, installed capacity would grow to 1,200 gigawatts (1.2 million megawatts).

Windforce 10, in its assessment of the number of jobs that might be generated over the next two decades, concludes that 17 job-years of employment are being created for every megawatt of wind energy capacity manufactured and an additional five job-years for every megawatt installed, or a total of 22 job-years per megawatt. As labor productivity rises, the per-megawatt job figures are expected to gradually decrease to 15.5 by 2010 and 12.3 by 2020.

Assuming these ratios hold, the study projected that total wind power employment will climb from something under 100,000 jobs today to almost 2 million over the next two decades, with most of the growth occurring in Europe, North America, and China.

This growth includes the "direct" jobs of manufacturing and installing wind turbines, as well as the "indirect" jobs in supplier industries. It does not include any jobs that may be produced by the still embryonic off-shore wind industry. Nor, significantly, does it include the work of maintaining wind installations once they are built.

Offshore installations, which would be placed in relatively shallow waters somewhat like offshore oil rigs, were not included in the *Windforce* study. But they are expected to play a growing role in coming years, particularly in Europe. A study released by the German Wind Energy Institute and Greenpeace in October 2000 ("North Sea Offshore Wind—A European Powerhouse") concludes that five North Sea countries—Germany, Britain, the Netherlands, Belgium and Denmark—have the potential to generate almost 2,000 terawatt hours of electricity per year from offshore wind, an amount that is more than triple their current combined demand for power. Tapping just one percent of this wind source in a year would provide electricity for 6.5 million homes and could employ 160,000 persons, according to Greenpeace.

Light Footprint

Wind power is not a "heavy" industry of the traditional kind; its fuel is weightless, and its plants don't require the massive structures of coal or nuclear plants. Here, construction workers for Enron Wind lower a section of a wind tower into place. Enron has erected over 4,300 towers so far.

Additional employment is generated through operating and maintaining wind turbines, though reliable numbers are unavailable. The European Wind Energy Association estimates that between 100 and 450 people are employed per year for every terawatt-hour of electricity produced, depending on the age and type of turbine used. In 1999, when about 29 terawatt-hours were generated, that would have meant anywhere from 3,000 to 13,000 additional jobs worldwide. As wind power capacity expands, obviously so will these numbers. Even at the lower end of this range, there may be some 3 million jobs in running and maintaining the world's wind energy turbines by the year 2020, if the *Windforce 10* projections hold up.

What Kinds of Jobs?

Wind power development opens up employment opportunities in a variety of fields. It requires meteorologists and surveyors to rate appropriate sites (to ensure that the areas with the greatest wind potential are selected); people trained in anemometry (measuring the force, speed, and direction of the wind); structural, electrical, and mechanical engineers to design turbines, generators, and other equipment and to supervise their assembly; workers to form advanced composite and metal parts; quality control personnel to monitor machining, casting, and forging processes; computer operators and software specialists to monitor the system, and mechanics and technicians to keep it in good working order. Many of these are highly skilled positions with good pay.

The lion's share of the world's wind power–generating capacity has been installed in Western Europe, and European companies are the leading manufacturers of wind turbines (accounting for about 90 percent of worldwide sales in 1997), so most of the world's wind power–related jobs are being generated there. In the United States, now the second-leading force in wind power, capacity is expected to almost double by the end of 2001.

As other regions with high wind power potential gear up, the picture will gradually change. India and China, especially, have the meteorological potential to greatly increase wind power production and employment. With roughly 1,000 megawatts of capacity, India is already among the five leading wind power nations. It currently has 14 domestic turbine manufacturers, and spare parts production and turbine maintenance are helping some of its regions and villages to generate needed income and employment.

Other developing countries, too, are showing rising interest. Although they currently have little wind generating capacity installed, wind companies in Argentina hope to create 15,000 permanent jobs over the next decade. Latin American and East European nations are able to manufacture nearly all needed components within their own regions. Imports will be needed for at least a portion of new installations in Asia, and for the bulk of installations in the Middle East and Africa.

Fossil Fuel Jobs—A Disappearing Act

The traditional energy sector, with its many millions of jobs once providing a large part of the industrial world's employment, is now a shrinking source of employment, even though the overall production of fossil fuels is still creeping upward. World coal output began stagnating in the mid-1980s, and the industry has become one of bigger and fewer companies, larger equipment, and less and less need for labor. In Europe, employment in this field has dropped particularly fast, since production is being driven down both by coal imports and by a shift to other sources of energy. During the past two decades, British coal employment has collapsed from 224,000 to just 10,000 miners, the result of mine closures and aggressive automation at remaining sites. About 50,000 jobs were lost in Germany during the 1990s. Even though the German coal industry continues to receive massive subsidies—with some $20 billion allocated for the 2000–2005 period alone—its cuts in employment are expected to continue.

China, which produces more coal than any other nation, has undertaken to deliberately cut its coal output by 20 percent over the next several years, in order to bring production more in line with declining demand, reduce pollution, and bring down the human toll of mining. (At least 10,000 people die in Chinese coal mines each year—80 percent of the global number of victims—and increasingly, these jobs are scorned by all but the most desperate workers.) To this end, China has reduced its subsidies to coal production, with the result that some 870,000 coal industry jobs have been cut since 1994 and another 400,000 workers are expected to be laid off.

In the United States, coal production increased 32 percent between 1980 and 1999, but coal-mining employment nevertheless declined 66 percent, from 242,000 to 83,000 workers. One reason is that production has shifted from more labor-intensive underground mines in the eastern United States to surface mines in the West. Ton for ton, strip-mining employs only about one-third to one-half the number of workers required in underground mines. Environmental considerations played a role in this shift, insofar as efforts to combat acid rain have led to a greater preference for lower-sulfur coal, and western coal is lower in sulfur content than eastern coal. Employment is expected to fall by another 36,000 workers between 1995 and

Wind Power in the Year 2020

Region	Installed Capacity	Electricity Generation Per Year	Share of Electricity Consumption	Employment
	(megawatts)	(terawatt hours)	(percent)	(number)
North America	300,000	735.8	12	325,000
OECD Europe	220,000	539.6	12	270,600
China	180,000	441.5	11	369,000
Eastern Europe[1]	140,000	343.2	10	270,600
Asia[2]	140,000	343.4	9	265,700
Japan, Australia, New Zealand	90,000	218.5	12	184,000
Latin America	90,000	218.5	11	184,500
Middle East	25,000	61.5	7	43,100
Africa	25,000	61.3	7	51,720
World	1,210,000	2,963.3	11	1,964,220

[1]Including Russia. [2]Excluding China and Japan.

Source: Adapted from European Wind Energy Association, Forum for Energy and Development, and Greenpeace International, *Windforce 10: A Blueprint to Achieve 10% of the World's Electricity from Wind Power by 2020* (London: 1999).

2020, even in the absence of any measures to address the threat of climate change.

Similar trends can be seen in other parts of the energy and utility industries, as increasing mechanization and automation have cut jobs even as output rises. In the United States, more than half of all oil and gas production jobs were lost between 1980 and 1999; during the same period of time, almost 40 percent of oil-refining jobs were cut. Today, petroleum refining and wholesale distribution accounts only for 0.3 percent of all U.S. employment. In EU countries, more than 150,000 utility and gas industry jobs have disappeared since the mid-1990s and another 200,000 jobs—one in five—are likely to be lost by 2004, as the new market liberalization program proceeds. In Germany alone, 60,000 utility sector jobs—one quarter of the total—were eliminated between 1990 and 1998.

The Labor Productivity Issue

Wind power is more labor-intensive than either coal- or nuclear-generated electricity. In Germany, currently the world leader with roughly 5,000 megawatts (roughly one-third of global capacity) installed, wind still contributes just 2 to 3 percent of the country's total electricity generation, while supporting about 35,000 jobs in manufacturing, installing, and operating wind machines. In comparison, nuclear power commands 33 percent of the electricity market, but supports a relatively meager 38,000 jobs; coal-fired power plants have a 26 percent market share and account for some 80,000 jobs.

Judging by the way the business press routinely describes companies that employ fewer people for a given level of

output as "lean" or "efficient," the high number of jobs in wind energy may seem to suggest that wind is a less economically efficient way of producing electricity. In today's globalizing economy, companies seem ever more intent on boosting labor productivity—the amount of goods and services produced per worker—and slashing labor costs as a means to stay competitive. Because wages and benefits are a major part of the cost of most businesses, the pursuit of greater labor productivity is an omnipresent concern.

In principle, however, a given industry—such as wind power—can become profitable while still remaining relatively more labor-intensive, by achieving superior efficiency in other major categories of cost—in its requirements for capital, materials, and energy. Unfortunately, in the calculus of most business executives, improving energy or materials productivity is given short shrift compared with improving labor productivity (or laying off employees). A key reason for this is that energy and materials appear to be cheaper than they really are, and therefore offer less incentive for pursuing increased efficiencies, because their production and use are subsidized and their environmental costs are "externalized"— meaning that those costs are not accounted for on a company's balance sheet. One of the costs of coal power, for example, is the acid rain that drifts over the eastern United States from Midwestern power plants and kills countless trees along the Appalachian Mountains. Because the power plants don't have to pay to restore those damaged forests, they have less incentive (than would be the case if the costs of restoration were included) to improve the efficiency of their fuel use than to cut the cost of their labor.

As industrial societies that pervasively allow such damage, not only to ecosystems but to human health and climate stability, we are deluding ourselves not only by not including the cost of such damage in assessing the overall productivity of a business, but also in thinking that by simply running the business with fewer workers we are truly being efficient in our ways. A society that widely exploits such accounting is not much different than those societies that wage repressive campaigns against workers and labor unions in order to keep wages low and the country's products "competitive."

The real news about wind-generated electricity is that it can be competitive—and can generate income that is not ill-gotten through overlooking human or environmental costs—even though it employs a comparatively larger number of people than a coal-fired plant. Unlike a conventional power plant, a wind turbine does not have to purchase fuel inputs, whether they be coal, oil, natural gas, or enriched uranium. At a wind power plant, the energy input comes for free. Wind power plants are less capital-intensive, as well: they require less investment in buildings and machinery than conventional power plants do. And, there are no worries about toxic mine tailings, radioactive wastes, and other problems or costs associated with fossil and nuclear energy.

The Wider Picture

Some widely quoted critics of the Kyoto climate treaty—some of them working for think tanks quietly funded by fossil fuel industries (see "Matters of Scale")—have declared that actions taken to substantially reduce carbon emissions would be terribly disruptive to the industrial economy. Their rhetoric echoes that of certain critics of U.S. policies aimed at saving Northwest rainforests, a decade ago, who displayed bumper stickers reading "Save a logger—kill an owl." Yet, just as environmental protection in Oregon and Washington have not brought the feared ravages of unemployment, it is now clear that environmental policies pose little threat to jobs in general—and that, in fact, the wind industry is far from alone in demonstrating that moving toward a more sustainable economy will bring abundant new jobs to replace the old. Wind power has been the fastest growing among alternative sources of energy, but others, such as solar photovoltaics and solar thermal energy, also have the potential to engage a growing portion of the public in meaningful and remunerative work. Additional opportunities will be found in the pursuit of such energy efficiency measures as retrofitting buildings to boost their thermal insulation.

The benefits to be gained by such shifts—the "double dividend" of a more protected environment and more jobs—will not just be one-to-one substitutions of beneficial investment dollars for destructive ones. The energy sector is a small employer relative to the size of the overall economy, yet it exerts enormous leverage because such large quantities of capital—much of it in the form of public subsidies for nuclear, oil, and coal—are bound up in it. Withdrawing some of the hundreds of billions of dollars that have been propping up these obsolescent industries could free up capital to invest in a wide range of more sustainable industries—not only the wind industry discussed here, but a phalanx of new enterprises aimed at achieving greater materials/energy efficiency and pollution prevention. These enterprises might include greatly ramped-up recycling and remanufacturing, as well as designing and redesigning of products (and of buildings, communities, and whole economies) to put greater emphasis on durability, repairability, and reusability. Like wind power, many of these new industries are still quite small, but with the right kinds of subsidy, tax, and research policies—they can be scaled up significantly. It is becoming clear that making it possible for people to work productively does not have to depend on destabilizing the natural world.

Michael Renner is a senior researcher at the Worldwatch Institute and author of Worldwatch Paper 152, *Working for the Environment: A Growing Source of Jobs.*

Power Struggle

California's Engineered Energy Crisis and the Potential of Public Power

By Harvey Wasserman

THE U.S. BARONS OF FOSSIL AND NUCLEAR FUEL have used a contrived energy crisis in California and the nation as a pretext to declare an all-out assault on environmental protection. George Bush, Dick Cheney and their cohorts from the oil industry claim the rolling West Coast blackouts justify rolling over a century of carefully crafted environmental law.

But this rationalization ignores a telling detail: California does not have an energy shortage. And according to some of the state's highest officials, the blackouts have been choreographed for massive price gouging by some of Bush's closest associates and contributors.

The blackouts, rising electricity prices, government subsidies, utility bankruptcies and near-bankruptcies have many causes. But neither a shortfall in supply nor a surge in demand for electricity is among them.

California is in fact among the most energy efficient of states. Though its population and economy have soared, its overall demand for electricity has risen only modestly in recent years. The amounts of electricity it can generate in state or has contracts to purchase out-of-state are more than sufficient to meet overall and peak demand, and have been throughout the state's crisis.

California's electricity crisis was precipitated by a botched deregulatory scheme pushed by the very utilities now screaming about alleged losses, a plan that has immensely profited both the utilities' parent corporations and a few pirate power generators close to George W. Bush.

California's deregulatory disaster is a "failure by design," prompted not by a real shortage, but by a corporate agenda, says Paul Fenn of the Oakland-based American Public Power Project.

In the view of the state's leading consumer and clean power advocates, California's consumers and taxpayers are victims of a massive, complex double-theft, first by the state's biggest electric power utilities, and then by huge oil and gas companies close to George W. Bush.

STRANDED COST CATASTROPHE

There was very little public debate leading up to the Golden State's decision to deregulate its electric utilities in 1996. The early battles were muddied and muzzled. The state legislature deliberated for a scant three weeks. The media barely covered those few hearings that were open to the public. Southern California Edison (SoCalEd) essentially wrote much of the legislation, AB1890, in its corporate offices.

When the legislature unanimously voted for the bill and watched its September signing by a beaming governor Pete Wilson, the utilities and their lobbyists gushed. This is "a great day for us," cheered John Bryson, president of Southern California Edison, widely regarded as the bill's chief architect. "We believe this plan is the best way to facilitate a smooth, timely transition to a competitive electricity market and maximize value for our shareholders and customers." It was "a large achievement and a sound achievement for the state in terms of giving customers choice," he said.

Ostensibly, AB1890 was meant to dismantle the regulated monopolies that had supplied California with electric power since the early twentieth century. Instead, consumers would be able to choose among competing suppliers. The bill presumed the electric market would fill with power companies vying to sell low-cost juice of all varieties, including "green power" from wind and solar.

But the state's three biggest utilities—SoCalEd, Pacific Gas & Electric (PG&E) and San Diego Gas & Electric—made sure that before competition could come they were first reimbursed for their investments in nuclear power, which they argued was inefficient and could not compete in an open market.

In exchange for this payback, they engineered consumer rate caps that were to remain in place until the utilities collected up to $28.5 billion in surcharges for their obsolete generating plants. Once they collected that money, they said, rates would fall as competing generators came into the open market.

MODEST DEMAND, WITHHELD SUPPLY

No one disputes that a mismatch between supply and demand underlies the California energy crisis. But consumer advocates point out that the data show California's energy demand to be growing slowly, not surging as many news reports suggest. And, they say, the electricity shortage reflects not any real limits on supply, but the market manipulations of the independent generators.

A comprehensive San Francisco Chronicle study in March showed that California's energy demand is in fact rising slowly. "The industry has painted the summer of 2000 as the equivalent of a 100-year storm in meteorology—an event so powerful and unexpected that the existing infrastructure was devastated by its force," the Chronicle reported. "The statistics show that 2000, taken in total, was nothing of the sort." Overall electricity usage in California rose approximately 2 percent a year in the 1990s.

Most importantly, peak use—the demand level that actually stresses the system—was lower at the end of 2000 than in the previous year. While peak demand was high in May 2000, in four out of the last six months of 2000—July, August, October and December—peak demand was lower than in 1999, according to an analysis of statistics from California's Independent System Operator (CAISO) by Public Citizen's Critical Mass Energy and Environment Project.

California's available capacity and electricity on contract vastly exceeds its peak demand. In total, California has 55,500 megawatts of power generating capacity and 4,500 megawatts of power on long-term out-of-state contracts—approximately 15,000 megawatts more than peak demand, Public Citizen reports, citing statistics from CAISO.

The problem California is facing isn't supply, say the consumer/green groups, it is that the power supply companies, now a separate industry segment from the utilities, simply refuse to make the supply available. The independent power generators' alleged market manipulation is now the source of numerous lawsuits and investigations.

—*Robert Weissman*

With abundant infusions of utility cash (the utilities spent more than $3.6 million lobbying to win the bill in 1996, and another $4.1 million to promote it in 1997), the state's energy interests promoted AB1890 as a way to save customers money through the magic of the marketplace. In 1995, Bryson trumpeted deregulation as "the best, soundest way to move to a desirable competitive market that will benefit all customers, large

and small." SoCalEd, he said, was "committed to a 25 percent rate reduction effective January 1, 2000. As near as we're able to tell, this [legislation] is consistent with our goal."

A broad coalition of consumer and environmental groups knew better. They saw the "crisis" coming right from the start, and bitterly opposed the original AB1890. In 1998, as deregulation was taking effect, Herbert Chao Gunther of the San Francisco-based Public Media Center, Harvey Rosenfield of the statewide Foundation for Taxpayer and Consumer Rights, Ed Maschke and Anna Aurilio of the California and U.S. Public Interest Research Groups, Fenn's American Public Power Project and others fought back. Operating on a shoestring budget, they gathered an astonishing 700,000-plus signatures to put on the fall ballot an initiative—Proposition 9—that would have nipped the crisis in the bud. Among other things, Prop 9 would have restored regulation to the electric power system and prevented the huge "stranded cost" bailout that was AB1890's central feature.

But the supporters of Proposition 9 ran into a hugely funded utility opposition that would not be denied. Still intoxicated by the promises of deregulation, William Hauck, chair of the Concerned Stockholders of California, a SoCalEd front group, spoke for the industry when he warned that returning to public regulation would dismantle "the competitive electricity market and customer choice, and will actually result in higher electric rates."

Big energy steamrolled the campaign with a $40 million counteroffensive. The greens had only $1 million. California voters rejected Proposition 9's proposal to repeal electricity deregulation by a 73-to-27 percent margin. (A parallel Massachusetts campaign was crushed on the same day, by a similar margin.)

It was a grim day for green and consumer advocates who had fought hard to avoid what's now happened. Eugene Coyle, one of the state's most respected energy analysts, and a host of others warned that the electric power industry was a natural monopoly that would never foster true competition. They showed that AB1890 was a cover for a massive bailout for the utilities' bad nuclear investments. They demanded public control. They predicted disaster right from the start.

One early opponent of deregulation was Dan Berman, an energy expert working to win public utility ownership for his hometown of Davis, California. With Boston-based activist John O'Connor, Berman wrote in the 1996 book, *Who Owns the Sun*, "Today deregulation, cheap electricity, and natural gas are all the rage."

"But few people are paying attention to what will happen when the price of natural gas and oil go up, as they most surely will, after falling by 75 percent in the last decade," Berman and O'Connor wrote. "What will happen when the new, unrelated 'independent power producers' of cheap electric power fired by combined-cycle gas turbines pass on whopping rate increases to the public as the price of natural gas soars? Will big industry come weeping to the public, hat in hand, as the savings-and-loan investors did? Are the energy corporations crippling American industry by reinforcing an addiction to cheap fossil fuels and electricity? Will there be a massive ratepayers' revolt

POWER PLAYS: WHERE DID THE CALIFORNIA UTILITIES MONEY GO?

When California lawmakers voted unanimously to deregulate the state's electricity market in 1996, California's three biggest utilities—PG&E, SoCalEdison, and San Diego Gas & Electric (SDGE)—owned most of the state's power generating plants, as well as most of the state's distribution grid and power lines. The companies operated as regional monopoly utilities, with strict regulatory oversight over retail electricity rates.

Once deregulation began, however, the companies rapidly began to restructure. First, they "unbundled" their assets—selling off some of their power generating plants while keeping their distribution grid and power lines and other assets such as nuclear and hydro power generation facilities.

The money from the power plant sales and high utility charges was siphoned off to the utilities' parent companies (e.g., PG&E restructured in 1997, creating a separate holding company called PG&E Corporation), which then passed the money on to other unregulated subsidiaries, which acquired new generating facilities in other states and (in the case of Edison International and SDGE) other countries. The parent companies built a diverse array of new assets and services, including natural gas pipelines, storage and processing plants.

Since the summer of 2000 price spikes, PG&E and SoCalEdison have claimed they are going broke by having to pay the difference between high electricity prices charged by wholesale power suppliers in the newly deregulated market and lower fixed retail rates that the utilities themselves originally negotiated with state regulators.

Claiming they needed relief or else they would be forced to declare bankruptcy, the utilities convinced the state to begin spending $50 million of taxpayer monies per day to purchase wholesale electricity on their behalf. The state began buying super-high-priced electricity on the wholesale market, including on the spot markets where prices were orders of magnitude higher than just a couple years before, and conveying the electricity to the utilities. By May, California taxpayers had spent more than $6 billion, with no end in sight. As the state's budget surplus quickly eroded, Governor Davis proposed floating $13.4 billion in bonds to cover the cost (the bonds are to be paid off over 15 years by ratepayers via higher electricity rates).

Meanwhile, federal regulators (FERC) agreed in January to allow the companies to organize their corporate structure so as to preemptively shield their assets during bankruptcy proceedings.

Less than two months later, on April 6, PG&E—the utility, not the holding company—filed for bankruptcy, leaving consumers and taxpayers in even more of a bind.

But was the company really bankrupt? While the utility was losing $300 million a month in wholesale power purchase costs, its parent company was pulling down a healthy profit. At the end of 2000, the PG&E Corporation (the parent company) reported $30 billion in assets and $20 billion in revenues for the year. It had an ownership interest in 30 independent operating plants in 10 states, including a number under development. In October 2000, it spent nearly $8 billion to acquire 44 new turbines.

Moreover, between 1997 and 1999, PG&E transferred $4 billion to its parent corporation. In the first nine months of 2000, PG&E transferred an additional $632 million. An audit sponsored by the California Public Utility Commission (CPUC) concluded that "historically, cash has flowed in only one direction, from PG&E to PG&E Corp. and then to the unregulated affiliates." Similarly, SoCalEdison transferred $4.8 billion to its parent corporation (Edison International) between 1995 and 2000.

CPUC's audit also found that PG&E Corporation (the parent) is expected to receive an additional federal tax refund of up to $1 billion "largely due to losses sustained by PG&E."

The San Francisco-based Utility Reform Network explains in its report "Cooking the Books" that the utilities' position was "akin to a situation in which one pocket is empty and another is full of cash. A reasonable person would check both pockets before assuming that they are penniless."

PG&E's Jon Tremayne responds that deregulation was set up "to pay the utility investors back money that they invested in power plants that they were now forced under law to sell. So the shareholders essentially recovered their investments and reinvested it. This was done under the direction of the Public Utilities Commission, in synch with the state law."

With support from the FERC, PG&E maintains that its California subsidiary is completely separate from the parent holding company, despite the fact that the two companies have virtually identical boards and file a joint annual report with the Securities and Exchange Commission.

The company may not be entirely in the clear. PG&E admitted in April in a quarterly filing with the U.S. Securities and Exchange Commission that, depending on the terms and conditions of the company's reorganization plan adopted by the company's Bankruptcy Court, creditors of the bankrupt PG&E utility unit may be able to grab some assets from PG&E National Energy Group, a subsidiary the parent had intended to insulate through its restructuring plan.

—Charlie Cray

RECENT PURCHASES OR COMMITMENTS OF PG&E CORPORATION (PARENT OF PG&E COMPANY) AND EDISON INTERNATIONAL (PARENT OF SOCAL EDISON)

Subsidiary	What they built/bought/where	Cost ($ billions)	Date
PG&E Corporation			
Nat'l Energy Group	810 MW Southhaven power plant in Mississippi	??	11/00
Nat'l Energy Group	44 turbines, 15 other projects from Societe General	7.8	10/00
PG&E Generating	Okeehobee County, FL, power plant to be completed 2003	0.2	9/00
Nat'l Energy Group	Madison Windpower in New York	0.02	9/00
Nat'l Energy Group	Attala 500 MW power plant, MS	??	9/00
Energy Trading	Tolling rights to peaking plant in suburban Indianapolis	??	9/00
Nat'l Energy Group	Tolling rights to Liberty power plant in suburban Philadelphia	??	6/00
PG&E	Stake in True Quote trading software	??	4/00
PG&E	Aerie broadband pipeline project	??	4/00
PG&E Generating	Pleasant Prairie, Wisconsin power plant	0.5	1999
Nat'l Energy Group	Lake Road power plant in Killingly, CT	0.5	1999
US Gen	New England Electrical Systems hydro, coal, oil, gas generating plants (4,800 MW total capacity; 18 plants)	1.8	9/98
PG&E Corp.	Valero (gas pipelines, processing)	1.5	1997
Edison International			
Citizens Power	P&L Cal Holdings/Boston	0.05	9/00
Edison Capital	Swisscom, telecom network	0.3	9/00
Mission Energy	Italian Vento Power Corp.	0.04	3/00
Mission Energy	Commonwealth Edison's 12 plants in Illinois	5.0	12/99
Mission Energy	Ferrybridge & Fiddler's Ferry power plants in England	2.0	7/99
Mission Energy	40 percent stake in New Zeland's Contact Energy	0.7	5/99
Mission Energy	Homer City power plant, PA	1.8	3/99
EME del Caribe	EcoElectrica co-gen facility in Puerto Rico	0.2	12/98
Mission Energy	1,230 MW coal-fired power plant Paiton, Indonesia (40% stake; under construction since 1994)	.3	1999

Sources: Public Citizen Critical Mass Energy and Environment Program, company annual reports, 10K filings.

when utilities try to stick consumers with doubled and even quadrupled utility bills?"

AB1890 did include measures that appeared to benefit ratepayers. The bill implemented an immediate 10 percent rate cut, and froze it into place for as many as four years. "The same critics who now say the bill was written by the utilities to benefit the utilities were there in Sacramento in 1996 when the legislation was drafted and passed," says Jon Tremayne, a PG&E spokesperson. "Their input, for instance, ensured that residential customers were included in the bill."

But the resulting rates were still 50 percent higher than the national average. And the fixed rate blocked what would have been a natural decline with the onset of new renewables.

Deregulation opponents emphasize that had the natural transition to cheaper and more desirable wind and solar power been allowed to proceed along with efficiency and conservation programs being mandated by the Public Utility Commission (PUC) rates would have drifted downward as green supply increased and demand was held steady.

In the early 1990s, a major program for building renewable-based capacity was eliminated by the utilities through a legal filing in front of the Federal Energy Regulatory Commission. Had that renewable building plan been allowed to proceed, rates would have been on their way down throughout California.

Furthermore, AB1890's so-called rate cut was financed by an elaborate bonding scheme that would force consumers to pay huge sums of long-term interest.

"In effect, the small customers are borrowing to give themselves this rate cut, which is like borrowing money to give yourself a raise," said Coyle at the time. This is a "hidden tax that Californians will have to pay to private utility owners."

GREEN POWER BLOCKED

In the midst of today's crisis, Vice President Dick Cheney, media pundits and others contend that the environmental movement has somehow blocked construction of new power plants that could have helped the state avoid the current crisis.

RECENT SEMPRA ENERGY PURCHASES OR COMMITMENTS

(Sempra was created by the 1998 merger of Pacific Enterprises, the parent of
Southern California Gas and Enova, the parent of San Diego Gas & Electric)

Subsidiary	Purchase/Location	Cost ($ millions)	Date
Sempra Energy International	Sodigas Pampeana S.A./Sodigas Sur S.A. (Argentina natural gas holding co.s)	180	10/00
SEI	Baja natural gas pipeline (joint venture w/PG&E)	230	6/00
SEI	Chilquinta Energia S.A., Chile's third largest electricity distributor; deal includes Energas S.A., gas distributor	830	6/99
SEI	Luz del Sur S.A.A., Peru. (included in above deal)	25	1997
SEI	ECOGAS Mexicali pipeline		
SEI	Termoelectica de Mexicali (Mexico) 600 MW gas-fired power plant (online 2003)	350	2001
SEI	ECOGAS Chihuahua, Mexico (privatized by PEMEX); gas distribution.	50	1997
SEI	DGN De La Laguna-Durango (Mexico) gas pipeline and distribution.	40	2000
SEI	Transportadora de Gas Natural de Baja California, Mexico, pipeline.	??	2000
SEI	Natural gas pipeline, Arizona to Tijuana, Mexico (PG&E will develop US portion).	38	2000
PE	Natural gas/propane distribution (Uruguay) (15 percent interest in Conecta consortium)	160	1998
SEI	Sempra Atlantic Gas (Nova Scotia) exclusive natural gas distribution rights.	800	1999
SEI	Bangor, Maine natural gas distribution	50	2000
SEI	Frontier Energy nat. gas distribution, NC	70	2001
Sempra Energy Trading (SET)	Trading subsidiary purchased from Consolidated Natural Gas, Stamford, CT	36	12/97
SET	Utility.com (Internet utility company)	4	4/00
Sempra Energy Resources (SER)	550 MW Elk Hills power plant Bakersfield, California (online 2003)	360	12/00
SER	1200 MW Mesquite Power plant, Phoenix, AZ	630	12/00
SER	500 MW combined cycle plant, Boulder City, Nevada	280	4/98
Sempra Energy Financial	1200 properties, including housing in Houston, Modesto (CA), Puerto Rico and the Virgin Islands	??	??
Sempra Communications	Aerie Networks nation broadband fiber optic network	??	4/00

Sources: Public Citizen Critical Mass Energy and Environment Program, company annual reports, 10K filings.

Consumer and environmental groups respond that the Federal Electrical Regulatory Commission (FERC) colluded with SoCalEdison to kill a clean generation initiative that would have provided enough additional generating capacity for the state to meet its own needs once deregulation began.

On February 23, 1995, responding to a SoCalEdison petition, FERC blocked a California Public Utilities Commission order that required the utilities to purchase more than 600 megawatts of renewable energy, primarily from wind and geothermal sources. Among other things, the FERC said—with what now seems terrible irony—"we have grave concerns about the need for this capacity," mostly because the state commission was relying on 1990 data, which FERC called "stale."

Throughout the 1990s, California's private utilities resisted the green demands for increased renewables and efficiency.

AB1890 further stifled competition to the benefit of the entrenched utilities that wrote the bill by making it more difficult

HURWITZ'S POWER GRAB

Charles Hurwitz, the man who critics say has made a fortune by pillaging old growth redwoods, busting unions, and engaging in shadowy stock market and savings and loan deals [See "Ravaging the Redwoods: Charles Hurwitz, Michael Milken and the Costs of Greed," *Multinational Monitor*, September 1994], can add another title to his infamous resume: power robber baron.

Hurwitz's Maxxam company controls Kaiser Aluminum, which has limited production at two smelters in Washington state since January 1999. Initially, Kaiser shut down its Washington aluminum smelters as part of the company's nationwide lockout of workers who belong to the United Steelworkers of America. The lockout ended last September, shortly after the federal government threatened to charge Kaiser with violating labor laws.

Then, Kaiser used a novel approach to turn a profit: it started selling the smelters' contracted allotments of electricity back to the Bonneville Power Administration (BPA), the agency that produced the power in the first place.

Under Kaiser's existing contract with the BPA, it purchased power at a rate of $22.50 per megawatt-hour. Kaiser resold the power to BPA for more than 20 times that amount. In late 2000, the spot market price for electricity in the western United States soared to as much as $750 per megawatt-hour.

Kaiser registered $135 million in power sales from October to December 2000.

At year's end, even with its entire U.S. production operations idle, the company's executives funneled some of this easy money into their own wallets. CEO Raymond Milchovich was paid a bonus of $978,000, on top of an annual salary of $630,000. Other executives received bonuses of $250,000 or more.

Aluminum companies have long benefited from generous relationships with the federal BPA, which operates 29 dams in the Columbia and Snake River basins. Smelters aggregate around sources of cheap energy because 45 percent of the cost of aluminum smelting is electricity.

Many Pacific Northwest smelters have been closed since last summer, when their owners found it more profitable to sell power earmarked for their operations to the spot energy market. In June 2000, Alcoa announced that it was halting production at its Troutdale, Oregon smelter. In western Montana, Columbia Falls Aluminum closed its smelter and is reselling the power. Golden Northwest is selling power from its allocation for smelter operations in The Dalles, Oregon, and Kilimat County, Washington.

Some smelter owners recognized the windfall nature of these sales, and pledged to give something back to the workers, local communities and the BPA.

Columbia Falls decided to maintain its 600 workers' salaries and benefits through the year 2001.

Golden Northwest said that it would pass along 25 percent of its estimated $400 million in power sales to the BPA. The company designated another 25 to 50 percent of its electricity revenue for the development of alternative energy, including a wind power plant. It is also converting from the heavily polluting production of primary aluminum to a secondary plant that makes aluminum from scrap, not alumina.

Kaiser has made no such pledges to channel energy resale profits to the BPA, its laid-off workers or alternative power plants. The sales could earn Kaiser a half-billion dollars until the current contract expires in October.

"It's difficult to conceive of a circumstance that would prevent them from coming to terms with the re-

gion's other ratepayers and their employees, given the amount of windfall profit," BRA spokesperson Ed Mosey said in January. "There's no way they should be profiteering from reselling federal power and then ask us to draw unemployment," says Steelworkers Local 329 steward Wayne Bentz. Over 900 workers are unemployed due to Kaiser's shutdown.

This winter, Kaiser Vice President Pete Forsyth said the company was keeping its profits to cushion the blow of the new electricity contract, which goes into effect later this year. Initially, BPA and Kaiser reached a deal in which the smelter's power purchase rate would rise by 20 percent. In April, however, BPA officials warned that, due to low water levels and high demand, the agency would not be able to supply enough energy to the region's smelters. BPA advised that the facilities should remain closed for another two years. Kaiser now acknowledges that it is looking to sell its Pacific Northwest smelters.

Like aluminum giants Alcoa and Alcan, Kaiser is slowly but surely moving out of the United States. As energy resources and environmental and labor regulations are tightening, primary aluminum production is shifting to the Third World. Powerful rivers in South America and Africa, coal mines in eastern India, and oil and gas fields of the Middle East are beginning to fuel the global aluminum market.

Ironically, U.S. governmental funding is helping to finance the shift. Kaiser's largest smelter, the 200,000 ton-per-year Valco operation in Ghana, owes its existence to a dam backed by the World Bank and the U.S. government's Overseas Private Investment Corporation (OPIC). The Akosombo dam, according to the International Rivers Network, "flooded more land than any other dam in the world—8,500 square kilometers." Valco consumes most of Akosombo's

(continued)

power. Drought and rising demand have led international financial institutions to back new sources of power for Kaiser's Ghanaian smelter, including a new oil-fired power plant in Takorade.

Hurwitz has focused Kaiser's investments in overseas locations since 1988, when he bought out the corporation with the assistance of fugitive commodities trader Marc Rich. Rich earned the nickname "Aluminum Finger" for his investments in Russia, Iran, and Jamaica, and his stubborn battle against the Steelworkers union

at a smelter in Ravenswood, West Virginia.

Under Hurwitz's control, Kaiser has invested in one of the world's largest smelters (Aluminum Bahrain), sold equipment to Russia's notoriously corrupt primary aluminum industry, and has contemplated or made bids to invest in smelters in Guinea, Azerbaijan and Ukraine.

Labor activists say that Kaiser's Pacific Northwest power grab represents a final swindling of the United States before Hurwitz waves goodbye. In April, the president of Steelworkers

Local No. 329, Dan Russell, lamented, "I see guys every day that are resigned to the fact that Kaiser is done here. A good number of them are just getting on with their lives... washing their hands of the whole thing."

—Jim Vallette

Jim Vallette is an investigative reporter based in Seawall, Maine. He is working on a report for the Institute of Policy Studies' Sustainable Energy and Economy Network that examines the global structure and social and environmental impacts of the aluminum industry.

---

for startups to lure new customers away from the established giants. But AB1890 included complex procedural roadblocks that made it virtually impossible for communities to band together to form buying groups that might circumvent the established monopolies.

As Paul Fenn puts it, AB1890 "raises serious questions about the ability of small consumers to exercise any market power. Unless residents and small businesses have the means to purchase power in aggregate through their local governments, 'consumer choice' will mean little more than paying higher rates to a middleman or to your current utility."

PG&E's Tremayne disagrees. "AB1890 provided the opportunity for customers to band together through aggregation and go out on the open market and seek better rates. That was not available prior to deregulation." People are not doing this now, Tremayne says, because "the market is all screwed up. Those lower rates are not out there. Some significant steps need to be taken to stop the out-of-state generators from gouging California customers. Whether it's short-term price caps at the federal level or those generators acting in a responsible fashion, we have to have changes that address the problems of the market."

There were some incentives for conservation and renewables written into AB1890, largely sponsored by the Natural Resources Defense Council, whose chief San Francisco-based energy advocate, Ralph Cavanagh, energetically supported the bill. But the provisions proved marginal at best.

Meanwhile, Cavanagh, with support from the Energy Foundation, supported the utilities at every turn, including acting as a key leader opposing the 1998 grassroots referendum aimed at repealing AB1890.

Cavanagh's role in helping to pass AB1890 and then defending it from repeal has earned him widespread outrage and contempt from the state's green/consumer groups. He still opposes legislation that would take mandated efficiency measures away from utility control and give it to municipalities.

The California experience, warns the state's green/consumer coalition, should stand as a warning to energy activists against accepting the marginal green provisions being tacked onto the Bush/Cheney energy plan.

TOO CHEAP TO METER?

AB 1890's driving force was a utility-sponsored provision to pay the utilities up to $28.5 billion in surcharges for investments in nuclear power, a technology once billed as "too cheap to meter."

In 1996 hearings, SoCalEd and PG&E branded their nuclear reactors at San Onofre and Diablo Canyon as too uneconomical to compete in the competitive free market that deregulation would allegedly bring.

They demanded that ratepayers compensate them for these and other bad investments before deregulation kicked in. That was the rationale for freezing rates at the 1996 established level.

The utility argument, echoed in states around the nation, was that since regulatory agencies had approved the nuclear investments, the public had an obligation to compensate the utilities for their nuclear expenses before opening the electricity market to competition.

Those watching closely argued that not only did this arrangement constitute an outrageous bailout for the utilities, but that the subsidies would be siphoned off for uses that would be of no benefit to Californians. Taken as a whole, warned Gene Coyle at an August 1996 press conference, deregulation and the torrent of cash it would generate for the utilities "will not build infrastructure in California. PG&E and Edison will likely invest it overseas, in places like Indonesia and Australia where both companies are already active. In fact, the entire $27 billion is a liquidation of California assets, with almost all of this ratepayer and taxpayer money likely to flow to foreign investments." That prediction proved prescient.

PG&E's Tremayne responds that the company was under no obligation to reinvest in California. "Under deregulation, the utility's investors—the shareholders of this company—were to get paid back for the investments they made years and years ago. [After the company sold its California generating plants] the shareholders either received the dividends by stock repurchase programs or by investing that money in other investments, including generation facilities outside of California. The way deregulation was set up, [it] intended to do exactly that—

pay the utility investors back money that they had invested in power plants that they were now forced under law to sell. So the shareholders essentially recovered their investments and reinvested them. This was done under the direction of the Public Utilities Commission, in synch with state law."

Both SoCalEd and PG&E say they have suffered huge losses in the last year or two, as the wholesale energy market has spun out of control, though these are losses for the utility subsidiaries, not necessarily for the parent companies which own generating companies as well as the utility subsidiaries' power distribution system. PG&E, the utility subsidiary, not the parent company, has run to bankruptcy protection (handing its executives huge bonuses the day before). In mid-May a federal judge barred consumer groups from participating in those bankruptcy proceedings, meaning the prime suitors in those hearings will be the very power generators who jacked up wholesale prices in the first place.

But by most accounts, the utilities' losses are billions less than the stranded cost bailouts they've laundered to their parent corporations. "It's a mafia operation," says Paul Fenn. "What happened to all that money?"

The Public Media Center's Gunther says the parent companies have spent much of this "rogue cash" as if they were "drunken sailors." Just as Coyle and others warned when AB1890 first became law, Pacific Gas & Electric's owner has made huge investments in power supply networks in New England and New York through its National Energy Group affiliate. SoCalEd's parent, operating through its Mission Energy subsidiary, has been deeply immersed in controversial speculations in Indonesia during the regime of the deposed dictator Suharto.

"The money has not gone to help things in California, that's for sure," says Gunther. "But where is it?"

FOLLOW THE MONEY

While an angry California public increasingly demands to know what its own utilities did with their money, they have also been forced to confront a second band of power magnates—the oil and gas companies close to George W. Bush.

To create competition, the AB1890 deregulation bill of 1996 established a complex scheme by which the utilities divested some, but not all, of their power plants. Those promoting the bill claimed the utilities would become pure distribution companies that would battle one another for the business of small customers.

The transmission wires that delivered the power would remain as regulated monopolies.

And then the generating facilities would, in theory, be bought by dozens of small, entrepreneurial power companies. The magic of the marketplace would drive prices down and service up.

The key to the utilities' deregulation scheme was the assumption that wholesale electric prices would stay low. SoCalEd and PG&E had devised its cap on consumer prices based on the idea that they could dominate supply.

In fact, the utilities did not divest themselves of all their power plants. And, for a series of complex reasons, they failed to enter into long-term contracts with the new generators, thus leaving the utilities dependent on spot markets, where short-term prices could shoot up without notice.

Whereas long-term contracts might have established stable prices over time, the spot market—where energy is sold daily, often in small quantities—is prone to price spikes. Those spikes are supposed to be moderated by the Federal Electrical Regulatory Commission. But over the past year demand on the spot market has regularly been higher than the supply generators make available at any given moment. Prices soar, leaving utilities little choice but to bid prices up dramatically in an effort to procure the electricity to supply their customers.

Thus California was put at the mercy of a handful of out-of-state energy speculators, most notably Duke Power of North Carolina, and Dynergy, Reliant and Enron, all of Texas. These are very big players, who more closely resemble the OPEC cartel than feisty Silicon Valley-type competitors that free market zealots envisioned.

According to Washington, D.C.-based Public Citizen, Enron, Dynergy and Reliant gave in excess of $1.5 million to Bush's campaign and inauguration committee, and to the Republican National Committee. In all, Public Citizen says nine power companies and a trade association with substantial interests in the California energy market gave more than $4 million to Republican candidates and party committees in the 2000 campaign. Bush's new Secretary of Energy, Spencer Abraham, was the energy industry's largest single campaign recipient during his failed U.S. Senate re-election bid in Michigan. Kenneth Lay, president of Enron, the largest U.S. natural gas supplier, is one of George W. Bush's key contributors, and very closest advisers. So is James Baker III, his father's former Secretary of State, and a principal at Reliant.

So while Governor Gray Davis and much of the California legislature received big campaign contributions from the state's utilities, the power companies that manipulated their power supply had their key connections in Washington, which they made good use of.

In a national radio address on May 19, Davis charged that "price gouging" sent the state's annual power bill soaring from $7 billion in 1999 to as much as $60 billion in 2001. Davis blamed it, "pure and simple," on "unconscionable price gouging by the big energy producers—most of them, incidentally, located in Texas."

While Bush has emphatically rejected price caps, widespread charges that his backers' price manipulations are illegal have been given new heft by Loretta Lynch, head of the California Public Utilities Commission. On May 19, the San Francisco Chronicle reported that whistleblower evidence indicated "that generators illegally manipulated prices by deliberately withholding electricity during shortages."

Lynch told a state Senate committee that in at least one instance three power plants simultaneously reduced output, causing prices to spike, and then restored output to cash in. "We certainly see a pattern," she said, while warning of possible criminal prosecutions to come.

By and large the targets of those charges have invested heavily in George Bush, and stand to gain billions if his national energy plan moves ahead. When push came to shove in the California case, they were vastly rewarded by the FERC's crucial refusal to cap prices at which they were selling power to California on the spot market. They also benefited from the FERC's prior opposition to investments in renewables and efficiency, which would have dampened the demand that helped fuel the crisis.

Lynch and the CPUC are not the only state officials warning of legal retribution. On May 2, Lt. Governor Cruz M. Bustamante filed a civil lawsuit against Dynergy, Duke Energy, Mirant, Reliant Energy and Williams Energy Services, alleging that they have systematically engaged in a price-fixing conspiracy to manipulate California's electricity market to "extract" unlawful profits that are draining the state's treasury.

"A cartel of five out-of-state generators has been holding us hostage through a practice of illegal and unfair price-fixing," Bustamante says.

"The energy crisis is not a problem of supply," says Democratic California Assemblywoman Barbara Matthews, who joined Bustamante in filing the suit on behalf of California taxpayers. "It's a problem of manipulation of our supply by out-of-state generators. Generators withhold power, create artificial shortages and play the 'Great American Shell Game' at the public's expense. We will not tolerate this any longer."

While Duke Energy and the others would not comment on the lawsuit, the company says it is "committed to continue playing a major role to help California address its electricity shortfall and the high prices many felt [last] summer." Duke officials say they are working to solve the problem by modernizing and bringing additional generating capacity online at existing power plants as well as signing long-term wholesale electricity contracts with Pacific Gas & Electric (PG&E). But the companies are keeping the terms of those contracts secret.

Public Citizen reports that Enron showed a 42 percent increase in profits last year, Reliant a 55 percent jump and Dynergy a 210 percent rise, all thanks to federal regulators' refusal to cap wholesale prices.

As energy analyst Eugene Coyle puts it: "We've been FERCed."

In April 2001, FERC did finally agree to a price cap scheme that is at best a mixed bag. But in the hot summer of 2000—with Bill Clinton in the White House—FERC stood by while wholesale prices soared.

San Diego Gas & Electric, having collected its final stranded cost money, was allowed by the Public Utilities Commission to unfreeze its consumer prices. The first shockwave of the deregulation disaster hit southern California consumers. SDG&E doubled and tripled its bills.

Consumer rates for SoCalEd and PG&E, however, remained capped. As wholesale prices soared, these large utilities claim to have lost more than $12 billion. When they—and Governor Davis—appealed to President Bush to recap wholesale rates, Bush refused, yielding spectacular profits for his friends at

Enron and Reliant, among others, not to mention the utilities' parent companies, Mission Energy and the National Energy Group.

It is this vise—between skyrocketing wholesale prices combined with frozen retail rates—that has prompted the contention that the problem in California is not a failure of deregulation, but rather that there simply hasn't been enough. The utilities have been desperate to end the rate freeze for consumers that they themselves invented.

In the breach, the utilities convinced Gray Davis to use state funds to buy power, continuing to deliver huge profits for the gas companies. But while escalating his rhetoric against the out-of-state suppliers, Governor Davis has refused widespread consumer demand that he use the state's leverage to take over the assets of the utilities that were still in the process of collecting more than $20 billion in the stranded cost cash bailout, money they promptly laundered to their parent companies.

In all, the double rip-off has yielded at least $20 billion for the utility parent corporations, and another $20 billion for the Bush-related gas companies, all for which California has nothing tangible to show. While PG&E has filed for bankruptcy protection, SoCalEd continues to pressure Davis for concessions, and the gas producers continue to demand that the taxpayers guarantee the purchase of power at rates that appear to fluctuate wildly based not on supply, but on their willingness to sell at uncapped prices.

FOSSIL & NUCLEAR VS. CLEAN AND PUBLIC

In the midst of a convoluted crisis that has so vastly enriched George Bush's supporters, the utility and gas industries are now furthering their agenda on other fronts. Bush and the industry's Congressional allies—most importantly Republican Senator Frank Murkowski of Alaska—want to lift environmental restrictions so that more power plants can be built, as Davis has already done within California. Alaska should be drilled, they say, as well as sensitive protected offshore eco-systems along both coasts and in the Gulf of Mexico, and virtually anywhere else the oil companies think they might make money.

But oil, however, has virtually nothing to do with solving an electricity crisis. Nationwide, less than 4 percent of U.S. electricity is generated from oil. The percentage is even smaller in California.

And then there's the push for more nuclear power plants. Around the April 26, 2001 anniversary of 15 years of fallout from the Chernobyl catastrophe, the mainstream media filled with talk of a revival of atomic energy. Not one of the major stories carried coverage of the huge stranded cost bailouts that had prompted the California crisis in the first place, or the fact that the builders of those nuke plants had labeled them "uncompetitive." Nor was there much mention of the February 3 fire at San Onofre that knocked out a turbine and may keep that nuclear plant off-line for months, at a cost of up to $100 million. In the midst of the state's worst energy crisis, this disaster knocked out, in a flash, fully 25 percent of SoCalEd's generating capacity, and 25 percent of the state's nuclear capacity, enough to supply more than 1.1 million homes.

GIVING IT TO CALIFORNIA CONSUMERS: A REVIEW OF THE DOUBLE RIP-OFF

Thanks to the opposition from utilities and FERC, California has been deprived of the natural-based energy sources the environmental movement tried to win for it.

Thanks to AB1890, the state has been forced to pay surcharges of $20 billion or more to bail the utilities out of their bad nuclear investments.

Thanks to the independent power generators, the state is being gouged for billions to buy electricity it might have otherwise gotten from in-state green sources and increased efficiency.

Thanks to Governor Davis's unwillingness to leverage state money or eminent domain into a takeover of utility assets, the electric grid and production facilities remain owned by utilities and the power producers who caused the crisis in the first place, leaving the state as vulnerable as ever to future manipulations.

And now, thanks to the hysteria whipped up by the rolling blackouts, the state is being inundated by new privately owned, fossil-fuel-burning generators that will escalate pollution levels by rolling over decades of hard-won environmental protections.

—H.W.

Little attention has been devoted to the fact that wind power is now far cheaper than all other sources of new generation except brown coal. At 2.5 cents/kilowatt hour, it is far cheaper than any projected costs for new nuclear capacity, and is even lower than natural gas in many instances. In 2000, Germany, which is moving to shut its 19 nuclear reactors, brought on line 1,300 megawatts of wind, 200 more than was lost at San Onofre. Great Britain has just committed $2.5 billion to offshore turbines. The Public Utilities Commission of Minnesota has deemed wind its "least cost" alternative and ordered at least 300 megawatts of capacity to be added to the 400 already in place. A major wind farm along the Oregon-Washington border began construction in February, and will go on line in December. It is widely known that the Great Plains between the Mississippi and the Rockies are the "Saudi Arabia of wind," with virtually infinite capacity. With current technology, and with additional offshore capacity, wind power could meet much if not all the nation's electric needs long before a new generation of nuclear reactors could be built.

Even atomic energy's staunchest supporters acknowledge it would take at least five years to bring new reactors on line, far more than it would take to bring on vast new supplies of still other alternative technologies, most importantly photovoltaic cells (PV), which transform sunlight directly to electricity. Within five years, PV is expected to plummet far below even the most optimistic projected costs of atomic power. Increased efficiency and conservation are already far cheaper.

PG&E's Tremayne says conservation is only part of the solution. "PG&E is a leader in conservation efforts and has been for 20 years. Our programs have been looked at from across the country as models of how to institute energy efficiency programs with high customer participation. As a result, Californians are the most energy efficient—in terms of per capita consumption—in the country."

"The problem is that over the past 10 years demand has grown and supply hasn't," Tremayne says. "We've been able to rely upon the Pacific Northwest and the Southwest to supply power to help California meet its needs: As much as 8,000 or 9,000 megawatts on any given summer day. But those two West Coast regions have experienced significant growth as well over the last six years, so they don't have the available power to export any longer. We haven't built any new power plants in California" in a period where 600,000 new residential customers have been added each year.

But as Paul Fenn puts it, the crisis "is not about supply. There's plenty of capacity around. It's a problem of who controls the supply, and the money that pays for it. That's why local control of electricity supply is critical to real solutions and why the idea of gutting environmental laws under the auspices of energy relief is such a horribly impotent gesture."

"We're so afraid to let these companies go under," Fenn complains. "But when all is said and done, the public would be better off letting them go down and using eminent domain to buy their assets. At least then we'll have gotten something tangible out of the deal."

While Governor Davis has talked about possibly buying the utilities' transmission lines, the green/consumer community is increasingly demanding public utility ownership. But Fenn and others are not particularly eager to have the state run a single public-owned utility. Rather, they look to a more local-based solution.

That point of view has gotten powerful backing from James McClatchy, publisher of the Sacramento Bee, and one of the few dissenting voices in the major media's coverage of the California crisis. Beyond the state's taking over the transmission lines, McClatchy wrote in a February 18 editorial, "the next step would be for the state to buy the associated generating facilities."

"Any final solution," says McClatchy, "would have to include public ownership of the generating plants that PG&E and Southern California Edison sold to speculators, as well as the facilities they still own."

The way to deliver the power, McClatchy adds, would be through "existing locally owned and managed utility districts," as well as through ones newly organized to handle the job. "With public ownership of these systems," he says, "would come increased public transparency on all aspects of the operations—where there is little now—and thus less opportunity for sweetheart deals with friendly financiers or broker," leaving less room for "the rapaciousness of speculators and selfish political partisanship."

PG&E's Tremayne denies that public ownership of the companies' generating plants will solve the current crisis.

"Our facilities continue to be regulated under a cost-of-service structure under the Public Utilities Commission. So our generation is being provided to customers here in California at cost and aren't part of the problem. The real problem comes from the out-of-state generators that have been gouging the market for a year. That's where people need to be focusing their efforts."

As for local ownership of the distribution or transmission system, Tremayne says that wouldn't solve the problem either. "Those [distribution] costs are regulated by the CPUC and the FERC. The problem is in the generation market. It doesn't matter who owns the distribution or transmission system—the problem will still exist if we don't get control of the generation market."

THE MUNICIPAL ALTERNATIVE

The greatest of the untold stories of the California crisis is the stunning success of the state's two municipal-owned utilities, the Sacramento Municipal Utility District (SMUD) and the Los Angeles Department of Water and Power (LADWP). Both have not only weathered the storm, but thrived in it.

McClatchy and others point to SMUD and the LADWP for keeping rates stable for their customers while reaping substantial profits selling power into the grid.

In June 1989, Sacramento voted to shut the district's one nuclear reactor, at Rancho Seco. S. David Freeman, who now runs the Los Angeles utility and previously ran the Tennessee Valley Authority, led SMUD into an era powered increasingly—though modestly—by wind and PV facilities, including enough rooftop solar panels for some 6,500 families.

SMUD also inaugurated an unprecedented campaign for increased efficiency. It offered its customers $100 rebates for retiring wasteful old refrigerators (refrigerators account for 20 percent of the average household's electricity consumption). It distributed energy-efficient light bulbs. It promoted solar water heaters and PV panels. It also planted thousands of shade trees to slash summer air conditioning demand.

SMUD's progress occurred amidst a state-wide deregulatory wave that headed in precisely the opposite direction. In the rest of the state, public conservation and efficiency programs were weakened with the deregulation of wholesale supply. The uncertainties of impending consumer price deregulation stalled strong statewide movements to build windmills and install solar panels. And AB1890 freed the utilities from the renewable and efficiency programs mandated by the Public Utilities Commission—which SoCalEd's Bryson had once headed.

This record confirmed public interest advocates' belief that deregulation was incompatible with environmental concerns. Meanwhile, Dan Berman says SMUD was doing "the kinds of things you expect a utility owned by the public to do. It's what we want to see all over California."

Berman has worked for years to win a municipal utility for his home town of Davis. Parallel campaigns have sprung up elsewhere throughout the state. The San Francisco Board of Supervisors has approved a fall referendum for municipal ownership, and five East Bay towns, including Oakland and Berkeley, will hold similar votes. San Diego may also vote on municipal power, and other cities, towns and counties are virtually certain to join in, though they will face massive utility resistance. In mid-May, San Francisco's board of supervisors announced it will take bids on at least 50 megawatts of solar power to be installed in the city and paid for with public bonds—an effort that may be joined by Sacramento.

"SMUD is a role model," says Berman. "If we're to see any progress at all," public power and municipal control is crucial.

Nuclear opponents have long argued that if the hundreds of billions that went into reactors like the ones at the core of the California crisis had gone instead into renewables and efficiency, nothing like California's current crisis would have ever happened. Critics who predicted the disaster that is AB1890 insist that electricity is a service, not a commodity, and that its production and distribution can never be left to an uncertain market that will always be subject to the whims of private utilities and the barons of fossil fuels.

"The last thing the utilities and the independent power producers want is public ownership," says Fenn. "But municipal power, controlled at the local level, is ultimately the key. Until we get it, things are only going to get worse."

"Deregulation of the electricity monopoly is a failure," adds McClatchy. "The monopoly should be returned to the tax-paying consumers who support it and depend on it."

Harvey Wasserman is the author of *The Last Energy War: The Battle Over Utility Deregulation* and a senior advisor to Greenpeace USA.

From *Multinational Monitor*, June 2001, pp. 9-20. © 2001 by Multinational Monitor. Reprinted by permission.

UNIT 4

Biosphere: Endangered Species

Unit Selections

Key Points to Consider

- How and why are alien plants and animals spreading so rapidly and what threats do they pose to native species in nearly all parts of the world? What gives the alien species advantages that native plants do not have?

- Assess the nature of biodiversity in the United States. Why are tropical forests so often mentioned as centers of biodiversity while midlatitude regions like the United States have been ignored as storehouses of biological complexity?

- Why do biologists have such a difficult time in being precise about the number of plants and animals that are going extinct? How does the current rate of mass extinction compare with past periods of species loss in terms of numbers and in terms of time?

- How can newer forms of ecotourism such as wildlife watching protect species against extinction? Does wildlife watching, as contrasted with hunting and poaching, pose similar kinds of problems for wild animals?

 **Links: www.dushkin.com/online/**
These sites are annotated in the World Wide Web pages.

Friends of the Earth
http://www.foe.co.uk/index.html

Smithsonian Institution Web Site
http://www.si.edu

World Wildlife Federation (WWF)
http://www.wwf.org

Tragically, the modern conservation movement began too late to save many species of plants and animals from extinction. In fact, even after concern for the biosphere developed among resource managers, their effectiveness in halting the decline of herds and flocks, packs and schools, or groves and grasslands has been limited by the ruthlessness and efficiency of the competition. Wild plants and animals compete directly with human beings and their domesticated livestock and crop plants for living space and for other resources such as sunlight, air, water, and soil. As the historical record of this competition in North America and other areas attests, since the seventeenth century human settlement has been responsible—either directly or indirectly—for the demise of many plant and wildlife species. It should be noted that extinction is a natural process—part of the evolutionary cycle—and not always created by human activity, but human actions have the capacity to accelerate a natural process that might otherwise take millennia.

In the opening article of this unit, the losses of Earth's plants and animals from human impact are tallied. Author David Quammen makes the point in "Planet of Weeds" that the nature of human impact on the biosphere reduces biodiversity and encourages the dominance of plant and animal species that can be termed "weeds": hardy, resilient, able to live almost anywhere, and capable of choking out native species almost everywhere. Among the weedy species, human beings are paramount and Quammen's pessimistic view of a future world is of humans and a few other survivors "picking through the rubble" of a once-fertile planet.

The theme of biodiversity is continued in the unit's second article in which Bruce Stein, vice-president of the Association for Biodiversity Information, discusses the susceptibility of the United States for disruption of its biological systems. In "A Fragile Cornucopia: Assessing the Status of U.S. Biodiversity," Stein notes that the United States has as rich an array of ecological conditions as can be found anywhere in the world, and a correspondingly high level of biodiversity. In environments stretching from the Alaskan Arctic to the Hawaiian tropics, the lush coastal swamps of Florida to the arid deserts of Nevada, plant and animal species of national and global significance exist. The task of researchers is to identify those species and to fit them into the global pattern of biological complexity in order to understand the implications of actions that would seem, at first glance, to have only local relevance.

The third article in the section also deals with the human impact on environmental systems that begins locally but stretches across the world. Environmental researcher Christopher Bright describes one of the least anticipated results of the growing global economy: the spread of "invasive species." In "Invasive Species: Pathogens of Globalization," Bright notes that thousands of alien species are traveling aboard ships, planes, and railroad

cars from one continent to another, often carried in commodities themselves. This consequence of the worldwide global trading network poses enormous hazards for native species in affected areas and confronting the problem may be as important a challenge to environmental quality as global carbon emissions. And as recent events that Bright could not have predicted have shown, when pathogens are used intentionally as weapons, the challenge becomes even greater.

The last two articles in this section deal more with wildlife than plants, although a clear distinction between human impact on plants and on animals is impossible. The prospects for the future of both plants and wildlife are not appreciably different. Land developers destroy habitats as cities encroach upon the countryside and have an impact on both plants and animals. Living space for all organisms is destroyed as river valleys are transformed into reservoirs for the generation of hydropower and as forests are removed for construction materials and for paper. Toxic wastes from urban areas work their way through the food chain of plants into animal populations, annually killing thousands of animals in the United States alone. Rural lands are sprayed with herbicides and pesticides, which, while eradicating unwanted plants, also kill bird life and small mammals. Most important, wild plants and animals are placed in jeopardy for the very simple reason that they compete with domesticated species and humans for the same resources. And in any instance in which the protection of other organisms comes at the expense of crops, livestock, or humans, the other organisms are going to lose.

One of the clearest and most recent examples of heightened competition between domesticated and nondomesticated species is currently being played out in virtually all of the world's environments where thousands of species are becoming extinct, most of them because of competition or competition-related processes. The rate of extinction produced through human activities is apparently between 100 to 10,000 times the normal biological rate of change. In "Mass Extinction," science writer David Hosansky catalogues this global process and warns that we can't even imagine the long-term consequences of the loss of biodiversity that extinctions pose. One hopeful sign in the ongoing territorial battle between humans and other organisms is the subject of the concluding article in the section. In "Watching vs. Taking," writer and ecotourist Howard Youth describes the shift in the relationship between humans and wildlife. He explains that millions of people are turning away from the hunting of species for food, pelts, or sport and turning to the simple pleasure of watching. The trend began with the "birders" who turned the hobby of bird-watching into an important economic enterprise in many areas. It continues with the increasing number of people willing to engage in this new kind of hunt.

PLANET OF WEEDS

Tallying the losses of Earth's animals and plants

By David Quammen

Hope is a duty from which paleontologists are exempt. Their job is to take the long view, the cold and stony view, of triumphs and catastrophes in the history of life. They study the fossil record, that erratic selection of petrified shells, carapaces, bones, teeth, tree trunks, leaves, pollen, and other biological relics, and from it they attempt to discern the lost secrets of time, the big patterns of stasis and change, the trends of innovation and adaptation and refinement and decline that have blown like sea winds among ancient creatures in ancient ecosystems. Although life is their subject, death and burial supply all their data. They're the coroners of biology. This gives to paleontologists a certain distance, a hyperopic perspective beyond the reach of anxiety over outcomes of the struggles they chronicle. If hope is the thing with feathers, as Emily Dickinson said, then it's good to remember that feathers don't generally fossilize well. In lieu of hope and despair, paleontologists have a highly developed sense of cyclicity. That's why I recently went to Chicago, with a handful of urgently grim questions, and called on a paleontologist named David Jablonski. I wanted answers unvarnished with obligatory hope.

Jablonski is a big-pattern man, a macro-evolutionist, who works fastidiously from the particular to the very broad. He's an expert on the morphology and distribution of marine bivalves and gastropods—or clams and snails, as he calls them when speaking casually. He sifts through the record of those mollusk lineages, preserved in rock and later harvested into museum drawers, to extract ideas about the origin of novelty. His attention roams back through 600 million years of time. His special skill involves framing large, resonant questions that can be answered with small, lithified clamshells. For instance: By what combinations of causal factor and sheer chance have the great evo-lutionary innovations arisen? How quickly have those innovations taken hold? How long have they abided? He's also interested in extinction, the converse of abidance, the yang to evolution's yin. Why do some species survive for a long time, he wonders, whereas others die out much sooner? And why has the rate of extinction—low throughout most of Earth's history—spiked upward cataclysmically on just a few occasions? How do these cataclysmic episodes, known in the trade as mass extinctions, differ in kind as well as degree from the gradual process of species extinction during the millions of years between? Can what struck in the past strike again?

The concept of mass extinction implies a biological crisis that spanned large parts of the planet and, in a relatively short time, eradicated a sizable number of species from a variety of groups. There's no absolute threshold of magnitude, and dozens of different episodes in geologic history might qualify, but five big ones stand out: Ordovician, Devonian, Permian, Triassic, Cretaceous. The Ordovician extinction, 439 million years ago, entailed the disappearance of roughly 85 percent of marine animal species—and that was before there were any animals *on land*. The Devonian extinction, 367 million years ago, seems to have been almost as severe. About 245 million years ago came the Permian extinction, the worst ever, claiming 95 percent of all known animal species and therefore almost wiping out the animal kingdom altogether. The Triassic, 208 million years ago, was bad again, though not nearly so bad as the Permian. The most recent was the Cretaceous extinction (sometimes called the K-T event because it defines the boundary between two geologic periods, with K for Cretaceous, never mind why, and T for Tertiary), familiar even to schoolchildren because it ended the age of dinosaurs. Less familiarly, the K-T event also brought extinction of the marine reptiles and the ammonites, as well as major losses of species among fish, mammals, amphibians, sea urchins, and other groups, totaling 76 percent of all species. In between these five episodes occurred some lesser mass extinctions, and throughout the intervening lulls extinction continued, too—but at a much slower pace, known as the background rate, claiming only about one species in any major group every million years. At the background rate, extinction is infrequent enough to be counterbalanced by the evolution of new species. Each of the five major episodes, in contrast, represents a drastic net loss of species diversity, a deep trough of biological impoverishment from which Earth only slowly recovered. How slowly? How long is the lag between a nadir of impoverishment and a recovery to ecological fullness? That's another of Jablonski's research interests. His rough estimates run to 5 or 10 million years. What drew me to this man's work, and then to his doorstep, were his special competence on mass extinctions and his willingness to discuss the notion that a sixth one is in progress now.

THE EARTH HAS UNDERGONE FIVE MAJOR EXTINCTION PERIODS, EACH REQUIRING MILLIONS OF YEARS OF RECOVERY

Some people will tell you that we as a species, *Homo sapiens*, the savvy ape, all 5.9 billion of us in our collective impact, are destroying the world. Me, I won't tell you that, because "the world" is so vague, whereas what we are or aren't destroying is quite specific. Some people will tell you that we are rampaging suicidally toward a degree of global wreckage that will result in our own extinction. I won't tell you that either. Some people say that the environment will be the paramount political and social concern of the twenty-first century, but what they mean by "the environment" is anyone's guess. Polluted air? Polluted water? Acid rain? A frayed skein of ozone over Antarctica? Greenhouse gases emitted by smokestacks and cars? Toxic wastes? None of these concerns is the big one, paleontological in scope, though some are more closely entangled with it than others. If the world's air is clean for humans to breathe but supports no birds or butterflies, if the world's waters are pure for humans to drink but contain no fish or crustaceans or diatoms, have we solved our environmental problems? Well, I suppose so, at least as environmentalism is commonly construed. That clumsy, confused, and presumptuous formulation "the environment" implies viewing air, water, soil, forests, rivers, swamps, deserts, and oceans as merely a milieu within which something important is set: human life, human history. But what's at issue in fact is not an environment; it's a living world.

Here instead is what I'd like to tell you: The consensus among conscientious biologists is that we're headed into another mass extinction, a vale of biological impoverishment commensurate with the big five. Many experts remain hopeful that we can brake that descent, but my own view is that we're likely to go all the way down. I visited David Jablonski to ask what we might see at the bottom.

O n a hot summer morning, Jablonski is busy in his office on the second floor of the Hinds Geophysical Laboratory at the University of Chicago. It's a large open room furnished in tall bookshelves, tables piled high with books, stacks of paper standing knee-high off the floor. The walls are mostly bare, aside from a chart of the geologic time scale, a clipped cartoon of dancing tyrannosaurs in red sneakers, and a poster from a Rodin exhibition, quietly appropriate to the overall theme of eloquent stone. Jablonski is a lean forty-five-year-old man with a dark full beard. Educated at Columbia and Yale, he came to Chicago in 1985 and has helped make its paleontology program perhaps the country's best. Although in not many hours he'll be leaving

on a trip to Alaska, he has been cordial about agreeing to this chat. Stepping carefully, we move among the piled journals, reprints, and photocopies. Every pile represents a different research question, he tells me. "I juggle a lot of these things all at once because they feed into one another." That's exactly why I've come: for a little rigorous intellectual synergy.

Let's talk about mass extinctions, I say. When did someone first realize that the concept might apply to current events, not just to the Permian or the Cretaceous?

He begins sorting through memory, back to the early 1970s, when the full scope of the current extinction problem was barely recognized. Before then, some writers warned about "vanishing wildlife" and "endangered species," but generally the warnings were framed around individual species with popular appeal, such as the whooping crane, the tiger, the blue whale, the peregrine falcon. During the 1970s a new form of concern broke forth—call it wholesale concern—from the awareness that unnumbered millions of narrowly endemic (that is, unique and localized) species inhabit the tropical forests and that those forests were quickly being cut. In 1976, a Nairobi-based biologist named Norman Myers published a paper in *Science* on that subject; in passing, he also compared current extinctions with the rate during what he loosely called "the 'great dying' of the dinosaurs." David Jablonski, then a graduate student, read Myers's paper and tucked a copy into his files. This was the first time, as Jablonski recalls, that anyone tried to quantify the rate of present-day extinctions. "Norman was a pretty lonely guy, for a long time, on that," he says. In 1979, Myers published *The Sinking Ark*, explaining the problem and offering some rough projections. Between the years 1600 and 1900, by his tally, humanity had caused the extinction of about 75 known species, almost all of them mammals and birds. Between 1900 and 1979, humans had extinguished about another 75 known species, representing a rate well above the rate of known losses during the Cretaceous extinction. But even more worrisome was the inferable rate of unrecorded extinctions, recent and now impending, among plants and animals still unidentified by science. Myers guessed that 25,000 plant species presently stood jeopardized, and maybe hundreds of thousands of insects. "By the time human communities establish ecologically sound life-styles, the fallout of species could total several million." Rereading that sentence now, I'm struck by the reck-

less optimism of his assumption that human communities eventually will establish "ecologically sound life-styles."

BIOLOGISTS BELIEVE THAT WE ARE ENTERING ANOTHER MASS EXTINCTION, A VALE OF BIOLOGICAL IMPOVERISHMENT

Although this early stab at quantification helped to galvanize public concern, it also became a target for a handful of critics, who used the inexactitude of the numbers to cast doubt on the reality of the problem. Most conspicuous of the naysayers was Julian Simon, a economist at the University of Maryland, who argued bullishly that human resourcefulness would solve all problems worth solving, of which a decline in diversity of tropical insects wasn't one.

In a 1986 issue of *New Scientist*, Simon rebutted Norman Myers, arguing from his own construal of select data that there was "no obvious recent downward trend in world forests—no obvious 'losses' at all, and certainly no 'near catastrophic' loss." He later co-authored an op-ed piece in the *New York Times* under the headline "Facts, Not Species, Are Periled." Again he went after Myers, asserting a "complete absence of evidence for the claim that the extinction of species is going up rapidly—or even going up at all." Simon's worst disservice to logic in that statement and others was the denial that *inferential* evidence of wholesale extinction counts for anything. Of inferential evidence there was an abundance—for example, from the Centinela Ridge in a cloud-forest zone of western Ecuador, where in 1978 the botanist Alwyn Gentry and a colleague found thirty-eight species of narrowly endemic plants; including several with mysteriously black leaves. Before Gentry could get back, Centinela Ridge had been completely deforested, the native plants replaced by cacao and other crops. As for inferential evidence generally, we might do well to remember what it contributes to our conviction that approximately 105,000 Japanese civilians died in the atomic bombing of Hiroshima. The city's population fell abruptly on August 6, 1945, but there was no one-by-one identification of 105,000 bodies.

Nowadays a few younger writers have taken Simon's line, pooh-poohing the concern over extinction. As for Simon himself, who died earlier this year, perhaps the truest sentence he left behind was, "We must also try to get more reliable information about the number of species that might be lost with various changes in the forests." No one could argue.

But it isn't easy to get such information. Field biologists tend to avoid investing their precious research time in doomed tracts of forest. Beyond that, our culture offers little institutional support for the study of narrowly endemic species in order to register their existence *before* their habitats are destroyed. Despite these obstacles, recent efforts to quantify rates of extinction have supplanted the old warnings. These new estimates use satellite imaging and improved on-the-ground data about deforestation, records of the many human-caused extinctions on islands, and a branch of ecological theory called island biogeography, which connects documented island cases with the mainland problem of forest fragmentation. These efforts differ in particulars, reflecting how much uncertainty is still involved, but their varied tones form a chorus of consensus. I'll mention three of the most credible.

EVEN BY CONSERVATIVE ESTIMATES, HUGE PERCENTAGES OF EARTH'S ANIMALS AND PLANTS WILL SIMPLY DISAPPEAR

W. V. Reid, of the World Resources Institute, in 1992 gathered numbers on the average annual deforestation in each of sixty-three tropical countries during the 1980s and from them charted three different scenarios (low, middle, high) of presumable forest loss by the year 2040. He chose a standard mathematical model of the relationship between decreasing habitat area and decreasing species diversity, made conservative assumptions about the crucial constant, and ran his various deforestation estimates through the model. Reid's calculations suggest that by the year 2040, between 17 and 35 percent of tropical forest species will be extinct or doomed to be. Either at the high or the low end of this range, it would amount to a bad loss, though not as bad as the K-T event. Then again 2040 won't mark the end of human pressures on biological diversity or landscape.

Robert M. May, an ecologist at Oxford, co-authored a similar effort in 1995. May

and his colleagues noted the five causal factors that account for most extinctions: habitat destruction, habitat fragmentation, overkill, invasive species, and secondary effects cascading through an ecosystem from other extinctions. Each of those five is more intricate than it sounds. For instance, habitat fragmentation dooms species by consigning them to small, island-like parcels of habitat surrounded by an ocean of human impact and by then subjecting them to the same jeopardies (small population size, acted upon by environmental fluctuation, catastrophe, inbreeding, bad luck, and cascading effects) that make island species especially vulnerable to extinction. May's team concluded that most extant bird and mammal species can expect average life spans of between 200 and 400 years. That's equivalent to saying that about a third of one percent will go extinct each year until some unimaginable end point is reached. "Much of the diversity we inherited," May and his co-authors wrote, "will be gone before humanity sorts itself out."

The most recent estimate comes from Stuart L. Pimm and Thomas M. Brooks, ecologists at the University of Tennessee. Using a combination of published data on bird species lost from forest fragments and field data gathered themselves, Pimm and Brooks concluded that 50 percent of the world's forest-bird species will be doomed to extinction by deforestation occurring over the next half century. And birds won't be the sole victims. "How many species will be lost if current trends continue?" the two scientists asked. "Somewhere between one third and two thirds of all species—easily making this event as large as the previous five mass extinctions the planet has experienced."

IN THE NEXT FIFTY YEARS, DEFORESTATION WILL DOOM ONE HALF OF THE WORLD'S FOREST-BIRD SPECIES

Jablonski, who started down this line of thought in 1978, offers me a reminder about the conceptual machinery behind such estimates. "All mathematical models," he says cheerily, "are wrong. They are approximations. And the question is: Are they usefully wrong, or are they meaninglessly wrong?" Models projecting present and future species loss are useful, he suggests, if they help people realize that *Homo sapiens* is perturbing Earth's biosphere to a

degree it hasn't often been perturbed before. In other words, that this is a drastic experiment in biological drawdown we're engaged in, not a continuation of routine.

Behind the projections of species loss lurk a number of crucial but hard-to-plot variables, among which two are especially weighty: continuing landscape conversion and the growth curve of human population.

Landscape conversion can mean many things: draining wetlands to build roads and airports, turning tallgrass prairies under the plow, fencing savanna and overgrazing it with domestic stock, cutting second-growth forest in Vermont and consigning the land to ski resorts or vacation suburbs, slash-and-burn clearing of Madagascar's rain forest to grow rice on wet hillsides, industrial logging in Borneo to meet Japanese plywood demands. The ecologist John Terborgh and a colleague, Carel P. van Schaik, have described a four-stage process of landscape conversion that they call the land-use cascade. The successive stages are: 1) *wildlands*, encompassing native floral and faunal communities altered little or not at all by human impact; 2) *extensively used areas*, such as natural grasslands lightly grazed, savanna kept open for prey animals by infrequent human-set fires, or forests sparsely worked by slash-and-burn farmers at low density; 3) *intensively used areas*, meaning crop fields, plantations, village commons, travel corridors, urban and industrial zones; and finally 4) *degraded land*, formerly useful but now abused beyond value to anybody. Madagascar, again, would be a good place to see all four stages, especially the terminal one. Along a thin road that leads inland from a town called Mahajanga, on the west coast, you can gaze out over a vista of degraded land—chalky red hills and gullies, bare of forest, burned too often by graziers wanting a short-term burst of pasturage, sparsely covered in dry grass and scrubby fan palms, eroded starkly, draining red mud into the Betsiboka River, supporting almost no human presence. Another showcase of degraded land—attributable to fuelwood gathering, overgrazing, population density, and decades of apartheid—is the Ciskei homeland in South Africa. Or you might look at overirrigated crop fields left ruinously salinized in the Central Valley of California.

Among all forms of landscape conversion, pushing tropical forest from the *wildlands* category to the *intensively used* category has the greatest impact on biological diversity. You can see it in western India, where a spectacular deciduous ecosystem known as the Gir forest (home to the last surviving population of the Asiatic lion, *Panthera leo persica*) is yielding along its ragged edges to new mango orchards, peanut fields, and lime quarries for cement. You can see it in the central Amazon, where big tracts of rain forest have been felled and burned, in a largely futile attempt (encouraged by misguided government incentives, now revoked) to pasture cattle on sun-hardened clay. According to the United Nations Food and Agriculture Organization, the rate of deforestation in tropical countries has increased (contrary to Julian Simon's claim) since the 1970s, when Myers made his estimates. During the 1980s, as the FAO reported in 1993, that rate reached 15.4 million hectares (a hectare being the metric equivalent of 2.5 acres) annually. South America was losing 6.2 million hectares a year. Southeast Asia was losing less in area but more proportionally: 1.6 percent of its forests yearly. In terms of cumulative loss, as reported by other observers, the Atlantic coastal forest of Brazil is at least 95 percent gone. The Philippines, once nearly covered with rain forest, has lost 92 percent. Costa Rica has continued to lose forest, despite that country's famous concern for its biological resources. The richest of old-growth lowland forests in West Africa, India, the Greater Antilles, Madagascar, and elsewhere have been reduced to less than a tenth of their original areas. By the middle of the next century, if those trends continue, tropical forest will exist virtually nowhere outside of protected areas—that is, national parks, wildlife refuges, and other official reserves.

How many protected areas will there be? The present worldwide total is about 9,800, encompassing 6.3 percent of the planet's land area. Will those parks and reserves retain their full biological diversity? No. Species with large territorial needs will be unable to maintain viable population levels within small reserves, and as those species die away their absence will affect others. The disappearance of big predators, for instance, can release limits on medium-size predators and scavengers, whose overabundance can drive still other species (such as ground-nesting birds) to extinction. This has already happened in some habitat fragments, such as Panama's Barro Colorado Island, and been well documented in the literature of island biogeography. The lesson of fragmented habitats is Yeatsian: Things fall apart.

Human population growth will make a bad situation worse by putting ever more pressure on all available land.

Population growth rates have declined in many countries within the past several decades, it's true. But world population is still increasing, and even if average fertility suddenly, magically, dropped to 2.0 children per female, population would continue to increase (on the momentum of birth rate exceeding death rate among a generally younger and healthier populace) for some time. The annual increase is now 80 million people, with most of that increment coming in less-developed countries. The latest long-range projections from the Population Division of the United Nations, released earlier this year, are slightly down from previous long-term projections in 1992 but still point toward a problematic future. According to the U.N.'s middle estimate (and most probable? hard to know) among seven fertility scenarios, human population will rise from the present 5.9 billion to 9.4 billion by the year 2050, then to 10.8 billion by 2150, before leveling off there at the end of the twenty-second century. If it happens that way, about 9.7 billion people will inhabit the countries included within Africa, Latin America, the Caribbean, and Asia. The total population of those countries—most of which are in the low latitudes, many of which are less developed, and which together encompass a large portion of Earth's remaining tropical forest—will be more than twice what it is today. Those 9.7 billion people, crowded together in hot places, forming the ocean within which tropical nature reserves are insularized, will constitute 90 percent of humanity. Anyone interested in the future of biological diversity needs to think about the pressures these people will face, and the pressures they will exert in return.

THE LESSON TO BE LEARNED FROM FRAGMENTED, ISOLATED HABITATS IS YEATSIAN: THINGS FALL APART

We also need to remember that the impact of *Homo sapiens* on the biosphere can't be measured simply in population figures. As the population expert Paul Harrison pointed out in his book *The Third Revolution*, that impact is a product of three variables: population size, consumption level, and technology. Although pop-

ulation growth is highest in less-developed countries, consumption levels are generally far higher in the developed world (for instance, the average American consumes about ten times as much energy as the average Chilean, and about a hundred times as much as the average Angolan), and also higher among the affluent minority in any country than among the rural poor. High consumption exacerbates the impact of a given population, whereas technological developments may either exacerbate it further (think of the automobile, the air conditioner, the chainsaw) or mitigate it (as when a technological innovation improves efficiency for an established function). All three variables play a role in every case, but a directional change in one form of human impact—upon air pollution from fossil-fuel burning, say, or fish harvest from the seas—can be mainly attributable to a change in one variable, with only minor influence from the other two. Sulfur-dioxide emissions in developed countries fell dramatically during the 1970s and '80s, due to technological improvements in papermaking and other industrial processes; those emissions would have fallen still farther if not for increased population (accounting for 25 percent of the upward vector) and increased consumption (accounting for 75 percent). Deforestation, in contrast, is a directional change that *has* been mostly attributable to population growth.

WE CONFRONT THE VISION OF A HUMAN POPULATION PRESSING SNUGLY AROUND WHATEVER NATURAL LANDSCAPE REMAINS

According to Harrison's calculations, population growth accounted for 79 percent of the deforestation in less-developed countries between 1973 and 1988. Some experts would argue with those calculations, no doubt, and insist on redirecting our concern toward the role that distant consumers, wood-products buyers among slow-growing but affluent populations of the developed nations, play in driving the destruction of Borneo's dipterocarp forests or the hardwoods of West Africa. Still, Harrison's figures point toward an undeniable reality; more total people will need more total land. By his estimate, the minimum land necessary for food growing and other human needs (such as water supply and waste dumping) amounts to one fifth of a hectare per person. Given the U.N.'s projected increase of 4.9 billion souls before the human population finally levels off, that comes to another billion hectares of human-claimed landscape, a billion hectares less forest—even without allowing for any further deforestation by the current human population, or for any further loss of agricultural land to degradation. A billion hectares—in other words, 10 million square kilometers—is, by a conservative estimate, well more than half the remaining forest area in Africa, Latin America, and Asia. This raises the vision of a very exigent human population pressing snuggly around whatever patches of natural landscape remain.

Add to that vision the extra, incendiary aggravation of poverty. According to a recent World Bank estimate, about 30 percent of the total population of less-developed countries lives in poverty. Alan Durning, in his 1992 book *How Much Is Enough? The Consumer Society and the Fate of the Earth*, puts it in a broader perspective when he says that the world's human population is divided among three "ecological classes": the consumers, the middle-income, and the poor. His consumer class includes those 1.1 billion fortunate people whose annual income per family member is more than $7,500. At the other extreme, the world's poor also number about 1.1 billion people—all from households with less than $700 annually per member. "They are mostly rural Africans, Indians, and other South Asians," Durning writes. "They eat almost exclusively grains, root crops, beans, and other legumes, and they drink mostly unclean water. They live in huts and shanties, they travel by foot, and most of their possessions are constructed of stone, wood, and other substances available from the local environment." He calls them the "absolute poor." It's only reasonable to assume that another billion people will be added to that class, mostly in what are now the less-developed countries, before population growth stabilizes. How will those additional billion, deprived of education and other advantages, interact with the tropical landscape? Not likely by entering information-intensive jobs in the service sector of the new global economy. Julian Simon argued that human ingenuity—and by extension, human population itself—is "the ultimate resource" for solving Earth's problems, transcending Earth's limits, and turning scarcity into abundance. But if all the bright ideas generated by a human population of 5.9 billion haven't yet relieved

the desperate needfulness of 1.1 billion absolute poor, why should we expect that human ingenuity will do any better for roughly 2 billion poor in the future?

Other writers besides Durning have warned about this deepening class rift. Tom Athanasiou, in *Divided Planet: The Ecology of Rich and Poor*, sees population growth only exacerbating the division, and notes that governments often promote destructive schemes of transmigration and rain-forest colonization as safety valves for the pressures of land hunger and discontent. A young Canadian policy analyst named Thomas F. Homer-Dixon, author of several calm-voiced but frightening articles on the linkage between what he terms "environmental scarcity" and global sociopolitical instability, reports that the amount of cropland available per person is falling in the less-developed countries because of population growth and because millions of hectares "are being lost each year to a combination of problems, including encroachment by cities, erosion, depletion of nutrients, acidification, compacting and salinization and waterlogging from overirrigation." In the cropland pinch and other forms of environmental scarcity, Homer-Dixon foresees potential for "a widening gap" of two sorts—between demands on the state and its ability to deliver, and more basically between rich and poor. In conversation with the journalist Robert D. Kaplan, as quoted in Kaplan's book *The Ends of the Earth*, Homer-Dixon said it more vividly: "Think of a stretch limo in the potholed streets of New York City, where homeless beggars live. Inside the limo are the air-conditioned post-industrial regions of North America, Europe, the emerging Pacific Rim, and a few other isolated places, with their trade summitry and computer information highways. Outside is the rest of mankind, going in a completely different direction."

EVEN NOAH'S ARK ONLY MANAGED TO RESCUE PAIRED ANIMALS, NOT LARGE PARCELS OF HABITAT

That direction, necessarily, will be toward ever more desperate exploitation of landscape. When you think of Homer-Dixon's stretch limo on those potholed urban streets, don't assume there will be room inside for tropical forests. Even Noah's ark only managed to rescue paired animals, not large parcels of habitat. The

jeopardy of the ecological fragments that we presently cherish as parks, refuges, and reserves is already severe, due to both internal and external forces: internal, because insularity itself leads to ecological unraveling; and external, because those areas are still under siege by needy and covetous people. Projected forward into a future of 10.8 billion humans, of which perhaps 2 billion are starving at the periphery of those areas, while another 2 billion are living in a fool's paradise maintained by unremitting exploitation of whatever resources remain, that jeopardy increases to the point of impossibility. In addition, any form of climate change in the midterm future, whether caused by greenhouse gases or by a natural flip-flop of climatic forces, is liable to change habitat conditions within a given protected area beyond the tolerance range for many species. If such creatures can't migrate beyond the park or reserve boundaries in order to chase their habitat needs, they may be "protected" from guns and chainsaws within their little island, but they'll still die.

We shouldn't take comfort in assuming that at least Yellowstone National Park will still harbor grizzly bears in the year 2150, that at least Royal Chitwan in Nepal will still harbor tigers, that at least Serengeti in Tanzania and Gir in India will still harbor lions. Those predator populations, and other species down the cascade, are likely to disappear. "Wildness" will be a word applicable only to urban turmoil. Lions, tigers, and bears will exist in zoos, period. Nature won't come to an end, but it will look very different.

MAN'S ACCIDENTAL RELOCATION OF CERTAIN SPECIES HAS LONG CREATED PROFOUND DISLOCATIONS IN NATURE

The most obvious differences will be those I've already mentioned: tropical forests and other terrestrial ecosystems will be drastically reduced in area, and the fragmented remnants will stand tiny and isolated. Because of those two factors, plus the cascading secondary effects, plus an additional dire factor I'll mention in a moment, much of Earth's biological diversity will be gone. How much? That's impossible to predict confidently, but the careful guesses of Robert May, Stuart Pimm, and other biologists suggest losses reaching half to two thirds of all species. In the oceans, deepwater fish and shellfish populations will be drastically depleted by overharvesting, if not to the point of extinction then at least enough to cause more cascading consequences. Coral reefs and other shallow-water ecosystems will be badly stressed, if not devastated, by erosion and chemical runoff from the land. The additional dire factor is invasive species, fifth of the five factors contributing to our current experiment in mass extinction.

That factor, even more than habitat destruction and fragmentation, is a symptom of modernity. Maybe you haven't heard much about invasive species, but in coming years you will. The ecologist Daniel Simberloff takes it so seriously that he recently committed himself to founding an institute on invasive biology at the University of Tennessee, and Interior Secretary Bruce Babbitt sounded the alarm last April in a speech to a weed-management symposium in Denver. The spectacle of a cabinet secretary denouncing an alien plant called purple loosestrife struck some observers as droll, but it wasn't as silly as it seemed. Forty years ago, the British ecologist Charles Elton warned prophetically in a little book titled *The Ecology of Invasions by Animals and Plants* that "we are living in a period of the world's history when the mingling of thousands of kinds of organisms from different parts of the world is setting up terrific dislocations in nature." Elton's word "dislocations" was nicely chosen to ring with a double meaning: species are being moved from one location to another, and as a result ecosystems are being thrown into disorder.

The problem dates back to when people began using ingenious new modes of conveyance (the horse, the camel, the canoe) to travel quickly across mountains, deserts, and oceans, bringing with them rats, lice, disease microbes, burrs, dogs, pigs, goats, cats, cows, and other forms of parasitic, commensal, or domesticated creature. One immediate result of those travels was a wave of island-bird extinctions, claiming more than a thousand species, that followed oceangoing canoes across the Pacific and elsewhere. Having evolved in insular ecosystems free of predators, many of those species were flightless, unequipped to defend themselves or their eggs against ravenous mammals. *Raphus cucullatus*, a giant cousin of the pigeon lineage, endemic to Mauritius in the Indian

Ocean and better known as the dodo, was only the most easily caricatured representative of this much larger pattern. Dutch sailors killed and ate dodos during the seventeenth century, but probably what guaranteed the extinction of *Raphus cucullatus* is that the European ships put ashore rats, pigs, and *Macaca fascicularis*, an opportunistic species of Asian monkey. Although commonly known as the crab-eating macaque, *M. fascicularis* will eat almost anything. The monkeys are still pestilential on Mauritius, hungry and daring and always ready to grab what they can, including raw eggs. But the dodo hasn't been seen since 1662.

The European age of discovery and conquest was also the great age of biogeography—that is, the study of what creatures live where, a branch of biology practiced by attentive travelers such as Carolus Linnaeus, Alexander von Humboldt, Charles Darwin, and Alfred Russel Wallace. Darwin and Wallace even made biogeography the basis of their discovery that species, rather than being created and plopped onto Earth by divine magic, evolve in particular locales by the process of natural selection. Ironically, the same trend of far-flung human travel that gave biogeographers their data also began to muddle and nullify those data, by transplanting the most ready and roguish species to new places and thereby delivering misery unto death for many other species. Rats and cats went everywhere, causing havoc in what for millions of years had been sheltered, less competitive ecosystems. The Asiatic chestnut blight and the European starling came to America; the American muskrat and the Chinese mitten crab got to Europe. Sometimes these human-mediated transfers were unintentional, sometimes merely shortsighted. Nostalgic sportsmen in New Zealand imported British red deer; European brown trout and Coastal rainbows were planted in disregard of the native cutthroats of Rocky Mountain rivers. Prickly-pear cactus, rabbits, and cane toads were inadvisedly welcomed to Australia. Goats went wild in the Galapagos. The bacterium that causes bubonic plague journeyed from China to California by way of a flea, a rat, and a ship. The Atlantic sea lamprey found its own way up into Lake Erie, but only after the Welland Canal gave it a bypass around Niagara Falls. Unintentional or otherwise, all these transfers had unforeseen consequences, which in many cases included the extinction of less competitive, less opportunistic native species. The rosy wolfsnail,

a small creature introduced onto Oahu for the purpose of controlling a larger and more obviously noxious species of snail, which was itself invasive, proved to be medicine worse than the disease; it became a fearsome predator upon native snails, of which twenty species are now gone. The Nile perch, a big predatory fish introduced into Lake Victoria in 1962 because it promised good eating, seems to have exterminated at least eighty species of smaller cichlid fishes that were native to the lake's Mwanza Gulf.

THE SPECIES THAT SURVIVE WILL BE LIKE WEEDS, REPRODUCING QUICKLY AND SURVIVING ALMOST ANYWHERE

The problem is vastly amplified by modern shipping and air transport, which are quick and capacious enough to allow many more kinds of organisms to get themselves transplanted into zones of habitat they never could have reached on their own. The brown tree snake, having hitchhiked aboard military planes from the New Guinea region near the end of World War II, has eaten most of the native forest birds of Guam. Hanta virus, first identified in Korea, burbles quietly in the deer mice of Arizona. Ebola will next appear who knows where. Apart from the frightening epidemiological possibilities, agricultural damages are the most conspicuous form of impact. One study, by the congressional Office of Technology Assessment, reports that in the United States 4,500 nonnative species have established free-living populations, of which about 15 percent cause severe harm; looking at just 79 of those species, the OTA documented $97 billion in damages. The lost value in Hawaiian snail species or cichlid diversity is harder to measure. But another report, from the U.N. Environmental Program, declares that almost 20 percent of the world's endangered vertebrates suffer from pressures (competition, predation, habitat transformation) created by exotic interlopers. Michael Soulé, a biologist much respected for his work on landscape conversion and extinction, has said that invasive species may soon surpass habitat loss and fragmentation as the major cause of "ecological disintegration." Having exterminated Guam's avifauna, the brown tree snake has lately been spotted in Hawaii.

Is there a larger pattern to these invasions? What do fire ants, zebra mussels, Asian gypsy moths, tamarisk trees, maleleuca trees, kudzu, Mediterranean fruit flies, boll weevils, and water hyacinths have in common with crab-eating macaques or Nile perch? Answers: They're *weedy* species, in the sense that animals as well as plants can be weedy. What that implies is a constellation of characteristics: They reproduce quickly, disperse widely when given a chance, tolerate a fairly broad range of habitat conditions, take hold in strange places, succeed especially in disturbed ecosystems, and resist eradication once they're established. They are scrappers, generalists, opportunists. They tend to thrive in human-dominated terrain because in crucial ways they resemble *Homo sapiens:* aggressive, versatile, prolific, and ready to travel. The city pigeon, a cosmopolitan creature derived from wild ancestry as a Eurasian rock dove (*Columba livia*) by way of centuries of pigeon fanciers whose coop-bred birds occasionally went AWOL, is a weed. So are those species that, benefiting from human impacts upon landscape, have increased grossly in abundance or expanded their geographical scope without having to cross an ocean by plane or by boat—for instance, the coyote in New York, the raccoon in Montana, the while-tailed deer in northern Wisconsin or western Connecticut. The brown-headed cowbird, also weedy, has enlarged its range from the Eastern United States into the agricultural Midwest at the expense of migratory songbirds. In gardening usage the word "weed" may be utterly subjective, indicating any plant you don't happen to like, but in ecological usage it has these firmer meanings. Biologists frequently talk of weedy species, meaning animals as well as plants.

WILDLIFE WILL CONSIST OF PIGEONS, COYOTES, RATS, ROACHES, HOUSE SPARROWS, CROWS, AND FERAL DOGS

Paleontologists, too, embrace the idea and even the term. Jablonski himself, in a 1991 paper published in *Science*, extrapolated from past mass extinctions to our current one and suggested that human activities are likely to take their heaviest toll on narrowly endemic species, while causing fewer extinctions among those species that are broadly adapted and broadly distributed. "In the face of ongoing habitat alteration and fragmentation," he wrote, "this implies a biota increasingly enriched in widespread, weedy species— rats, ragweed, and cockroaches—relative to the large number of species that are more vulnerable and potentially more useful to humans as food, medicines, and genetic resources." Now, as we sit in his office, he repeats: "It's just a question of how much the world becomes enriched in these weedy species." Both in print and in talk he uses "enriched" somewhat caustically, knowing that the actual direction of the trend is toward impoverishment.

Regarding impoverishment, let's note another dark, interesting irony: that the two converse trends I've described—partitioning the world's landscape by habitat fragmentation, and unifying the world's landscape by global transport for weedy species—produce not converse results but one redoubled result, the further loss of biological diversity. Immersing myself in the literature of extinctions, and making dilettantish excursions across India, Madagascar, New Guinea, Indonesia, Brazil, Guam, Australia, New Zealand, Wyoming, the hills of Burbank, and other semi-wild places over the past decade, I've seen those redoubling trends everywhere, portending a near-term future in which Earth's landscape is threadbare, leached of diversity, heavy with humans, and "enriched" in weedy species. That's an ugly vision, but I find it vivid. Wildlife will consist of the pigeons and the coyotes and the white-tails, the black rats (*Rattus rattus*) and the brown rats (*Rattus norvegicus*) and a few other species of worldly rodent, the crab-eating macaques and the cockroaches (though, as with the rats, not *every* species—some are narrowly endemic, like the giant Madagascar hissing cockroach) and the mongooses, the house sparrows and the house geckos and the houseflies and the barn cats and the skinny brown feral dogs and a short list of additional species that play by our rules. Forests will be tiny insular patches existing on bare sufferance, much of their biological diversity (the big predators, the migratory birds, the shy creatures that can't tolerate edges, and many other species linked inextricably with those) long since decayed away. They'll essentially be tall woody gardens, not forests in the richer sense. Elsewhere the landscape will have its strips and swatches of green, but except on much-poisoned lawns and golf courses the foliage will be infested with cheatgrass and European buckthorn and spotted knapweed and Russian thistle and leafy spurge and salt meadow cordgrass and Bruce Babbitt's purple loosestrife. Having recently passed the great age of biogeography, we will have entered the age *after* biogeography, in that virtually everything will live virtually everywhere, though the list of species that constitute "everything" will be small. I see this world implicitly foretold in the U.N. population projections, the FAO reports on deforestation, the northward advance into Texas of Africanized honeybees, the rhesus monkeys that haunt the parapets of public buildings in New Delhi, and every fat gray squirrel on a bird feeder in England. Earth will be a different sort of place—soon, in just five or six human generations. My label for that place, that time, that apparently unavoidable prospect, is the Planet of Weeds. Its main consoling felicity, as far as I can imagine, is that there will be no shortage of crows.

HOMO SAPIENS— REMARKABLY WIDESPREAD, PROLIFIC, AND ADAPTABLE—IS THE CONSUMMATE WEED

Now we come to the question of human survival, a matter of some interest to many. We come to a certain fretful leap of logic that otherwise thoughtful observers seem willing, even eager, to make: that the ultimate consequence will be the extinction of us. By seizing such a huge share of Earth's landscape, by imposing so wantonly on its providence and presuming so recklessly on its forgiveness, by killing off so many species, they say, we will doom our own species to extinction. My quibbles with the idea are that it seems ecologically improbable and too optimistic. But it bears examining, because it's frequently offered as the ultimate argument against proceeding as we are.

Jablonski also has his doubts. Do you see *Homo sapiens* as a likely survivor, I ask him, or as a casualty? "Oh, we've got to be one of the most bomb-proof species on the planet," he says "We're geographically widespread, we have a pretty remarkable reproductive rate, we're incredibly good at co-opting and monopolizing resources. I think it would take really serious, concerted effort to wipe out the human species." The point he's making is one that has probably already dawned on you: *Homo sapiens* itself is the consummate weed. Why shouldn't we survive, then, on the Planet of Weeds? But there's a wide range of possible circumstances, Jablonski reminds me, between the extinction of our species and the continued growth of human population, consumptions, and comfort. "I think we'll be one of the survivors," he says, "sort of picking through the rubble." Besides losing all the pharmaceutical and generic resources that lay hidden within those extinguished species, and all the spiritual and aesthetic values they offered, he foresees unpredictable levels of loss in many physical and biochemical functions that ordinarily come as benefits from diverse, robust ecosystems—functions such as cleaning and recirculating air and water, mitigating droughts and floods, decomposing wastes, controlling erosion, creating new soil, pollinating crops, capturing and transporting nutrients, damping short-term temperature extremes and longer-term fluctuations of climate, restraining outbreaks of pestiferous species, and shielding Earth's surface from the full brunt of ultraviolet radiation. Strip away the ecosystems that perform those services, Jablonski says, and you can expect grievous detriment to the reality we inhabit. "A lot of things are going to happen that will make this a crummier place to live—a more stressful place to live, a more difficult place to live, a less resilient place to live—before the human species is at any risk at all." And maybe some of the new difficulties, he adds, will serve as incentive for major changes in the trajectory along which we pursue our aggregate self-interests. Maybe we'll pull back before our current episode matches the Triassic extinction or the K-T event. Maybe it will turn out to be no worse than the Eocene extinction, with a 35 percent loss of species.

"Are you hopeful?" I ask.

Given that hope is a duty from which paleontologists are exempt, I'm surprised when he answers, "Yes, I am."

I'm not. My own guess about the midterm future, excused by no exemption, is that our Planet of Weeds will indeed be a crummier place, a lonelier and uglier place, and a particularly wretched place for the 2 billion people comprising Alan Durning's absolute poor. What will increase most dramatically as time proceeds, I suspect won't be generalized misery or futuristic modes of consumption but the gulf between two global classes experiencing those extremes. Progressive failure of ecosystem functions? Yes, but human resourcefulness of the sort Julian Simon so admired will probably find stopgap technological remedies, to be available for a price. So the world's privileged class—that's your class and my class—will probably still manage to maintain themselves inside Homer-Dixon's stretch limo, drinking bottled water and breathing bottled air and eating reasonably healthy food that has become incredibly precious, while the potholes on the road outside grow ever deeper. Eventually the limo will look more like a lunar rover. Ragtag mobs of desperate souls will cling to its bumpers, like groupies on Elvis's final Cadillac. The absolute poor will suffer their lack of ecological privilege in the form of lowered life expectancy, bad health, absence of education, corrosive want, and anger. Maybe in time they'll find ways to gather themselves in localized revolt against the affluent class. Not likely, though, as long as affluence buys guns. In any case, well before that they will have burned the last stick of Bornean dipterocarp for firewood and roasted the last lemur, the last grizzly bear, the last elephant left unprotected outside a zoo.

WHAT WILL HAPPEN AFTER THIS MASS EXTINCTION, AFTER WE DESTROY TWO THIRDS OF ALL LIVING SPECIES?

Jablonski has a hundred things to do before leaving for Alaska, so after two hours I clear out. The heat on the sidewalk is fierce, though not nearly as fierce as this summer's heat in New Delhi or Dallas, where people are dying. Since my flight doesn't leave until early evening, I cab downtown and take refuge in a nouveau-Cajun restaurant near the river. Over a beer and jambalaya, I glance again at Jablonski's 1991 *Science* paper, titled "Extinctions: A Paleontological Perspective." I also play back the tape of our conversation, pressing my ear against the little recorder to hear it over the lunch-crowd noise.

Among the last questions I asked Jablonski was, What will happen *after* this mass extinction, assuming it proceeds to a worst-case scenario? If we destroy half or two thirds of all living species, how long will it take for evolution to fill the planet back up? "I don't know the answer to that," he said. "I'd rather not bottom out and see what happens next." In the journal paper he had hazarded that, based on fossil evidence in rock laid down atop the K-T event and others, the time required for full recovery might be 5 or 10 million years. From a paleontological perspective, that's fast. "Biotic recoveries after mass extinctions are geologically rapid but immensely prolonged on human time scales," he wrote. There was also the proviso, cited from another expert, that recovery might not begin until *after* the extinction-causing circumstances have disappeared. But in this case, of course, the circumstances won't likely disappear until *we* do.

Still, evolution never rests. It's happening right now, in weed patches all over the planet. I'm not presuming to alert you to the end of the world, the end of evolution, or the end of nature. What I've tried to describe here is not an absolute end but a very deep dip, a repeat point within a long, violent cycle. Species die, species arise. The relative pace of those two processes is what matters. Even rats and cockroaches are capable—given the requisite conditions, namely, habitat diversity and time—of speciation. And speciation brings new diversity. So we might reasonably imagine an Earth upon which, 10 million years after the extinction (or, alternatively, the drastic transformation) of *Homo sapiens*, wondrous forests are again filled with wondrous beasts. That's the good news.

David Quammen is the author of eight books, including The Song of the Dodo *and, most recently,* Wild Thoughts from Wild Places. *His last article for* Harper's Magazine, *"Brazil's Jungle Blackboard," appeared in the March 1998 issue.*

A Fragile Cornucopia: Assessing the Status of U.S. Biodiversity

Bruce A. Stein

A vast land of contrasts, the United States stretches from above the Arctic Circle to below the tropic of Cancer and spans nearly a third of the globe from eastern Maine to the tip of the Aleutian chain. This enormous expanse harbors a wide array of ecological conditions—from the lush forests of the Appalachians to California's thorny chaparral and from Alaska's frigid tundra to the parched desert of Death Valley. Although tropical rainforests come to mind when most people think of biodiversity—the variety of life—the United States itself contains a surprising diversity.

The Death Valley region, for instance, known for its scorching temperatures and austere landscapes, also holds some watery secrets. Devils Hole, an abrupt fissure in the desert floor, shelters the remnant of a huge lake that formed during the Ice Ages. In this 70-by-10-foot pool dwells the entire population of the Devils Hole pupfish (*Cyprinodon diabolis*), a tiny fish that descended from the lake's original inhabitants and has been isolated for more than 20,000 years. Subsisting on the algae that grow on the pool's single sunlit ledge, this fish has the distinction of

being the world's most narrowly distributed vertebrate species.

Such extreme rarity also makes the fish susceptible to even slight changes in its tiny oasis. Lowering the pool's water below the level of the algal-covered ledge would eliminate the fish's sole food source, leading inevitably to its extinction. Indeed, such a threat occurred in the late 1960s when owners of a nearby ranch began pumping irrigation water from the same aquifer that maintains Devils Hole. In a landmark decision concerning the 1973 Endangered Species Act (ESA), the U.S. Supreme Court ruled in favor of protecting the Devils Hole pupfish and limited the ground water extractions.[1]

Since that time, ESA has become a far-reaching legal tool for protecting the nation's threatened living resources. Conversely, ESA has been and continues to be a lightning rod for those concerned about private property rights and the scope and role of government. One of the few things that all sides in this contentious debate can agree on, however, is the need for sound scientific information on which to base decisions.

A National Status Assessment

What is the full scope of the nation's biological inheritance, and how is it faring? Which species and ecosystems are at greatest risk, and what is threatening them? Where are the most biologically significant places, and where should society direct its conservation efforts? To address these questions, the Association for Biodiversity Information (ABI) and The Nature Conservancy carried out a major assessment of the nation's species and ecosystems, recently published as the book *Precious Heritage: The Status of Biodiversity in the United States*.[2] By drawing together a quarter century of information gathered by the state-based natural heritage programs, this study provides the first-ever comprehensive view of the state of the nation's biota. (For information on ABI and its web site, see the boxes "Knowledge to Protect the Diversity of Life" and "NatureServe: An Online Resource for More Information".)

Biodiversity is a word that gained great currency during the 1990s, yet the concept is still foreign to most people. In its most elemental form, biodiversity refers to the full array of life on Earth. Although plant and animal species are the most tangible manifestations of biodiversity, the concept covers the full hierarchy of biological organization recognized by scientists, including genetic, species, and ecosystem levels. The Devils Hole pupfish is just one example of the extraordinary diversity of species in the United States. A host of other biological superlatives can also be found within the 50 states. The redwood of coastal northern California, for example, is the world's tallest tree, with individual specimens rising as high as 35-story office buildings. What is currently thought to be the world's most massive organism is a 107-acre aspen grove in Utah. Because each trunk is connected by a common root system, the entire grove represents a single, genetic individual.[3] Other oddities and peculiarities abound among the national flora and fauna, including yard-long salamanders, frogs with "antifreeze" in their blood that enables them to survive freezing and thawing, and birds that travel 25,000 miles roundtrip on their annual migration.

A Catalog of U.S. Biodiversity

In an age when exploration of distant planets seems routine, many take it for granted that we have done an adequate job of discovering what exists here on Earth. Nothing could be farther from the truth. The mysteries of life on our own living planet are still profound and none more so than determining the most fundamental question of how many species of plants, animals, and microorganisms exist. A more complete knowledge of the diversity of life is essential to understanding how ecosystems function and evaluating the health of our environment. Only about 1.75 million species have been studied well enough to receive a scientific name. Yet estimates of the total number of living species span more than an order of magnitude, from around 3 million species to more than 100 million, although a working estimate hovers around 14 million.[4] (This enormous range in estimates is due to emerging research that points to previously under-appreciated reservoirs of life such as tropical tree canopies, deep sea floors, and soil generally.)

The United States is one of the best studied nations on Earth, but, even here, producing a tally of life is extremely difficult. Surveying the scientific literature and querying taxonomic specialists reveals that in excess of 200,000 native species are currently known to inhabit the United States.[5] This figure is conservative, including only formally named species and leaving out such poorly known groups as bacteria, protists, or viruses. Still, this represents more than one-tenth of all scientifically documented species on Earth. Although these figures accentuate the richness of the U.S. biota, they also call attention to the disparity between cataloging efforts in the temperate regions, where most biologists live, and the tropics, where most of the world's species are thought to reside.

Even taking into account the disparity in inventory efforts, the United States emerges as an exceptionally rich country biologically and a world leader in the diversity of certain groups of organisms (see Table 1). Four out of every ten salamander species, for example, are found in the United States, the most of any country on Earth. Almost one-

Knowledge to Protect the Diversity of Life

The Association for Biodiversity Information (ABI) is a new nonprofit organization working in partnership with the network of state natural heritage programs. Dedicated to providing the knowledge necessary to protect the diversity of life, ABI reflects a continuation of The Nature Conservancy's 25-year commitment to science-based conservation. Representing a unique institutional and scientific collaboration, ABI and its natural heritage program members are the leading source for detailed information about the condition and location of rare and endangered spe-

cies and threatened ecosystems. Network programs operate in all 50 U.S. states, across Canada, and in a dozen countries of Latin America and the Caribbean, with each center maintaining detailed maps and computer records about the species and ecosystems that are of greatest conservation concern within their state. This information and expertise is provided to thousands of users annually to assist in project planning, environmental review, and conservation efforts. ABI serves as the network's coordinating body, providing scientific and technical support to ensure consistency of

information across state boundaries. ABI also provides a single access point for heritage data and expertise, particularly at regional and national scales. To improve society's understanding of and ability to protect biodiversity, ABI provides policy makers and the public with a variety of objective and credible information products, such as *Precious Heritage* and the recently launched NatureServe web site (see "NatureServe: An On-Line Resource for More Information"). (For more information about ABI or natural heritage programs, visit http://www.abi.org.)

NatureServe: An On-line Resource for More Information

A new web site launched by the Association for Biodiversity Information (ABI) provides an easily accessible source for information about the thousands of U.S. plants and animals that are summarized in this article. NatureServe: An On-line Encyclopedia of Life (http://www.natureserve.org) offers access to ABI's comprehensive databases on more than 50,000 U.S. and Canadian species and ecological communities. The site provides in-depth information about rare and endangered species as well as information about common plants and animals. NatureServe can be used to learn about a particular species or can be queried to learn, for instance, which endangered mammals or butterflies are found in a given state. The site includes scientific and common names, conservation and legal status, color-coded distribution maps, and summaries of life histories and conservation needs. This new web site puts details about our rich natural heritage within reach of everyone.

---

third of the world's freshwater mussel species (about 300) reside in the United States. By comparison, all the rivers of Europe have just 10 mussel species. Of all the nations in the world, the United States has the most species of freshwater crayfishes (61 percent), freshwater turtles (22 percent), and freshwater snails (17 percent). Several other groups of organisms are well represented in the United States. U.S. mammals rank sixth in global diversity, due largely to the diverse fauna of the arid lands in the Southwest. Although the tropics have the greatest diversity of freshwater fishes, the United States has by far the highest diversity of these important organisms among temperate countries. Among plants, the number of U.S. gymnosperms—including conifers such as pines, firs, and cypress—is second only to China.

The United States is also remarkably diverse at the scale of ecosystems. Several systems exist for dividing the globe into major ecological zones. To assist the World Conservation Union in targeting the placement of protected areas, in the 1970s, the now late University of California professor Miklos Udvardy mapped out biogeographic provinces worldwide, depicting distinctive floral and faunal assemblages. Each of these provinces was classed according to one of 14 biomes, representing major ecosystem groups such as temperate grasslands or tropical dry forests.[6] In another effort to classify large-scale global ecosystems, U.S. Forest Service ecologist Robert Bailey identified large ecologically defined areas that share similar vegetation and climate—termed ecoregions.[7] Interestingly, the United States has a larger number of both biomes and ecoregions than any other country on Earth.[8] Certain globally significant large-scale ecosystems are especially well represented in the United States, including temperate broadleaf forest and prairies. Coastal California's shrubby vegetation is particularly distinctive, representing one of only five examples worldwide of Mediterranean-climate vegetation.

The Condition of U.S. Species

How is this extraordinary array of species faring? The Association for Biodiversity Information and its affiliated natural heritage programs assess the conservation status of species based on several factors linked to increases in risk of extinction.[9] These criteria include the total number of individuals of a species, the number of distinct populations across which these individuals are distributed, the viability of these populations, and the short- and long-term trends for the species. Conservation status ranks are assigned on a scale from one to five, ranging from critically imperiled (G1) to demonstrably secure (G5) (see Table 2 for more details on the ranking system). In general, species classified as vulnerable (G3) or rarer may be considered to be "at risk."

Given the paucity of information about most of the nation's more than 200,000 species, any attempt to characterize the status of the biota overall will necessarily be incomplete. Nonetheless, ABI and natural heritage program scientists have assessed the status of more than 30,000 U.S. species and subspecies. These assessments are comprehensive for 14 of the best-known groups of plants and animals—that is, the conservation status has been evaluated for each and every species in these groups. These 14 groups, which include all vertebrates and vascular plants, represent about 20,900 species (see Table 3). These assessments begin to paint an overall picture of the condition of wildlife in America. And that picture is not particularly pretty.

Overall, a surprisingly high one-third of the native U.S. flora and fauna is at risk, with 16 percent considered vulnerable, 8 percent imperiled, 7 percent critically imperiled, and I percent missing or extinct. The proportion of species at risk varies greatly from one group of plants and animals to another (see Figure 1). Freshwater mussels in the United States show the highest risk levels, with nearly 70 percent in trouble. Indeed, one in ten mussels are already missing or extinct (listed as GX (presumed extinct) or GH (possibly ex-

Table 1. Global significance of selected U.S. plant and animal groups				
Taxonomic group	Number of U.S. species	Number of species worldwide	% of global species in U.S.	U.S. ranking worldwide
Mammals	416	4,600	9	6
Birds	768	9,700	8	27
Reptiles	283	6,600	4	14
Amphibians	231	4,400	5	12
Freshwater fishes	799	8,400	10	7
Freshwater mussels	292	1,000	29	1
Freshwater snails	661	4,000	17	1
Crayfishes	322	525	61	1
Tiger beetles	114	2,000	6	7
Dragonflies/ Damselflies	456	5,800	8	?
Butterflies/ Skippers	620	17,500	4	?
Flowering plants	15,320	235,000	7	>10
Gymnosperms	114	760	15	2
Ferns	556	12,000	5	>15

NOTE: Several species have their highest levels of diversity in the United States, Including Freshwater Mussels, Freshwater Snails, And Crayfishes: Several Other Taxonomic Groups, Such As Freshwater Fishes And Gymnosperms, Are Also Well Represented In The United States.

SOURCE: B. A. Stein, L. S. Kutner, and J. S. Adams, eds., Precious Heritage: The Status of Biodiversity in the United States (New York: Oxford University Press, 2000), Table 3.2, 67.

tinct)). Several other freshwater groups have exceptionally high risk levels, including crayfish (51 percent), stoneflies (43 percent), and freshwater fish (37 percent). In contrast, several vertebrate groups appear to be on relatively secure footing, at least at the full species level measured here. This includes the two groups of animals—birds and mammals—that receive the majority of public conservation attention. Of all groups considered, birds are best off, with only 14 percent of their species at risk.

Focusing on the number of species at risk in each group tells a different story. On a proportional basis, species depending on freshwater habitats appear to be at greatest risk. Based on the actual number of species at risk, however, the largely terrestrial flowering plants dominate by a huge margin, with 33 percent of the more than 15,300 native U.S. species—a sobering 5,090 species—at risk.

The Legacy of Extinctions

The Devils Hole pupfish's brush with extinction was staved off thanks to actions of the nation's highest court in 1976. Another small fish living not far from Devils Hole was not so lucky. The Ash Meadows poolfish (Empetrichthys merriami) was first discovered in 1891. Never abundant, the fish lived in just five isolated desert springs. Probably due to the introduction of two voracious predators—bullfrogs and crayfish—the species slipped into oblivion sometime between 1948, when it was last seen, and 1953, when researchers were unable to locate it. Like the Ash Meadows poolfish, a number of declining U.S. species never had friends in high places to guard over them and consequently have suffered the ultimate demise. Overall, at least 100 U.s. species have gone extinct since Euro-

pean colonization of North America and Hawaii and nearly 440 more are missing and may be extinct.[10]

These extinctions span the gamut of organisms, including vertebrates such as the great auk (Pinguinus impennis), plants like the Santa Catalina monkeyflower (Mimulus traskiae), and invertebrates such as the Wabash rifleshell mussel (Epioblasmna sampsonii). Snails have been particularly hard hit by extinctions; with 26 species presumed extinct and another 106 species missing and possibly extinct, gastropods lead all other groups in this unenviable category. Among vertebrates, birds have been most severely affected by extinctions, with 22 species of birds presumed extinct and another 3 missing. Although several bird species have disappeared from the mainland United States—the passenger pigeon, Carolina parakeet, Labrador duck, and great auk—most extinct birds are Hawaiian. A considerable number of plants—11 extinct and 130 missing—are also gone from the U.S. landscape.

Table 2. Definition of conservation status ranks

GX	Presumed Extinct: not located despite intensive searches
GH	Possibly Extinct: of historical occurrence; missing but still some hope of rediscovery
G1	Critically Imperiled: typically 5 or fewer occurrences or 1,000 or fewer individuals
G2	Imperiled: typically 6 to 20 occurrences or 1,000 to 3,000 individuals
G3	Vulnerable: rare; typically 21 to 100 occurrences and 10,000 individuals
G4	Apparently Secure: uncommon but not rare, some cause for long-term concern; usually more than 100 occurrences and 10,000 individuals
G5	Secure: common; widespread, and abundant

NOTE: The conservation status of a given species is assessed based on several factors using a one to five scale. This table summarizes key assessment criteria. "G" refers to the global or rangewide status of a species. Both national (N) and state (S) status ranks are also assessed.

SOURCE: B. A. Stein, L. S. Kutner, and J. S. Adams, eds., Precious Heritage: The Status of Biodiversity in the United States (New York: Oxford University Press, 2000), Table 4.2, 97.

Although extinctions have touched every state in the nation, certain regions have lost disproportionate numbers of species. States with large numbers of extinctions tend to have either high overall species numbers, an inherently fragile flora and fauna, or intense human alteration of the landscape. Hawaii, not surprisingly, has suffered the gravest losses, with 249 extinct species (29 presumed and 220 possibly extinct). The oceanic isolation of the Hawaiian Islands has produced one of the world's most distinctive biotas and one that evolved largely in the

absence of aggressive continental predators. The native Hawaiian species have proven to be extremely susceptible to the kinds of outside influences introduced first by the Polynesian immigration (between 200 and 500 A.D.) and later by European colonists.

On the mainland, Alabama tops the list of extinction-prone states, with 96 species gone (22 presumed and 74 possibly extinct). Home to an exceptionally rich variety of freshwater fauna, many of the waterways in Alabama have been dammed, dredged, or diverted, leading to the loss of numerous snails, mussels, and fish. California ranks third in the nation with 35 extinctions (11 presumed and 24 possibly extinct). The intensive conversion of the state's lands and waters for agriculture, urbanization, and other uses has had a severe impact on the many restricted-range species that have evolved in this ecologically unique state.

State of the States

The ecological complexion of the 50 states varies dramatically, reflected by the composition and character of the plants and animals inhabiting each state. The East is dominated by wide coastal plains and the ancient remains of the north-south trending Appalachian chain. Stretching across the Midwest and the Great Plains, the continent's vast interior exhibits relatively little topographic relief. In contrast, the western third of the country is a welter of mountains and valleys resulting from relatively recent (from a geological perspective) flurries of mountain-building. Topography interacts with and influences climatic factors, which in turn combine with regional evolutionary histories to produce distinctive assemblages of plants and animals.

The areas with the greatest species diversity are found in the topographically and climatically diverse Southwest, with California and Texas leading the nation in the number of species.[11] This is due in part to

the enormous size of these two states, in part to their southern location (a general ecological principle is that species diversity increases as one moves toward the equator), and partly because of their ecological complexity. Because of California's benign climate and extraordinary diversity of unique habitats, the state is often referred to as an ecological island, attached to but discrete from the rest of the continent. Centrally situated, Texas straddles several major ecological zones—the Great Plains, the southwestern deserts, the humid southeastern coastal plain, and even a touch of the Mexican subtropics—each contributing their own distinctive species to the state's mix. (For a map of state patterns of diversity, see Figure 2.)

Another way to look at the biological significance of different states is to consider those species that are restricted—or endemic—to a single state (see Figure 2). California, with almost 1,500 endemic species, leads the nation in this category, highlighting the ecological distinctiveness of the Golden State. Hawaii is the other state that stands out for its number of endemic species. Because of the island chain's extreme isolation, most plants and animals native to the archipelago are descended from a relatively few colonists. A mere 15 original colonists, for example, may account for all 53 species of endemic birds, including the famous honeycreepers.[12] As a result, Hawaii has some of the highest levels of endemism of any place on Earth. Forty-three percent of Hawaii's native vertebrates are endemic to the state, as are 87 percent of its vascular plants and an astonishing 97 percent of its insects.

Risk patterns among states—as reflected by the proportion of a state's species that are considered imperiled or vulnerable—also highlight Hawaii and California (see Figure 2). Secondary centers of rarity are found in several other western and southeastern states. The upper Midwest, in turn, shows relatively low levels of rarity, a condition due in part to the lingering effects of Ice

Table 3. Species report card: Status of U.S. plants and animals

		Presumed Extinct (GX)	Possibly Extinct (GH)	Critically Imperiled (G1)	Imperiled (G2)	Vulnerable (G3)	Apparently Secure (G4)	Secure (G5)	Other	Total
Vertebrate animals										
	Mammals	1	0	8	21	36	96	253	1	416
	Birds	22	3	27	20	36	87	572	1	768
	Reptiles	0	0	7	14	30	45	186	1	283
	Amphibians	1	1	21	28	33	42	104	1	231
	Freshwater fishes	16	1	91	87	105	147	351	1	799
Vertebrate totals		**40**	**5**	**154**	**170**	**240**	**417**	**1,466**	**5**	**2,497**
Invertebrate animals										
	Butterflies/Skippers	0	0	8	31	78	101	388	14	620
	Crayfishes	1	2	54	53	55	87	69	1	322
	Freshwater mussels	17	20	73	42	50	44	42	4	292
	Dragonflies/Damselflies	0	2	10	24	45	91	268	16	456
	Stoneflies	1	11	13	71	164	168	178	0	606
	Tiger beetles	0	0	3	5	14	19	62	11	114
Invertebrate totals		**19**	**35**	**161**	**226**	**406**	**510**	**1,007**	**46**	**2,410**
Vascular plants										
	Ferns/Fern allies	0	4	32	24	65	168	242	21	556
	Gymnosperms	0	0	7	8	12	27	60	0	114
	Flowering plants	11	126	1,031	1,309	2,615	4,469	5,618	141	15,320
Vascular plant totals		**11**	**130**	**1,070**	**1,341**	**2,692**	**4,664**	**5,920**	**162**	**15,990**
Totals		**70**	**170**	**1,385**	**1,737**	**3,338**	**5,591**	**8,393**	**213**	**20,897**

SOURCE: B. A. Stein, L. S. Kutner and J. S. Adams, eds., Precious Heritage: The Status of Biodiversity in the United States (New York: Oxford University Press, 2000), Table 4.4, 104.

Age glaciations and the resulting wide ranges of most midwestern species. Rare plants and animals in this region tend to be widespread but spotty in their distribution, often as a result of large-scale agricultural conversion of their habitat. Mead's milkweed (*Asciepias meadii*), for instance, formerly ranged throughout the tallgrass prairie but is now restricted to about 100 sites.

State patterns for specific groups of organisms can be very revealing and may differ significantly from the general patterns. The Southeast has been a center of evolution for freshwater fish, giving rise to a plethora of species like darters and mad toms. Maps of fish diversity reflect the large number of species in states such as Tennessee, Alabama, and Georgia (see Figure 3). Patterns of rarity, however, accentuate the arid states of the Southwest (see Figure 3). Although this region has far fewer rivers, streams, and lakes, their very scarcity has contributed to the evolution of numerous narrowly restricted species in groups such as pupfish and suckers. Most of the major rivers in the Southwest have also been extensively dammed or otherwise altered, leading to serious declines in even once abundant and wide-ranging fish species. The Colorado pikeminnow (*Ptycliocheilus lucius*), a "minnow" capable of reaching six feet in length, was once abundant throughout the major riv-

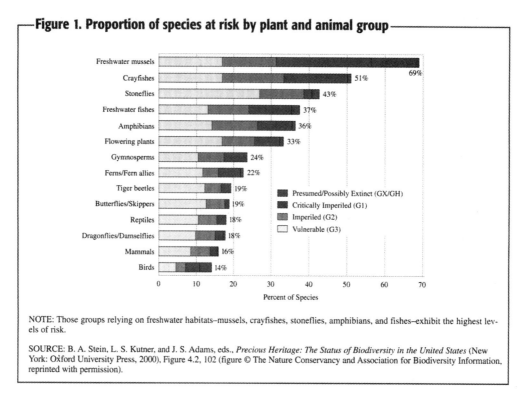

Figure 1. Proportion of species at risk by plant and animal group

NOTE: Those groups relying on freshwater habitats–mussels, crayfishes, stoneflies, amphibians, and fishes–exhibit the highest levels of risk.

SOURCE: B. A. Stein, L. S. Kutner, and J. S. Adams, eds., *Precious Heritage: The Status of Biodiversity in the United States* (New York: Oxford University Press, 2000), Figure 4.2, 102 (figure © The Nature Conservancy and Association for Biodiversity Information, reprinted with permission).

ers of the Colorado River basin. Due to changes in water flow and temperature brought about by dam operations, this magnificent fish now occupies barely a quarter of its former habitat.

Hot Spots of Imperilment

State-level assessments are useful for broadly identifying the magnitude of the conservation challenge in different regions, but to accomplish on-the-ground conservation and to minimize inadvertent damage to sensitive resources, it is necessary to have a much finer level of knowledge about the distribution of these species. Conducting field inventories and mapping the precise localities for rare and endangered species is a hallmark of the Association for Biodiversity Information (ABI) and its natural heritage program members. Collectively, the natural heritage programs maintain databases with nearly half a million detailed locality records for rare and endangered species and threatened ecological communities. This information is routinely used by government agencies, industry, landowners, and

conservationists to improve the environmental sensitivity of development projects and to target conservation activities. *Precious Heritage* is the first project to bring together this detailed information to create a truly national view of imperiled species in the United States.

A striking picture of biodiversity hot spots in United States emerges from mapping imperiled species against a uniform grid (see Figure 4).[13] Charting the rarest of the rare—approximately 2,800 imperiled and critically imperiled species—this map depicts the number of different imperiled species in each 640,000-acre hexagonal cell, an area about half the size of the typical eastern U.S. county. Most previous maps showing the distribution of U.S. endangered species have been based on county distributions, and therefore suffer distortions from the huge disparity in size among counties, especially in the West.[14] By using this equal-area hexagon grid, size differences arid edge effects are eliminated, leading to a more accurate picture of rarity.

Concentrations of imperiled species are particularly prominent in four major regions: Hawaii, Califor-

nia, the southern Appalachians, and Florida. Hawaii, in particular, has extraordinarily high concentrations of imperiled species, including a single hexagon centered on the Alakai Swamp that contains 128 different imperiled species. The highest concentration of imperiled species on the mainland is a hexagon centering on the Clinch and Powell Rivers in southwestern Virginia that contains 27 imperiled species, a large number of which are freshwater mussels.

Mapping these imperiled species by ecoregion provides a complementary view of biodiversity hot spots (see Figure 5).[15] Ecoregions consist of large, ecologically defined zones that share similar climate, vegetation, ecological processes, and suites of species. This view of imperiled species confirms the importance of coastal California and the Appalachian region but also highlights another region—the Great Basin, which stretches across Nevada and into Utah. Many of this region's isolated mountain ranges harbor at least a few highly localized species. Thus, even though few places have large concentrations of imperiled species, in combination these mountain range-restricted rarities elevate

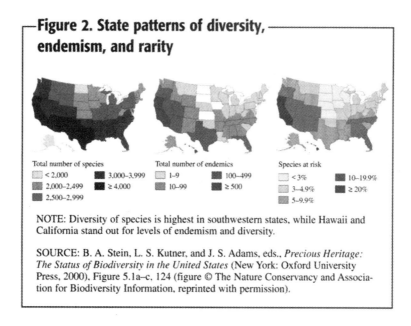

Figure 2. State patterns of diversity, endemism, and rarity

Total number of species
< 2,000   3,000–3,999
2,000–2,499   ≥ 4,000
2,500–2,999

Total number of endemics
1–9   100–499
10–99   ≥ 500

Species at risk
< 3%   10–19.9%
3–4.9%   ≥ 20%
5–9.9%

NOTE: Diversity of species is highest in southwestern states, while Hawaii and California stand out for levels of endemism and diversity.

SOURCE: B. A. Stein, L. S. Kutner, and J. S. Adams, eds., *Precious Heritage: The Status of Biodiversity in the United States* (New York: Oxford University Press, 2000), Figure 5.1a–c, 124 (figure © The Nature Conservancy and Association for Biodiversity Information, reprinted with permission).

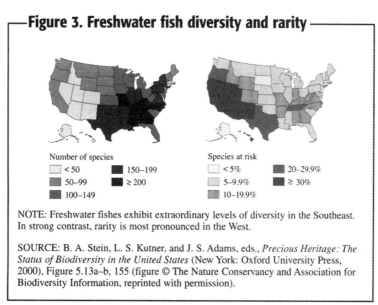

Figure 3. Freshwater fish diversity and rarity

Number of species
< 50   150–199
50–99   ≥ 200
100–149

Species at risk
< 5%   20–29.9%
5–9.9%   ≥ 30%
10–19.9%

NOTE: Freshwater fishes exhibit extraordinary levels of diversity in the Southeast. In strong contrast, rarity is most pronounced in the West.

SOURCE: B. A. Stein, L. S. Kutner, and J. S. Adams, eds., *Precious Heritage: The Status of Biodiversity in the United States* (New York: Oxford University Press, 2000), Figure 5.13a–b, 155 (figure © The Nature Conservancy and Association for Biodiversity Information, reprinted with permission).

the Great Basin as an ecoregional hot spot.

Implications for the Endangered Species Act

Many of the listed threatened and endangered species that command the highest public attention—and consume the major share of Endangered Species Act funding—are relatively wide-ranging species, such as the grizzly bear, the northern spotted owl, and the red-cockaded woodpecker. Some listed animals, like the desert tortoise, are also still quite abundant and listed primarily because of steep declines in their populations. The process for listing species as threatened or endangered has been criticized by some as not being based on sufficiently clear guidelines, with the potential for poor listing decisions.[16] A legitimate question, then, is to what extent such wide-ranging and relatively abundant species are representative of the endangered species list as a whole?

This relates to a current controversy regarding ESA: how to determine listing priorities. Specifically, how can the listing priorities of the implementing agencies be balanced with court-ordered priorities generated by lawsuits? The environmental community has been very effective at using the courts to focus listing activities on certain species, often with extensive habitat requirements, while both the Clinton and current Bush administrations have argued in favor of allowing the U.S. Fish and Wildlife Service greater flexibility in allocating listing funds towards internally derived priorities.

One approach for determining how rare a species needs to be before making it onto the federal endan-

Figure 4. Hot spots of rarity in the United States

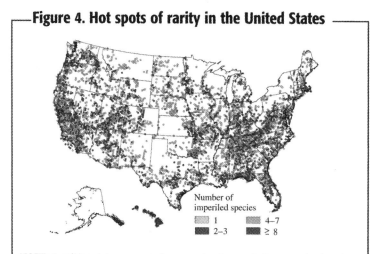

Number of
imperiled species

1 | 4–7
2–3 | ≥ 8

NOTE: A striking picture emerges from mapping the rarest of rare species, based on field data gathered by state natural heritage programs. Particular concentrations of imperiled species occur in California, Hawaii, Florida, and the Southern Appalachians.

SOURCE: B. A. Stein, L. S. Kutner, and J. S. Adams, eds., *Precious Heritage: The Status of Biodiversity in the United States* (New York: Oxford University Press, 2000), Figure 6.6, 169 (figure © The Nature Conservancy and Association for Biodiversity Information, reprinted with permission).

Figure 5. Rarity and ecological regions

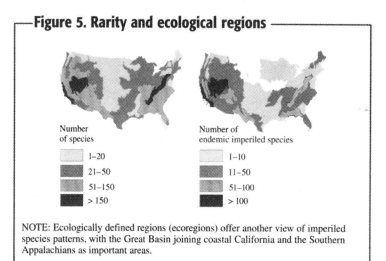

Number of species

1–20
21–50
51–150
> 150

Number of endemic imperiled species

1–10
11–50
51–100
> 100

NOTE: Ecologically defined regions (ecoregions) offer another view of imperiled species patterns, with the Great Basin joining coastal California and the Southern Appalachians as important areas.

SOURCE: B. A. Stein, L. S. Kutner, and J. S. Adams, eds., *Precious Heritage: The Status of Biodiversity in the United States* (New York: Oxford University Press, 2000), Figure 6.5a–b, 168 (figure © The Nature Conservancy and Association for Biodiversity Information, reprinted with permission).

gered species list is to consider how listed species correspond to ABI conservation status assessments (see Table 2). The vast majority (92 percent) of all federally listed species (including subspecies and distinct populations) are categorized by ABI as imperiled, critically imperiled, or historical (G1, G2, or GH, respectively).[17] Only 6 percent of listed species fall into the vulnerable category (G3), with fewer than 2 percent

in the more secure categories of G4 and G5. Those few species that are regarded by ABI as "secure or apparently secure" (G4 or G5) are for the most part wide-ranging vertebrates, for which the listing affects only specific populations. This analysis confirms that most species are already exceptionally rare and restricted in their distributions before making it onto the federal endangered species list.

The entry bar for inclusion on the list, however, seems to be more stringent for plants and invertebrates than for vertebrate animals. Whereas 5 percent of listed vertebrates are ranked as secure or apparently secure (G4 or G5) and 13 percent as vulnerable (G3), no listed plants or invertebrates fall into the secure categories and very few (5 percent and 0.7 percent, respectively) are regarded as vulnerable (G3).[18] These figures are consistent with earlier studies that have found that plants and invertebrates tend to be considerably more rare than vertebrates before they are considered for listing.[19]

This analysis also confirms that a biological basis underlies whether species are listed as endangered (formally defined as "in danger of extinction within the foreseeable future") or in the lesser category of threatened (defined as "likely to become endangered within the foreseeable future").[20] From these definitions one would predict that species listed as endangered should receive higher ABI ranks than those listed as threatened. In fact, this is the case: Almost three-quarters (74 percent) of listed endangered species are regarded as critically imperiled by ABI, while only 37 percent of listed threatened species are in that category.[21]

The distribution of federally listed species relative to different land ownership patterns also has important implications for how ESA is administered. For example, vertebrate animals receive protection from being killed or harmed even on private lands, while plants and invertebrates are fully protected only on federal lands. In documenting the locations for imperiled and endangered species, natural heritage programs keep track of the type of landowner on whose property the species occurs. Analyzing this data from a national perspective indicates that federal lands support examples of about three-fifths (59 percent) of federally listed species. Interestingly, Department of Defense (DOD) lands contain the most federally listed species of any federal agency, supporting examples of about one-fifth (21 percent) of all listed species. This finding is particularly striking, given that DOD lands represent just 3 percent of the overall federal estate. Many military bases turn out to be strategically placed, not just from a military standpoint, but also from a biological perspective. Often found in coastal areas with fast growing human populations, many of DOD land holdings, such as southern Cal-

ifornia's Camp Pendelton Marine Base, are becoming islands of natural habitat in rapidly urbanizing regions.

Federal land management agencies have special responsibilities for protecting federally listed species, and there are numerous legal mechanisms in place to promote such public stewardship. However, given that about 40 percent of federally listed species are nowhere found on the federal lands, protection for these species will necessarily need to focus on private lands and lands managed by state and local governments.

Beyond Fences

Historically, conservation efforts in America have targeted a fairly narrow set of the nation's biological wealth. These efforts were often opportunistic and tended to focus on scenic lands, often of limited economic value. As a result, the nation is well endowed with "rock and ice" parks that offer spectacular vistas and inspiring recreational experiences. These parks are not, however, representative of the biota as a whole. Yellowstone National Park, for instance, was originally set aside for its spectacular geysers and other geological features rather than its equally impressive wildlife and ecosystem functions. Current understanding of biodiversity and particularly the large-scale ecological processes necessary to sustain it, suggests the need for a more biologically rational strategy for targeting conservation efforts and for strategies that address larger geographic scales.

A chain-link fence encircles the tiny oasis sheltering the Devils Hole pupfish. While providing some protection against vandals, this fence is as much psychological as practical, because the most serious threats to the species' existence come from offsite. Fences, no matter how high, are ineffective against regional-scale threats, like the drawdown of aquifers. Conservationists now under-

stand that even for the most rare and localized species—and none are more so than this little fish—protection efforts must go well beyond the fences, whether symbolic or physical, that bound our parks and nature preserves. Indeed, to sustain and restore the nation's biodiversity over the long term, protection efforts will need to be planned for and undertaken at the scale of whole landscapes and regions.

Twenty-five years ago The Nature Conservancy began creating a mechanism to improve its own ability to rationally identify and protect ecologically significant lands. The organization's efforts to establish natural heritage programs in each of the 50 states set in motion a process for informing not just the Nature Conservancy's conservation priorities, but those of society at large. Building on this foundation, the conservation community is now focusing on the need to get ahead of the extinction curve by considering the scale of habitat conservation that will be needed to protect not just individual species but whole ecological systems. Although the Endangered Species Act is a powerful legal tool and a formal testament to U.S. society's concern for its living inheritance, it is designed to function as a safety net rather than the much broader support scaffolding that will be required to protect the full array of the nation's biodiversity.

Building such a robust conservation infrastructure will require identifying those lands and waters that will be most important for maintaining and restoring the nation's species and ecosystems. Such activities are already under way in numerous organizations working at local, state, and ecoregional scales. One thing is already clear though: Approaching biodiversity conservation through land acquisition alone would be extraordinarily costly and probably not possible. Fortunately, many of the lands important to biodiversity are currently in uses that could be compatible with conservation values if they were managed responsibly.

Across the nation, innovative collaborations involving farmers, ranchers, and loggers are demonstrating that land can be managed in a way that produces economic benefits as well as maintains ecosystem values. For example, in the Malpais borderlands region of New Mexico, local ranchers have banded together and are working with conservationists and government agencies to improve the quality of the regional ecosystem so that they can continue to draw their livelihood from ranching. Indeed, the future of biodiversity in the United States may well rely as much on how we manage the working landscape (those lands in some form of economic production) as on how much land is strictly protected. Developing effective incentives for private landowners to manage their lands in biodiversity-friendly ways must be a major policy focus in the next few years. In the debate over endangered species protection, incentive-based approaches are at least one thing on which fairly broad consensus exists.

As the scientific community learns more about how natural ecosystems work, conservationists inevitably must think on broader scales of both space and time. To be successful, though, these grand visions must be translated into action at particular places inhabited by real people. One of the great challenges is to link local biodiversity concerns with national and global priorities. The assessment of U.S. species and ecosystems contained in *Precious Heritage: The Status of Biodiversity in the United States* confirms that there is globally significant biodiversity in our backyard and that protecting this inheritance is an obligation that the nation cannot afford to take lightly. Yet as the Devils Hole pupfish and its tiny range reminds us, the implications of even our most local actions can and do have global resonance.

NOTES:

1. J. E. Deacon and C. D. Williams, "Ash Meadows and the Legacy of the Devils Hole Pupfish," in W. L. Minckley and J. E. Deacon, eds., *Battle against Extinction: Native Fish Management in the American West* (Tucson: University of Arizona Press, 1991), 69–87.

2. B. A. Stein, L. S. Kutner, and J. S. Adams, eds., *Precious Heritage: The Status of Biodiversity in the United States* (New York: Oxford University Press, 2000).

3. M. C. Grant, J. B. Mitton, and Y. B. Linhart, "Even Larger Organisms," *Nature* 360 (1992): 216.

4. P. M. Hammond, "The Current Magnitude of Biodiversity," in V. Heywood. ed., *Global Biodiversity Assessment* (Cambridge: Cambridge University Press, 1995). 113–38.

5. Stein, Kutner, Adams, eds., note 2 above, pages 62–3.

6. M. D. F. Udvardy, "A Classification of the Biogeographic Provinces of the World," *IUCN Occasional Paper*, no. 18 (Gland, Switzerland: IUCN, 1975).

7. R. G. Bailey, *Ecoregions of tire Continents*, Map 1:30,000,000 (Washington D.C.: U.S. Department of Agriculture, Forest Service, 1989).

8. Stein, Kutner, Adams, eds., note 2 above, page 207.

9. For a discussion of species status assessments, see B. A. Stein and S. R. Flack, "Conservation Priorities: The State of U.S. Plants and Animals." *Environment* May 1997, 6–11, 34–9.

10. Stein, Kutner, Adams, eds., note 2 above, pages 112–6.

11. Ibid., pages 123–5. Figures for overall species diversity, endemism, and rarity are based on an analysis of a total of 19,279 species, which includes all native vascular plants, vertebrates, mussels, and crayfish.

12. A. J. Berger. Hawaiian Birdlife (Honolulu: University Press of Hawaii, 1981), 15.

13. Stein, Kutner, Adams, eds., note 2 above, pages 169–70.

14. For two recent county-level analyses of endangered species distributions, see A. P. Dobson, J. P. Rodriguez, W. M. Roberts, and D. S. Wilcove, "Geographic Distribution of Endangered Species in the United States," *Science* 275, 24 January 1997, 550–53; and C. H. Flather, M. S. Knowles, and I. A. Kendall, "Threatened and Endangered Species: Geography Characteristics of Hot Spots in the Conterminous United States," *BioScience* 48, no. 5 (1998): 365–76.

15. Stein, Kutner, Adams, eds., note 2 above, pages 166–8. The ecoregions displayed here are those recognized by The Nature Conservancy, which are based initially on those defined by R. Bailey of the U.S. Forest Service (see note 7 above).

16. A. Easter-Pilcher, "Implementing the Endangered Species Act: Assessing the Listing of Species as Endangered or Threatened," *BioScience* 46, no. 5 (1996): 355–63.

17. Stein, Kutner, Adams, eds., note 2 above, page 109, This analysis is based on the 1,048 U.S. species listed an of April 1998.

18. Ibid., pages 109–10.

19. D. S. Wilcove, M. McMillan, and K. C. Winston, "What Exactly Is an Endangered Species? An Analysis of the U.S. Endangered Species List, 1985–1991," *Conservation Biology* 7, no. 1 (1993): 87–93.

20. U.S. Fish and Wildlife Service, *Endangered Species Act of 1973 as Amended through the 100th Congress* (Washington, D.C., 1988).

21. Stein, Kutner, Adams, eds., note 2 above, pages 109–10.

Bruce A. Stein is vice president for programs with the Association for Biodiversity Information and was formerly a senior scientist with The Nature Conservancy. He was a coauthor and lead editor of *Precious Heritage: The Status of Biodiversity in the United States* (Oxford University Press. 2000), from which this article draws. A tropical botanist by training, Stein's current work focuses on evaluating the condition of the nation's ecosystems and working with the network of natural heritage programs to make biological and ecological information more accessible to the public and to environmental decision makers. He may be reached at the Association for Biodiversity Information, 1101 Wilson Boulevard, 15th floor, Arlington, VA 22209 (telephone: 703-908-1800; email: bruce_stein@abi.org).

From *Environment,* September 2001. Reprinted with permission of the Helen Dwight Reid Educational Foundation. Published by Heldref Publications, 1319 Eighteenth St., NW, Washington, DC. 20036-1802. © 2001.

Invasive Species:
Pathogens of Globalization

by Christopher Bright

World trade has become the primary driver of one of the most dangerous and least visible forms of environmental decline: Thousands of foreign, invasive species are hitch-hiking through the global trading network aboard ships, planes, and railroad cars, while hundreds of others are traveling as commodities. The impact of these bioinvasions can now be seen on every landmass, in nearly all coastal waters (which comprise the most biologically productive parts of the oceans), and probably in most major rivers and lakes. This "biological pollution" is degrading ecosystems, threatening public health, and costing billions of dollars annually. Confronting the problem may now be as critical an environmental challenge as reducing global carbon emissions.

Despite such dangers, policies aimed at stopping the spread of invasive "exotic" species have so far been largely ineffective. Not only do they run up against far more powerful policies and interests that in one way or another encourage invasion, but the national and international mechanisms needed to control the spread of non-native species are still relatively undeveloped. Unlike chemical pollution, for instance, bioinvasion is not yet a working category of environmental decline within the legal culture of most countries and international institutions.

In part, this conceptual blindness can be explained by the fact that even badly invaded landscapes can still look healthy. It is also a consequence of the ancient and widespread practice of introducing exotic species for some tangible benefit: A bigger fish makes for better fishing, a faster-growing tree means more wood. It can be difficult to think of these activities as a form of ecological corrosion—even if the fish or the tree ends up demolishing the original natural community.

The increasing integration of the world's economies is rapidly making a bad situation even worse. The continual expansion of world trade—in ways that are not shaped by any real understanding of their environmental effects—is causing a degree of ecological mixing that appears to have no evolutionary precedent. Under more or less natural conditions, the arrival of an entirely new organism was a rare event in most times and places. Today it can happen any time a ship comes into port or an airplane lands. The real problem, in other words, does not lie with the exotic species themselves, but with the economic system that is continually showering them over the Earth's surface. Bioinvasion has become a kind of globalization disease.

They Came, They Bred, They Conquered

Bioinvasion occurs when a species finds its way into an ecosystem where it did not evolve. Most of the time when this happens, conditions are not suitable for the new arrival, and it enjoys only a brief career. But in a small percentage of cases, the exotic finds everything it needs—and nothing capable of controlling it. At the very least, the invading organism is liable to suppress some native species by consuming resources that they would have used instead. At worst, the invader may rewrite some basic ecosystem "rules"—checks and balances that have developed between native species, usually over many millennia.

Although it is not always easy to discern the full extent of havoc that invasive species can wreak upon an ecosystem, the resulting financial damage is becoming increasingly difficult to ignore. Worldwide, the losses to agriculture might be anywhere from $55 billion to nearly $248 billion annually. Researchers at Cornell University recently concluded that bioinvasion might be costing the United States alone as much as $123 billion per year. In South and Central America, the growth of specialty export crops—upscale vegetables and fruits—has spurred the spread of whiteflies, which are capable of transmitting at least 60 plant viruses. The spread of these viruses has forced the abandonment of more than 1 million hectares of cropland in South America. In the wetlands of northern Nigeria, an exotic cattail is strangling rice paddies, ruining fish habitats, and slowly choking off the Hadejia-Nguru river system. In southern India, a tropical American shrub, the bush morning glory, is causing similar chaos throughout the basin of the Cauvery, one of the region's biggest rivers. In the late 1980s, the accidental release into the Black Sea of *Mnemiopsis leidyi*—a comb jelly native to the east coast of the Americas—provoked the collapse of the already highly stressed Black Sea fisheries, with estimated financial losses as high as $350 million.

Controlling invasive species is difficult enough, but the bigger problem is preventing the machinery of the world trading system from releasing them in the first place.

Controlling such exotics in the field is difficult enough, but the bigger problem is preventing the machinery of the world trading system from releasing them in the first place. That task is becoming steadily more formidable as the trading system continues to grow. Since 1950, world trade has expanded sixfold in terms of value. More important in terms of potential invasions is the vast increase in the volume of goods traded. Look, for instance, at the ship, the primary mechanism of trade—80 percent of the world's goods travel by ship for at least part of their journey from manufacturer to consumer. From 1970 to 1996, the volume of seaborne trade nearly doubled.

Ships, of course, have always carried species from place to place. In the days of sail, shipworms bored into the wooden hulls, while barnacles and seaweeds attached themselves to the sides. A small menagerie of other creatures usually took up residence within these "fouling communities." Today, special paints and rapid crossing times have greatly reduced hull fouling, but each of the 28,700 ships in the world's major merchant fleets represents a honeycomb of potential habitats for all sorts of life, both terrestrial and aquatic.

The most important of these habitats lies deep within a modern ship's plumbing, in the ballast tanks. The ballast tanks of a really big ship—say, a supertanker—may contain more than 200,000 cubic meters of water—equivalent to 2,000 Olympic-sized swimming pools. When those tanks are filled, any little creatures in the nearby water or sediment may suddenly become inadvertent passengers. A few days or weeks later, when the tanks are discharged at journey's end, they may

become residents of a coastal community on the other side of the world. Every year, these artificial ballast currents move some 10 billion cubic meters of water from port to port. Every day, some 3,000 to 10,000 different species are thought to be riding the ballast currents. The result is a creeping homogenization of estuary and bay life. The same creatures come to dominate one coastline after another, eroding the biological diversity of the planet's coastal zones—and jeopardizing their ecological stability.

Some pathways of invasion extend far beyond ships. Another prime mechanism of trade is the container: the metal box that has revolutionized the transportation of just about every good not shipped in bulk. The container's effect on invasion ecology has been just as profound. For centuries, shipborne exotics were largely confined to port areas—but no longer. Containers move from ship to harbor crane to the flatbed of a truck or railroad car and then on to wherever a road or railroad leads. As a result, all sorts of stowaways that creep aboard containers often wind up far inland. Take the Asian tiger mosquito, for example, which can carry dengue fever, yellow fever, and encephalitis. The huge global trade in containers of used tires—which are, under the right conditions, an ideal mosquito habitat—has dispersed this species from Asia and the Indo-Pacific into Australia, Brazil, the eastern United States, Mozambique, New Zealand, Nigeria, and southern Europe. Even packing material within containers can be a conduit for exotic species. Untreated wood pallets, for example, are to forest pests what tires are to mosquitoes. One creature currently moving along this pathway is the Asian long-horn beetle, a wood-boring insect from China with a lethal appetite for deciduous trees. It has turned up at more than 30 locations around the United States and has also been detected in Great Britain. The only known way to eradicate it is to cut every tree suspected of harboring it, chip all the wood, and burn all the chips.

As other conduits for global trade expand, so does the potential for new invasions. Air cargo service, for example, is building a global network of virtual canals that have great potential for transporting tiny, short-lived creatures such as microbes and insects. In 1989, only three airports received more than 1 million tons of cargo; by 1996, there were 13 such airports. Virtually everywhere you look, the newly constructed infrastructure of the global economy is forming the groundwork for an ever-greater volume of biological pollution.

The Global Supermarket

Bioinvasion cannot simply be attributed to trade in general, since not all trade is "biologically dirty." The natural resource industries—especially agriculture, aquaculture, and forestry—are causing a disproportionate share of the problem. Certain trends within each of these industries are liable to exacerbate the invasion pressure. The migration of crop pests can be attributed, in part, to a global agricultural system that has become increasingly uniform and integrated. (In China, for example, there were about 10,000 varieties of wheat being grown at mid-century; by

1970 there were only about 1,000.) Any new pest—or any new form of an old pest—that emerges in one field may eventually wind up in another.

The key reason that South America has suffered so badly from white-flies, for instance, is because a pesticide-resistant biotype of that fly emerged in California in the 1980s and rapidly became one of the world's most virulent crop pests. The fly's career illustrates a common dynamic: A pest can enter the system, disperse throughout it, and then develop new strains that reinvade other parts of the system. The displacement of traditional developing-world crop varieties by commercial, homogenous varieties that require more pesticide, and the increasing development of pesticide resistance among all the major pest categories—insects, weeds, and fungi—are likely to boost this trend.

Similar problems pertain to aquaculture—the farming and exporting of fish, shellfish, and shrimp. Partly because of the progressive depletion of the world's most productive fishing grounds, aquaculture is a booming business. Farmed fish production exceeded 23 million tons by 1996, more than triple the volume just 12 years before. Developing countries in particular see aquaculture as a way of increasing protein supply.

But many aquaculture "crops" have proved very invasive. In much of the developing world, it is still common to release exotic fish directly into natural waterways. It is hardly surprising, then, that some of the most popular aquaculture fish have become true cosmopolitans. The Mozambique tilapia, for example, is now established in virtually every tropical and subtropical country. Many of these introductions—not just tilapia, but bass, carp, trout, and other types of fish—are implicated in the decline of native species. The constant flow of new introductions catalogued with such enthusiasm in the industry's publications are a virtual guarantee that tropical freshwater ecosystems are unraveling beneath the surface.

Aquaculture is also a spectacularly efficient conduit of disease. Perhaps the most virulent set of wildlife epidemics circling the Earth today involves shrimp production in the developing world. Unlike fish, shrimp are not a subsistence crop: They are an extremely lucrative export business that has led to the bulldozing of many tropical coasts to make way for shrimp ponds. One of the biggest current developments is an Indonesian operation that may eventually cover 200,000 hectares. A horde of shrimp pathogens—everything from viruses to protozoa—is chasing these operations, knocking out ponds, and occasionally ruining entire national shrimp industries: in Taiwan in 1987, in China in 1993, and in India in 1994. Shrimp farming has become, in effect, a form of "managed invasion." Since shrimp are important components of both marine and freshwater ecosystems worldwide, it is anybody's guess at this point what impact shrimpborne pathogens will ultimately have.

Managed invasion is an increasingly common procedure in another big biopolluting industry: forestry. Industrial roundwood production (basically, the cutting of logs for uses other than fuel) currently hovers at around 1.5 billion cubic meters annually, which is more than twice the level of the 1950s. An increasing amount of wood and wood pulp is coming out of tree plantations (not inherently a bad idea, given the rate at which the world is losing natural forests). In North America and Europe, plantation forestry generally uses native species, so the gradation from natural forest to plantation is not usually as stark as it is in developing countries, where exotics are the rule in industrial-plantation development.

For the most part, these developing-country plantations bear about as much resemblance to natural forests as corn fields do to undisturbed prairies. And like corn fields, they are maintained with heavy doses of pesticides and subjected to a level of disturbance—in particular, the use of heavy equipment to harvest the trees—that tends to degrade soil. Some plantation trees have launched careers as king-sized weeds. At least 19 species of exotic pine, for example, have invaded various regions in the Southern Hemisphere, where they have displaced native vegetation and, in some areas, apparently lowered the water tables by "drinking" more water than the native vegetation would consume. Even where the trees have not proved invasive, the exotic plantations themselves are displacing natural forest and traditional forest peoples. This type of tree plantation is almost entirely designed to feed wood to the industrialized world, where 77 percent of industrial roundwood is consumed. As with shrimp production, local ecological health is being sacrificed for foreign currency.

There is another, more poignant motive for the introduction of large numbers of exotic trees into the developing world. In many countries severely affected by forest loss, reforestation is recognized as an important social imperative. But the goal is often nothing more than increasing tree cover. Little distinction is made between plantation and forest or between foreign and native species. Surayya Khatoon, a botanist at the University of Karachi, observes that "awareness of the dangers associated with invasive species is almost nonexistent in Pakistan, where alien species are being planted on a large scale in so-called afforestation drives."

Even international agreements that focus specifically on ecological problems have generally given bioinvasion short shrift.

The industrial sources of biological pollution are very diverse, but they reflect a common mindset. Whether it is a tree plantation, a shrimp farm, or even a bit of landscaping in the back yard, the Earth has become a sort of "species supermarket"; if a species looks good for whatever it is that you have in mind, pull it off the shelf and take it home. The problem is that many of the traits you want the most—adaptability, rapid growth, and easy reproduction—also tend to make the organism a good candidate for invasion.

Launching a Counter-Attack

Since the processes of invasion are deeply embedded in the globalizing economy, any serious effort to root them out will run the risk of exhausting itself. Most industries and policymakers are striving to open borders, not erect new barriers to trade. Moreover, because bioinvasion is not yet an established policy category, jurisdiction over it is generally badly fragmented—or even absent—on both the national and international levels. Most countries have some relevant legislation—laws intended to discourage the movement of crop pests, for example—but very few have any overall legislative authority for dealing with the problem. (New Zealand is the noteworthy exception: Its Biosecurity Act of 1993 and its Hazardous Substances and New Organisms Act of 1996 do establish such an authority.) Although it is true that there are many treaties that bear on the problem in one way or another—23 at least count—there is no such thing as a bioinvasion treaty.

Even agreements that focus specifically on ecological problems have generally given bioinvasion short shrift. Agenda 21, for example—the blueprint for sustainable development that emerged from the 1992 Earth Summit in Rio de Janeiro—reflects little awareness of the dangers of exotic forestry and aquaculture. Among international agencies, only certain types of invasion seem to get much attention. There are treaties—such as the 1951 International Plant Protection Convention—that limit the movement of agricultural pests, but there is currently no clear international mechanism for dealing with ballast water releases. Obviously, in such a context, you need to pick your fights carefully. They have to be important, winnable, and capable of yielding major opportunities elsewhere. The following three-point agenda offers some hope of slowing invasion over the near term.

The first item: Plug the ballast water pathway. As a technical problem, this objective is probably just on the horizon of feasibility, making it an excellent policy target. Strong national and international action could push technologies ahead rapidly. At present, the most effective technique is ballast water exchange, in which the tanks of a ship are pumped out and refilled in the open sea. (Coastal organisms, pumped into the tanks at the ship's last port of call, usually will not survive in the open ocean; organisms that enter the tanks in mid-ocean probably will not survive in the next port of call.) But it can take several days to exchange the water in all of a ship's ballast tanks, so the procedure may not be feasible for every leg of a journey, and the tanks never empty completely. In bad weather, the process can be too dangerous to perform at all. Consequently, other options will be necessary—filters or even toxins (that may not sound very appealing, but some common water treatment compounds may be environmentally sound). It might even be possible to build port-side ballast water treatment plants. Such a mixture of

technologies already exists as the standard means of controlling chemical pollution.

This objective is drifting into the realm of legal possibility as well. As of July 1 this year, all ships entering U.S. waters must keep a record of their ballast water management. The United States has also issued voluntary guidelines on where those ships can release ballast water. These measures are a loose extension of the regulations that the United States and Canada have imposed on ship traffic in the Great Lakes, where foreign ballast water release is now explicitly forbidden. In California, the State Water Resources Control Board has declared San Francisco Bay "impaired" because it is so badly invaded—a move that may allow authorities to use regulations written for chemical pollution as a way of controlling ballast water. Australia now levies a small tax on all incoming ships to support ballast water research.

Internationally, the problem has acquired a high profile at the UN International Maritime Organization (IMO), which is studying the possibility of developing a ballast management protocol that would have the force of international law. No decision has been made on the legal mechanism for such an agreement, although the most likely possibility is an annex to MARPOL, the International Convention for the Prevention of Pollution from Ships.

Within the shipping industry, the responses to such proposals have been mixed. Although industry officials concede the problem in the abstract, the prospect of specific regulations has tended to provoke unfavorable comment. After an IMO meeting last year on ballast water management, a spokesperson for the International Chamber of Shipping argued that rigorous ballast exchange would cost the industry millions of dollars a year—and that internationally binding regulations should be avoided in favor of local regulation, wherever particular jurisdictions decide to address the problem. Earlier this year in California, a proposed bill that would have essentially prohibited foreign ballast water release in the state's ports provoked outcries from local port representatives, who argued that such regulations might encourage ship traffic to bypass California ports in favor of the Pacific Northwest or Mexico. Of course, any management strategy is bound to cost something, but the important question is: What impact will this additional cost have? It may not have much impact at all. In Canada, for example, the Vancouver Port Authority reported that its ballast water program has had no detectable effect on port revenues.

The second item on the agenda: Fix the World Trade Organization (WTO) Agreement on the Application of Sanitary and Phytosanitary Measures. This agreement, known as the SPS, was part of the diplomatic package that created the WTO in 1994. The SPS is supposed to promote a common set of procedures for evaluating risks of contamination in internationally traded commodities. The contaminants can be chemical (pesticide residues in food) or they can be living things (Asian longhorn beetles in raw wood).

One of the procedures required by the SPS is a risk assessment, which is supposed to be done before any trade-constricting barriers are imposed to prevent a contaminated good from entering a country. If you want to understand the fundamental flaw in this approach as it applies to bioinvasion, all you have to do is recall the famous observation by the eminent biologist E. O. Wilson: "We dwell on a largely unexplored planet." When it comes to the largest categories of living things—insects, fungi, bacteria, and so on—we have managed to name only a tiny fraction of them, let alone figure out what damage they can cause. Consider, for example, the rough, aggregate risk assessments done by the United States Department of Agriculture (USDA) for wood imported into the United States from Chile, Mexico, and New Zealand. The USDA found dozens of "moderate" and "high" risk pests and pathogens that have the potential for doing economic damage on the order of hundreds of millions of dollars at least—and ecological damage that is incalculable. But even with wide-open thoroughfares of invasion such as these, the SPS requirement in its current form is likely to make preemptive action vulnerable to trade complaints before the WTO.

Another SPS requirement intended to insure a consistent application of standards is that a country must not set up barriers against an organism that is already living within its borders unless it has an "official control program" for that species. This approach is unrealistic for both biological and financial reasons. Thousands of exotic species are likely to have invaded most of the world's countries and not even the wealthiest country could possibly afford to fight them all. Yet it certainly is possible to exacerbate a problem by introducing additional infestations of a pest, or by boosting the size of existing infestations, or even by increasing the genetic vigor of a pest population by adding more "breeding stock." The SPS does not like "inconsistencies"—if you are not controlling a pest, you have no right to object to more of it; if you try to block one pathway of invasion, you had better be trying to block all the equivalent pathways. Such an approach may be theoretically neat, but in the practical matter of dealing with exotics, it is a prescription for paralysis.

In the near term, however, any effort to repair the SPS is likely to be difficult. The support of the United States, a key member of the WTO, will be critical for such reforms. And although the United States has demonstrated a heightened awareness of the problem—as evidenced by President Bill Clinton's executive order to create an Invasive Species Council—it is not clear whether that commitment will be reflected in the administration's trade policy. During recent testimony before Congress, the U.S. Trade Representative's special trade negotiator for agricultural issues warned that the United States was becoming impatient with the "increasing use of SPS barriers as the 'trade barrier of choice.'" In the developing world, it is reasonable to assume that any country with a strong export sector in a natural resource industry would not welcome tougher regulations. Some developed countries, however, may be sympathetic to change. The European Union (EU) has sought very strict standards in its disputes with the United States over bans on beef from cattle fed with growth hormones and on genetically altered foods. It is possible that the EU might be willing to entertain a stricter SPS. The same might be true of Japan, which has attempted to secure stricter testing of U.S. fruit imports.

The third item: Build a global invasion database. Currently, the study of bioinvasion is an obscure and rather fractured en-

terprise. It can be difficult to locate critical information or relevant expertise. The full magnitude of the issue is still not registering on the public radar screen. A global database would consolidate existing information, presumably into some sort of central research institution with a major presence on the World Wide Web. One could "go" to such a place—either physically or through cyberspace—to learn about everything from the National Ballast Water Information Clearinghouse that the U.S. Coast Guard is setting up, to the database on invasive woody plants in the tropics that is being assembled at the University of Wales. The database would also stimulate the production of new media to encourage additional research and synthesis. It is a telling indication of how fragmented this field is that, after more than 40 years of formal study, it is just now getting its first comprehensive journal: *Biological Invasions.*

Better information should have a number of practical effects. The best way to control an invasion—when it cannot be prevented outright—is to go after the exotic as soon as it is detected. An emergency response capability will only work if officials know what to look for and what to do when they find it. But beyond such obvious applications, the database could help bring the big picture into focus. In the struggle with exotics, you can see the free-trade ideal colliding with some hard ecological realities. Put simply: It may never be safe to ship certain goods to certain places—raw wood from Siberia, for instance, to North America. The notion of real, permanent limits to economic activity will for many politicians (and probably some economists) come as a strange and unpalatable idea. But the global economy is badly in need of a large dose of ecological realism. Ecosystems are very diverse and very different from each other. They need to stay that way if they are going to continue to function.

WANT TO KNOW MORE?

Although the scientific literature on bioinvasion is enormous and growing rapidly, most of it is too technical to attract a readership outside the field. For a nontechnical, broad overview of the problem, readers should consult Robert Devine's *Alien Invasion: America's Battle with Non-Native Animals and Plants* (Washington: National Geographic Society, 1998) or Christopher Bright's *Life Out of Bounds: Bioinvasion in a Borderless World* (New York: W.W. Norton & Company, 1998).

If you have a long-term interest in bioinvasion, you will want to get acquainted with the book that founded the field: Charles Elton's *The Ecology of Invasions by Animals and Plants* (London: Methuen, 1958). A historical overview of bioinvasions can be found in Alfred Crosby's book *Ecological Imperialism: The Biological Expansion of Europe, 900–1900* (Cambridge: Cambridge University Press, 1986).

Many studies focus on invasion of particular regions. The focus can be very broad, as in P. S. Ramakrishnan, ed., *Ecology of Biological Invasions in the Tropics*, proceedings of an international workshop held at Nainital, India, (New Delhi: International Scientific Publications, 1989). Generally, however, the coverage is much narrower, as in Daniel Simberloff, Don Schmitz, and Tom Brown, eds., *Strangers in Paradise: Impact and Management of Nonindigenous Species in Florida* (Washington: Island Press, 1997). The other standard research tack has been to look at a particular type of invader. The most accessible results of this exercise are encyclopedic surveys such as Christopher Lever's *Naturalized Mammals of the World* (London: Longman, 1985) and his companion volumes on naturalized birds and fish. In the plant kingdom, the genre is represented by Leroy Holm, et al., *World Weeds: Natural Histories and Distribution* (New York: John Wiley and Sons, 1997).

There are many worthwhile documents available for anyone who is interested not just in the ecology of invasion, but also in its economic, social, and epidemiological implications. Just about every aspect of the problem is discussed in Odd Terje Sandlund, Peter Johan Schei, and Aslaug Viken, eds., *Proceedings of the Norway/UN Conference on Alien Species* (Trondheim: Directorate for Nature Management and Norwegian Institute for Nature Research, 1996). A groundbreaking study of invasion in the United States, with particular emphasis on economic effects, is *Harmful Nonindigenous Species in the United States* (Washington: Office of Technology Assessment, September 1993). An assessment of the ballast water problem is available from the National Research Council's Commission on Ships' Ballast Operations' *Stemming the Tide: Controlling Introductions of Nonindigenous Species by Ships' Ballast Water* (Washington: National Academy Press, 1996). Readers who are interested in exotic tree plantations as a form of "managed invasion" might look through Ricardo Carrere and Larry Lohmann's *Pulping the South: Industrial Tree Plantations and the World Paper Economy* (London: Zed Books, 1996) and the World Rainforest Movement's *Tree Plantations: Impacts and Struggles* (Montevideo: WRM, 1999). Unfortunately, there are no analogous studies of shrimp farms.

For links to relevant Web sites, as well as a comprehensive index of related FOREIGN POLICY articles, access **www.foreignpolicy.com.**

Christopher Bright is a research associate at the Worldwatch Institute in Washington, DC, and author of Life Out of Bounds: Bioinvasion in a Borderless World (New York: W.W. Norton & Company, 1998).

Mass Extinction

BY DAVID HOSANSKY

THE ISSUES

Zoo biologist Edward J. Maruska can remember exploring the rain forests of Costa Rica in the late 1970s, when thousands of shimmering golden toads gathered in ponds of the mist-shrouded Monteverde Cloud Forest Reserve to breed.

But all the toads and vanished by the 1980s. Scientists believe that the spectacular toads, easily recognized by the males' bright orange color, fell victim to disease, changing climate patterns or pollution.

"They were so unique," recalls Maruska, executive director of the Cincinnati Zoo and Botanical Garden. "Then they were gone."

Last sighted by scientists in 1989, the golden toad is among thousands of species that have become extinct in recent years. Humans are wiping out much of the Earth's plant and animal live by paving over open space for homes and factories; clearing forests for cultivation and grazing; polluting the air and water; and introducing non-native species into fragile ecological areas.

Even primates, which belong to the taxonomic group that includes human beings, are not immune. Scientists recently concluded that a West African monkey known as Miss Waldron's red colobus has been wiped out because of deforestation and hunting. The red-cheeked monkey lived in the rain forest canopy of Ghana and the Ivory Coast. It was the first time in several centuries that a primate had become extinct.[1]

In fact, civilization's unrelenting march across unspoiled lands has had such a profound effect on nature that scientists warn we have entered an age of mass extinction the likes of which have not been seen since the demise of dinosaurs some 65 million years ago.

The Earth, scientists say, has experienced wholesale loss of life on such a colossal scale only five times before. Some contend it will bring irrevocable changes for the planet's dominant species—humans—altering everything from food supplies to medical breakthroughs to the weather.

"There's scientific debate about the rate and the extent of species loss, but I don't think there's much remaining debate that we're in a period of mass extinction," says Eleanor Sterling, director of the Center for Biodiversity and Conservation at the American Museum of Natural History in New York City. "It's absolutely one of the most critical issues that's facing us today."

Biologists cannot quantify the rate of extinction because they do not know the total number of species that exist on Earth—let alone the numbers of mostly unknown animals, plants, fungi and other organisms that are vanishing. But leading scientists believe that based on the pace of destruction of the richest habitats, such as tropical rain forests and coral reefs, the world is losing species at a rate of 100 to 10,000 times the normal, or "background," rate. They warn that 50 percent or more of all species will be gone by the end of the current century.[2]

In the United States alone, the U.S. Fish and Wildlife Service lists 1,233 plants and animals as threatened or endangered. That probably greatly understates the full number of vanishing species because the government lacks the resources to search for all types of endangered organisms.[3] Around the globe, human activities are threatening countless species of frogs, tropical beetles, freshwater fish, birds, flowers and trees as well as familiar mammals such as tigers, gorillas and giant pandas.

"We anticipate we've lost a whole host of species, sometimes before we even documented them," says David Olsen, a conservation biologist with the World Wildlife Fund (WWF). "We know enough about patterns of biodiversity around the world and the loss of natural habitats to say that we are in the midst of a very serious event."

Although not all biologists agree on the extent of the loss, they generally regard mass extinction as one of the gravest issues facing humanity. According to a 1998 poll by the American Museum of Natural History, most scientists in the United States believe that the world is in the midst of a mass extinction. Moreover, they rate it a greater threat to society than more publicized problems such as pollution, global warming and the thinning of the ozone layer.

Scientists believe the loss of so many species of animals, plants and microorganisms could have profound and unpredictable effects on the United States and every other nation. Many plants and animals provide food, fibers and building materials, as well as new medicines. Others regulate the flow of water, influence weather patterns, fertilize crops, prevent topsoil erosion and reduce the amount of carbon dioxide in the atmosphere. Even tiny creatures such as insects, regarded by many as pests, are probably essential for the survival of Homo sapiens by filling critical niches in the global ecosystem.

"So important are insects and other land-dwelling arthropods that if all were to disappear, humanity probably could not last more than a few months," Harvard bi-

Vast Regions of Earth Are 'In Danger'

*In vast portions of the world, including Australia, India, China and much of South America, 7 percent or more of the mammal and bird species are in danger, the World Wildlife Fund says. * According to the WWF's Living Planet Index, the Earth has lost 30 percent of its plant and animal species since 1970, based on the loss of forest cover and the decline in abundance of marine and freshwater species.*

Mammal and bird species in danger

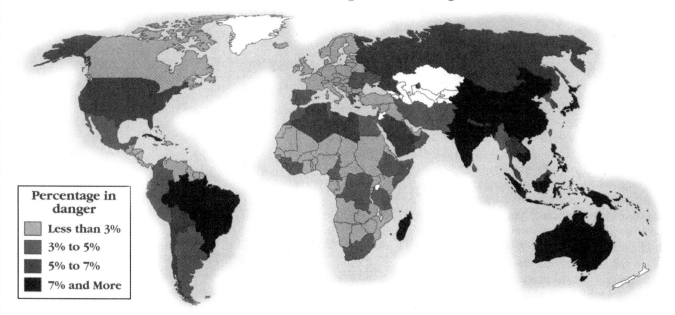

Percentage in danger
- Less than 3%
- 3% to 5%
- 5% to 7%
- 7% and More

Many places where losses are believed to be severe, such as Ivory Coast and other areas in Africa, don't appear on the map because of counting anomalies.

Source: "Living Planet Report," World Wildlife Fund, September 2000.

Reuters Graphics

ologist Edward O. Wilson wrote in his influential 1992 book, *The Diversity of Life*.[4]

World leaders have responded to the threat of biodiversity loss with a series of international agreements. The 1973 Convention on International Trade in Endangered Species of Wild Fauna and Flora (CITES) has helped preserve well-known species such as elephants and sea turtles. In the United States, the Endangered Species Act (ESA), along with other government measures such as pesticide regulations and passage of the Clean Water Act, are credited with fostering the recovery of several species on the brink of extinction, including the bald eagle.

A series of high-profile conferences, including the 1992 Earth Summit in Rio de Janeiro, Brazil, has helped focus attention on the worldwide loss of biodiversity. In recent years, powerful organizations such as the World Bank have begun to provide grants to

developing countries for projects that help restore the environment, and industrialized nations have provided incentives, such as debt forgiveness, to developing countries to preserve biologically important habitat. Some environmental groups promote ecotourism as a way to help local communities profit by limiting harmful development.

However, ecosystems are threatened by such a multitude of factors that even optimistic conservationists warn that numerous species are doomed. "We're at the beginning of the [extinction] curve, and the question is whether we'll get it together soon enough to avoid a lot of the loss," says Thomas Lovejoy, chief biodiversity adviser at the World Bank.

The No. 1 cause of extinction, both in the United States and worldwide, is habitat destruction. As people clear forests, drain wetlands and dam rivers, they destroy the homes of countless organisms. The United

States, for example, has lost almost half its wetlands since the 18th century, and tropical countries have lost more than half of their rain forests.[5]

The introduction of invasive species such as fire ants and kudzu, which proliferate rapidly and overcome native species, also has increased the pace of extinctions. Island ecosystems are particularly sensitive. In Guam and Hawaii, for example, newly introduced rats, cats and snakes have wiped out many species of native birds.

Pollution, overfishing and hunting also threaten wildlife habitat. And more potential threats loom. Environmentalists now fear global climatic changes could decimate habitats in natural parks before they can migrate to more suitable surroundings.

"We have quite static ways of protecting biodiversity at the moment, drawing boundaries about a population and saying, 'This population is now conserved,'" says

THE TEN MOST WANTED

The World Wildlife Fund has identified 10 species of plants and animals that are particularly threatened by illegal and unsustainable trade. The list was presented at the Convention on International Trade in Endangered Species of Wild Flora and Fauna (CITIES) annual conference in Nairobi, Kenya, last April.

Tiger
- **Habitat:** Russian Far East, Southeast Asia
- **Threat:** Poaching for skin, bones and body parts
- **Population:** under 6,000

Whale Shark
- **Habitat:** Highly migratory, coastal
- **Threat:** Overfishing for fins, meat
- **Population:** Unknown, decline in sightings

Javan Pangolin
- **Habitat:** Indonesia
- **Threat:** Skin, meat and scales, used in traditional Asian medicine
- **Population:** Unknown

Asian Box Turtles
- **Habitat:** South east Asia
- **Threat:** Hunted for shell, which is used in traditional medicine
- **Population:** Nearing extinction

Horned Parakeet
- **Habitat:** New Caledonia
- **Threat:** Habitat loss, illegal capture for the pet trade
- **Population:** under 1,700

Tibetan Antelope
- **Habitat:** East Asia
- **Threat:** Wool used to make highly priced Shahtoosh shawls. Up to five killed to make one shawl
- **Population:** under 75,000

Asian Ginseng
- **Habitat:** Russia, China, Korean Peninsula
- **Threat:** High prices paid for wild roots ($3,000 per ounce)
- **Population:** Isolated areas

Sumatran Rhinoceros
- **Habitat:** Sumatra, Borneo, Peninsular Malaysia
- **Threat:** Hunted for use in traditional medicine
- **Population:** under 300

Sumatran Rhinoceros

Hawksbill Sea Turtle

Hawksbill Sea Turtle
- **Habitat:** Caribbean
- **Threat:** Global market for tortoise shell, exploitation for eggs and meat
- **Population:** Unknown

Giant Panda
- **Habitat:** China
- **Threat:** Habitat loss and poaching. Demand from overseas zoos
- **Population:** 1,000

Source: World Wildlife Fund

Reuters Graphics

the American Museum of Natural History's Sterling. "As the temperature changes, you're going to have an empty preserve and a species that has migrated out."

But while scientists have expressed growing alarm about the rapid disappearance of more and more species, the Museum of Natural History poll indicates the issue of mass extinction has not yet resonated with the general public. Unlike more visible environmental problems, such as air pollution and contaminated drinking water, the loss of obscure beetles and salamanders does not affect the daily lives of most people.

Furthermore, the business community is reluctant to support government regulations that restrict development to protect plants and creatures that seem to have little significance. That was dramatically illustrated in the late 1980s and early '90s, when loggers heatedly protested plans to set aside forests that were habitat to the rare northern spotted owl.

"Let's prioritize what the real costs are, because our resources are limited," says William L. Kovacs, vice president of environmental and regulatory affairs for the U.S. Chamber of Commerce. "We can spend tens of billions of dollars in trying to protect something that has very little benefit to man."

Complicating the issue, scientists disagree about both the extent and the implications of mass extinction. Biologists, who have identified about 1.75 million species worldwide, estimate the actual total ranges anywhere from 3.6 million to more than 100 million.

Harvard's Wilson believes that more than half of these will be gone by the end of the century unless strong conservation measures are undertaken. He and other biologists have arrived at such estimates by calculating the number of species in biologically diverse habitats—especially coral reefs and tropical rain forests—and the rate at which those habitats are being destroyed by human actions.

"Biologists who explore biodiversity see it vanishing before their eyes," he wrote in a special Earth Day edition of *Time* magazine earlier this year.[6]

But some contend that concerns about extinction are overstated. Michael Gilpin, a University of San Diego biology professor, predicts that future extinctions will be largely confined to obscure organisms in developing countries. "It's not that we're going to lose zebras and wildebeests," he says. "Beetles we're going to lose, but we're never going to know what they are."

Others go as far as to argue that human disruptions of nature actually may be good in the long run. They contend that past extinctions and other disturbances that alter ecosystems created openings for new organisms, often increasing overall diversity. "Without extinction, without a loss of current variety, future variation diminishes," two professors argued in a controversial 1992 article.[7]

However, few in the scientific community take such a sanguine view. Even a skeptic such as Gilpin warns the loss of species deprives the world of genetic diversity, making it harder for scientists to develop new medicines and more disease-resistant strains of crops. "We're burning our genetic library," he says.

Others believe that a mass extinction will result in a natural world dominated by disease-carrying animals that have learned to adapt to human society. What sorts of animals are likely to proliferate if the pace of extinction continues? Sterling predicts "rats and cockroaches," among others.

As policy-makers confront the specter of mass extinction, here are key questions being asked:

Should we preserve endangered plants and animals?

As an aspiring botanist growing up in the San Francisco area, Peter H. Raven enjoyed tromping about and identifying numerous species of local plants. Now he finds many formerly wild areas paved over, and the plants either gone or dying out.

Raven, who is director of the Missouri Botanical Garden and a leading advocate for protecting biodiversity, warns that the same pattern is being repeated around the globe, threatening to impoverish human society on a vast scale because scientists constantly turn to nature to develop new foods, medicines and other products.

"As we lose biodiversity, we lose many opportunities for a rich and sustainable and healthy life," Raven says. "It's biodiversity that makes this a living planet. Most of the species that are likely to be lost in the coming century have never been seen by anybody. We will never be able to exploit their potential."

Raven and other conservationists cite three major reasons why it is critical for world leaders to do everything possible to preserve fast-disappearing species:

• Plants and animals provide humans with essentials such as food, clothing, shelter and medicine. Food supplies could dwindle if not for wild plants that can be crossbred with cultivated varieties to provide hardier crops. For example, a wild species of Mexican maize, nearly extinct when found in the 1970s, has been used to develop disease-resistant corn.

Similarly, endangered plants have helped to spur pharmaceutical breakthroughs. The potent anti-cancer drug Taxol, for example, is derived from the bark of the Pacific yew tree. A study by the National Institutes of Health and other agencies concluded that about 40 percent of the most commonly prescribed drugs are developed from natural sources.[8]

• Living organisms stabilize the environment. Scientists are finding that forests, wetlands and other ecosystems play a major role in regulating water drainage, preventing landslides and soil erosion, and even influencing rain patterns and other types of weather. When the environment is degraded, catastrophe may occur. Environmentalists point to Haiti, now virtually deforested and gradually turning into a desert.

• Society has ethical and aesthetic obligations to preserve the environment. In essence, conservationists—and some religious leaders—argue, every organism has an intrinsic beauty and a place in the world that should not be disturbed by human actions. In fact, some go so far as to say that human beings can never feel at home in a world scarred by environmental degradation.

"Our brains are formed around biodiversity," says Raven of the Missouri Botanical Garden. "A lot of our art is based on it. We are related to it."

"If we don't pay more attention to these issues, then the consequences down the road for ourselves and our future generations will be very significant," says Mark Schaefer, president of the Association for Biodiversity Information, a conservation research organization.

Perhaps surprisingly, some skeptics argue that preservation efforts, by disrupting natural processes, actually can damage the environment. They point out that extinction, after all, is a vital part of evolution: Certain species vanish, and others take their place. "I would make the argument that species live and die and evolve," says the Chamber of Commerce's Kovacs. "Whether you like it or not, you and I are part of evolution. At some point of time, our ancestors died out and changed."

Some also question whether the government should continue to stress the protection of species in an age when technological breakthroughs are allowing scientists to develop genetically modified crops that are resistant to disease and adverse weather. Scientists are even beginning to try to recover species through cloning.

Business leaders also are concerned about the cost of protecting endangered species. They point out that the construction of roads, hospitals, housing developments and other structures have been blocked for relatively trivial environmental concerns, sometimes at the cost of human health.

In 1998, for example, senators battled over building a single-lane gravel road through Alaska's Izembek National Wildlife Refuge, disrupting a pristine habitat but giving residents of Cold Bay, an isolated fishing village, year-round access to a healthcare facility. In the end, lawmakers decided to leave the preserve intact while agreeing to spend $37.5 million on an all-weather airport building and other facilities for the community.[9]

Kovacs says the government should pay more attention to the needs of society, even if environmentalists protest. "Human beings are species too," he says, "and they have some rights on the planet."

Are environmental laws helping rare species?

A boater on the Potomac River just outside Washington, D.C., can expect to see something that would have been remarkable just three decades ago: pairs of nesting bald eagles. The majestic birds, which nearly vanished from the continental United States in the 1960s, have staged a strong comeback thanks to the 1970 federal ban on the highly toxic pesticide DDT and the 1973 Endangered Species Act.

"The return of the bald eagle is a fitting cap to a century of environmental stewardship," President Clinton declared at a White House ceremony celebrating the comeback of the bird that has symbolized the nation for 200 years.[10]

Although the bald eagle remains comparatively rare because of the destruction of its habitat, it nevertheless symbolizes the success of environmental laws. Other successes include the grizzly bear, the gray whale and the gray wolf.

The Endangered Species Act is part of a network of environmental laws that preserve habitats and rare species. The 1972 Marine Mammal Protection Act restricts the hunting of seals, polar bears and other marine mammals; the 1972 Clean Water Act helps rehabilitate waterways and restore many species of fish, and the 1964 Wilderness Act protects remote areas from road building and other development that can fragment habitat and isolate species.

Congress in 1990 took a step toward stopping the introduction of non-native species, which can wipe out native species by preying upon them or outcompeting with them for food, by passing the Non-Indigenous Aquatic Nuisance Act. The law established a task force that coordinates federal efforts to keep out non-native aquatic species, such as the zebra mussel, which is notorious for proliferating in the pipes of drinking water systems, hydroelectric plants and industrial facilities, constricting water flow and affecting heating and cooling systems.

The Endangered Species Act, however, is the only U.S. law targeted specifically at helping rare animals and plants. Policy-makers are divided on whether it has been successful.

Since its passage, more than 1,200 species in the United States have been listed as either endangered or threatened. Of that number, just 11 species have recovered

sufficiently to be taken off the list; nine were removed because of improved data, such as the discovery of additional populations; and seven have become extinct.[11]

Based on these results, the law's critics brand it a failure. "It hasn't really been effective in achieving its goals," says Duane Desiderio, assistant staff vice president of the National Association of Home Builders. "It's become little more than an act with a list."

He contends that environmentalists are more intent on stopping development by "adding more and more species to the list" than they are on fostering the recovery of rare organisms.

Environmentalists, however, say that many more animals and plants would be extinct today if it were not for the law. For example, the number of black-footed ferrets—a Western weasel that feeds on prairie dogs—had dwindled to 18 in the mid-1980s before the government stepped in and helped nurture the population back to several thousand.

"If you look at it in terms of preventing extinctions and getting species to a point where at least there is a chance to bring them back, then on that level it's been effective," says Christopher Williams, senior program officer for wildlife conservation policy at the WWF.

The Endangered Species Act is controversial because it imposes strict constraints on property owners who want to develop land that is home to rare plants and animals. As a result, it can have the perverse effect of encouraging property owners to destroy habitat on their land to make sure that it never becomes home to an endangered species.

In Austin, Texas, for example, some landowners in the 1990s bulldozed juniper trees on their property as a pre-emptive strike to prevent their land from being listed by the federal government as critical habitat for two endangered songbirds, the golden-cheeked warbler and black-capped vireo.[12] The rare birds nest only in junipers.

"The incentives are all wrong. The more wildlife habitat that a landowner leaves on his land, the more likely it is that he's going to be prevented from using his own land," says R. J. Smith, senior environmental scholar at the Competitive Enterprise Institute, a free-market-oriented think tank. "It punishes landowners for being good stewards."

Critics also contend that environmentalists are more interested in using the ESA to stop development than to protect rare

species. They trace the tactic back to the late 1970s, when residents along the Little Tennessee River who were trying to halt construction of a $31 million dam discovered that the river was home to a tiny endangered fish, the snail darter trout. By invoking the Endangered Species Act, the Tennessee residents won court rulings that temporarily blocked the project. Congress eventually overruled the courts and let construction proceed.

The law originally was conceived to help majestic animals such as the bald eagle. But, to the exasperation of developers, it often is imposed instead to protect rodents and insects such as the Delhi fly and the Indiana bat—even though few people appear concerned about such species.[13]

"Should you stop an entire road or an entire hospital that serves a region just because you've found some flies that scientists say they can't find any use for?" asks Kovach of the Chamber of Commerce. "The program has gotten ridiculous. The environmentalists have really lost control of their common sense."

But environmentalists say it is essential to protect all species, even the most obscure, to preserve an ecosystem. "Each plant or animal, of course, has its own unique set of genes. Once that organism is lost, so is that unique genome," says Schaffer of the Association of Biodiversity Information. "We simply don't know where the next critical bacterium or fungi or plant may be found that contains some unique chemical substance encoded in its gene."

U.S. laws aside, policy-makers debate the effectiveness of international efforts to protect endangered species, including the Convention on International Trade in Endangered Species of Wild Fauna and Flora, which restricts or even bans trading in rare animals such as tigers and rhinoceroses. Like the Endangered Species Act, CITES has produced mixed results. Its ban on ivory trading, for example, inadvertently may have encouraged ivory poachers to instead hunt rhinoceroses for their horns.

Still, says WWF Vice President Richard N. Mott, "It's been instrumental in averting the decline of marine turtles, elephants" and other species.

Will humans survive the current wave of extinctions?

Ever since biologists began studying the extinction of species, they have pondered how long humans will endure. Since animal species typically survive for a few million years, and Homo sapiens evolved

When Mass Extinctions Ruled the Earth

Extinction is a natural event that usually occurs at a low level, periodically claiming a relative handful of species. A mass extinction, by contrast, wipes out large numbers of plants and animals in a comparatively short time period in geologic terms—perhaps "only" a few hundred thousand years.

Scientists believe that there have been at least five mass extinctions since complex life evolved on Earth a half-billion years ago. In each case, at least one-quarter of all species are thought to have perished, and the toll often was more than one-half. It may have taken millions of years for the diversity of life to recover each time.

Scientists are not sure what caused most of the mass die-offs. They have speculated about giant asteroids striking the planet or massive volcanic eruptions that could have caused devastating climate shifts and changes in the earth's atmosphere. Some have even raised the possibility of deadly cosmic radiation.[1]

The five previous mass extinctions occurred during the following geological eras:

• **Ordovician (440 million years ago)**—Glaciation may have wiped out an estimated 25 percent of Earth's plant and animal families, some of which may have had thousands of species. Freezing temperatures lowered sea levels and eliminated much marine habitat. Groups of marine organisms, such as trilobites, suffered especially heavy losses.

• **Devonian (370 million years ago)**—A massive loss of biodiversity may have occurred within a period of 500,000 years, or there may have been several smaller extinction episodes over a period as long as 15 million years. As many as 70 percent of all species disappeared, with marine organisms more affected than those living in freshwater. Scientists speculate that climate change may have played a role, as well as a drop in oxygen levels in shallow waters.

• **Permian (250 million years ago)**—The most catastrophic of all mass extinctions may have wiped out 96 percent of all marine species and more than three-fourths of the vertebrate families on land. Scientists speculate that the cause may have been volcanic activity, a change in ocean salinity or climate shifts.

• **Triassic (210 million years ago)**—Although far less devastating than the Permian extinction, the Triassic is important because it opened a niche for dinosaurs and early mammals. The cause of this extinction, which wiped out many sponges, insects and vertebrate groups, is poorly understood. But some scientists speculate it was caused by dramatic climate change.

• **Cretaceous (65 million years ago)**—An estimated 85 percent of Earth's species, including all dinosaurs and flying lizards (pterosaurs) were wiped out during this much studied period. However, mammals, birds and many reptiles, including crocodiles, survived virtually unscathed. Since 1980, scientists have found evidence that a large asteroid crashing into the Earth caused the Cretaceous extinction, although some speculate that volcanic eruptions or other causes played a role.

1. Background taken from *National Geographic*, February 1999, pp. 48–49 and British Broadcasting Corporation, www.bbc.co.uk/education/darwin/esfiles/massintro.htm.

---

only about 100,000 years ago, the odds would appear to be good that humans will be around for quite a while.

However, if extinctions occur at up to 1,000 times the normal rate in the 21st century, could human beings disappear as well? The reassuring answer from most biologists: Not likely. To be sure, humans have the capability of destroying themselves through nuclear or biological warfare, and some scientists even speculate that machines or genetically engineered versions of humans could take over the world.

But, apart from such extraordinary scenarios, the laws of nature suggest that humans are well-positioned to survive a mass extinction. "We are by far the most widely distributed... species on the planet and, with our technology, I believe the most unassailable," University of Washington geologist and paleontologist Peter D. Ward wrote in a 1997 book on extinction, *The Call of Distant Mammoths*. "We are the least endangered species on the planet."[14]

Plant and animal species most vulnerable to extinction usually are few in number, live in a limited area and lack the ability to adjust to change. In contrast, there are more than 6 billion people who live throughout the world and eat all types of food. It is hard to picture a natural scenario—even repeated volcanic eruptions, massive flooding or global climate change—in which the Earth's environment is so altered that people can no longer survive, paleontologists say.

However, some experts warn that the loss of biodiversity could undermine the well-being of society because it may become harder to grow crops and develop new medicines and other products.

"I can't see a scenario where the destruction of biodiversity is going to drive human beings to extinction," says Raven of the Missouri Botanical Garden, "but I see a world that is dull, gray, homogenized and bleak, with many fewer possibilities for developing new products and with many fewer interesting things to do. As we lose biodiversity, we lose many opportunities for rich and sustainable and healthy lives."

Raven and other scientists also believe that the Earth may not be able to sustain its population, which is expected to reach anywhere from 7.3 billion to 10.7 billion by 2050, according to the United Nations projections.[15] In particular, they say that people in wealthy countries like the United States have to consume less or risk depleting the world's resources. "If everybody in the world were consuming at the level of the United States, we would need about three planets like Earth to support them," he says.

Others, however, reject such bleak scenarios. They believe that, thanks to the advances in genetics, scientists will be able to develop better crops and more effective medicines despite the loss of wild species. Indeed, they speculate, genetically modified organisms may even help speed up habitat restoration.

The debate may be moot. Scientists are not certain of what prompted previous mass extinctions, but they know that the causes must have been cataclysmic. If a massive asteroid were to strike the Earth—which may have wiped out the dinosaurs 65 million years ago—humans could perish along with most other plants and animals.

University of Chicago statistical paleontologist David M. Raup estimated the odds of a significant asteroid or comet striking Earth during a person's 75-year lifespan at 1 in 4,000. Although that may indicate that humans are pretty safe, Raup writes that experiencing a "civilization-destroying impact" would appear to be a far greater possibility than dying in an airplane crash.

"We don't know," Raup concludes, "whether we chose a safe planet."[16]

Notes

1. Andrew C. Revkin, "A West African Monkey Is Extinct, Scientists Say," *The New York Times*, Sept. 12, 2000, p. A20.

2. For more detail, see Edward O. Wilson, "Vanishing Before Our Eyes," *Time*, special Spring 2000 Earth Day edition, p. 29; M. Lynne Corn, "Endangered Species: Continuing Controversy," Congressional Research Service Report, May 8, 2000.

3. See U.S. Fish and Wildlife Service, "General Statistics on Endangered Species," http://endangered.fws.gov/stats/gen-stats.html.

4. Edward O. Wilson, *The Diversity of Life* (1992). p. 133.

5. For background, see Jeffrey A. Zinn and Claudia Copeland, "Wetland Issues," Congressional Research Service Report, May 1, 2000, viewed at http://www.cnie.org/nle/wet-5.html#_1_3; David Hosansky, "Saving the Rain Forests," *The CQ Researcher*, June 11, 1999, pp. 497–520.

6. Wilson, "Vanishing Before Our Eyes," *op. cit.*, p. 30.

7. Julian L. Simon and Aaron Wildavsky, cited in Brenda Stalcup, ed., *Endangered Species Opposing Viewpoints* (1996), p. 24. Simon, now deceased, was a University of Maryland business administration professor; Wildavsky is a University of California, Berkeley, political science and public policy professor.

8. Hosansky, *op. cit.*, p. 505.

9. "Congress Compiles a Modest Record in a Session Sidetracked by Scandal," *CQ Weekly*, Nov. 14, 1998, p. 3987.

10. Edwin Chen, "Eagle Lauded With an Eye Toward Future Environment," *Los Angeles Times*, July 3, 1999, p. A17.

11. "Endangered Species: Continuing Controversy," Congressional Research Service Report, May 8, 1999, www.cnie.org/nle/biodv-1.html#_1_5.

12. Geneva Overholser, "A New Tool to Help the Environment," *Austin American-Statesman*, Feb. 4, 2000, p. A15.

13. H. Josef Herbert, "On 25th Anniversary, Endangered Species Act Elicits Admiration, Ire," *Los Angeles Times*, March 14, 1999, p. B4.

14. Peter D. Ward, *The Call of Distant Mammoths* (1997), p. 95.

15. Jeffrey Kluger, "The Big Crunch," *Time*, special Spring 2000 Earth Day edition, p. 46.

16. David M. Raup, *Extinction: Bad Genes or Bad Luck?* (1991), p. 199.

David Hosansky is a freelance writer in Denver who specializes in environmental issues. He previously was a senior writer at *CQ Weekly* and a reporter at the *Florida Times Union* in Jacksonville, where he was twice nominated for a Pulitzer Prize.

Watching vs. Taking

We are seeing a shift in human relationships with wildlife, as millions turn from taking other species for furs, food, or sport to just watching. In a way, it's a new kind of hunt.

by Howard Youth; Illustrations by Mark Geyer

"Get over here... It's back," whispers a stone-faced Philadelphia man into his walkie-talkie. He's crouched at a campsite in the Bentsen-Rio Grande Valley State Park in southernmost Texas. A short distance away, two forty-something men launch into a fervent but silent racewalk. The site, festooned with bird feeders, backs up to a dense curtain of granjeno, catclaw, and other native brush that harbors some of North America's rarest birds. Eight binocular-toting bird watchers (or "birders," as they call themselves) already ring the site, peering through their optics at a brown, sparrow-sized bird—a female blue bunting. Unaccompanied by a colorful male companion, this Mexican bird is the only member of her species known to be visiting the United States.

"Where's the blue?" asks a woman, a native Texan who is camping nearby.

"You won't find any. That's the female," says a Maryland birder who, tipped off by the local rare bird alert, pulled up in his rental car just a few minutes before.

"Yes!" whispers one of the just-arrived racewalkers, his teeth and fist clenched.

"That's a big tick for you," comments his friend—the "tick" referring to a check on his lifetime list of birds seen.

"Look at the warm brown tone all over the bird," says another watcher, perhaps talking herself out of any disappointment at not seeing the bright blue of the missing male. **"She's a real beauty in her own right."**

The blue bunting encounter, which I witnessed during a December 1999 trip to South Texas, could have taken place in any of a thousand locations across North America. The species of bird might vary—it might be a piping plover or an elegant trogon or a northern hawk owl—but the intensity of the fascination would be much the same. Birding has become one of the continent's fastest growing outdoor pastimes, and it's leading a whole parade of newly popular wildlife-watching avocations: there are also people (and organizations) devoted to sighting butterflies, wildflowers, wolves, mountain lions, and whales. Nor is this growing fascination with wildlife confined to North America; it appears to be a global phenomenon, with large economic and ecological implications.

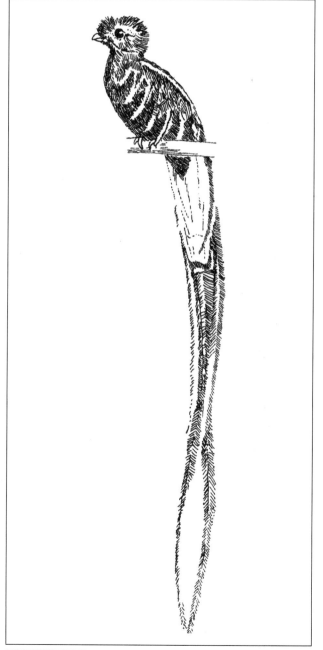

RESPLENDENT QUETZAL

GOLDEN TOAD

It has come on fast. Today, some species are worth as much—if not more—in their natural habitat, alive and free, than they bring as game. In parts of Africa where tourists pay well to see lions, for example, a lion that remains alive has been estimated to be worth $575,000. In many cases, where wildlife-watching tourism has grown, poaching has declined. In Belize, as manatee-watching tours have brought a growing income, illegal hunting of the big mammals has become less of a problem. And in South Texas, there is now hope that the remaining habitats of the blue bunting, the small wild cats called ocelots, and other rare species won't fall to bulldozers.

BLUE BUNTING Only one has been seen in the United States during the past year.

The growing interest in wildlife watching draws attention to a growing perceptual chasm, or difference of fundamental values and sensibilities, between those who view wildlife as a resource to be exploited, and those who

share a resurgent awareness of wild animals as our co-inhabitants on a fragile planet. The interest in watching may also reflect a new kind of frontier for human curiosity. Over the past millennium, the goals of exploration were geographical—and strange beasts or trees encountered along the way were seen mainly as curiosities. Now, the goals of exploration are increasingly those of understanding the nature of the world we have conquered. The geographical mysteries have given way to new ecological and biological ones. Close watching of wildlife has led to the realization that many animals have surprising levels of intelligence and social organization, and that many plants play powerful and complex ecological roles as well. In an era of dangerously unsustainable human domination, it is of growing interest that many of the animals we have considered our inferiors have in fact thrived for millions of years longer than we have, and have proven to be masters of adaptability and survival. Whatever our reasons for watching, we are less inclined now to take wildlife for granted.

Down on the Rio Grande

Once best known for citrus, cotton, cabbage, and as a haven for winter-fleeing "snowbirds" (people in Winnebagos or Airstreams) from the north, towns in the southern tip of Texas now herald themselves as wildlife-watching paradises. Birders, butterfly watchers, and other nature lovers arrive to take in subtropical sights that can be found nowhere else north of the U.S.-Mexican border. The new tourism may have arrived just in time. By 1999, after decades of conversion to cattle pastures, farm fields, and housing subdivisions, more than 95 percent of the natural environment of the region had disappeared. Today, Bentsen-Rio Grande Valley State Park and other protected areas account for most of the remaining natural habitat of the Lower Rio Grande Valley.

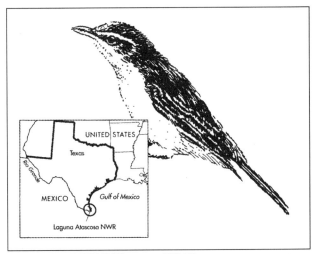

YELLOW-GREEN VIREO Was worth $150,000 a year, untouched, to businesses in South Texas.

Although birders have been coming to southern Texas for more than 30 years, their growing numbers only recently caught the attention of many local businesses. Now, visitors to the Best Western Inn in Harlingen are greeted by paintings *not* of cowboys, bucking broncos, and ten gallon hats, but of the red-crowned parrot and ringed kingfisher. A Harlingen Area Chamber of Commerce brochure, titled "Hooters, Hawks, and Hummingbirds," contains alluring photos of the green jay, chachalaca, kiskadee flycatcher, buff-bellied hummingbird, groove-billed ani, and other local birds. Three South Texas cities—Harlingen, McAllen, and Raymondville—now hold annual birding festivals that lure crowds of out-of-state bird lovers for field trips, seminars, and specialty sales, where they snap up the latest binoculars, books, and spotting scopes. These were among the first such events on the continent; now more than 200 annual nature-oriented festivals are scheduled throughout the United States and Canada. "A lot of things coalesced about five years ago," says Frank Judd, a 30-year Texas resident and biology professor at the University of Texas-Pan American in Edinburg. "Ecotourism is being championed in the local press and other media, and this has made local people aware that they can derive income from it. And it doesn't hurt anything—it just brings in money."

Just how much money? One measure is the worth of the only known U.S. nesting pair of yellow-green vireos, which resided in South Texas's Laguna Atascosa National Wildlife Refuge for several years in the early 1990s. These chickadee-sized songbirds generated an estimated $150,000 per year for local businesses near the refuge. About a three-hour drive to the north, 200 wintering whooping cranes attract an annual $1.2 million in tourist dollars for the small town of Rockport. The cranes arrive on the central Texas coast each fall, after breeding in Canada's Wood Buffalo National Park. Most of the cranes spend the winter at nearby Aransas National Wildlife Refuge. Visitors staying in Rockport can buy tickets for boat rides that ferry them past spots where the birds can be seen searching for crabs, fish, and frogs in the shallow water. Down on the Rio Grande, the Santa Ana National Wildlife Refuge draws 100,000 birders each year, who annually contribute about $14 million to the local economy.

While the full economic impact of wildlife-related activities can't always be so easily quantified, a study by the U.S. Departments of the Interior and Commerce, the "National Survey of Fishing, Hunting, and Wildlife-associated Recreation," suggests that watching has become more than a fringe industry. In 1996, according to the report, 77 million adults—about 40 percent of the U.S. adult population—participated in some form of wildlife-related recreation. Their activities generated $100 billion from sales of equipment, transportation, permits, lodging, food, and other expenses relating to their outdoor interests. Of course, these figures include those for whom wildlife-related recreation means hunting or fishing, and

a lot of that $100 billion was spent on guns, bullets, and lures. But if those who *watch* wildlife are broken out, the number still comes to 63 million people, who generated $29 billion. Between 1991 and 1996, wildlife watching trip and equipment expenditures rose 21 percent. Almost equal numbers of men and women reported participating in these activities.

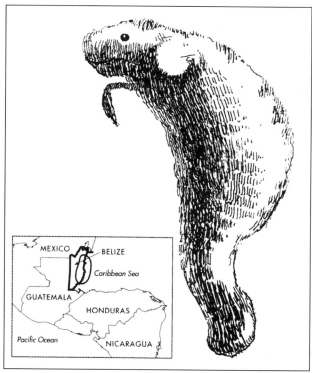

MANATEE THE MORE IT'S WATCHED, THE MORE PROTECTED IT IS.

Wildlife watching has gained popularity in other regions of the world as well. In 1994, an Australian government report indicated that 53 percent of adult Australians planned to take nature-based trips within the following year. And according to a survey conducted by the Royal Society for the Protection of Birds (RSPB), more than 1 million people in Britain, out of a total population of some 59 million, are regular birders. "Birdwatching in this country is the third most popular leisure pursuit, only beaten by angling and golf," says Graham Madge, an RSPB spokesman. Madge adds that an interest in birds often evolves into broader interests in butterflies, small mammals, and wildflowers that share bird habitats.

Why Watch?

Undoubtedly, at least some of the millions of people who watch birds or whales do so for reasons that have little to do with raising ecological awareness. Keeping personal lists of species sighted can be like keeping score in a game, and weekend outings may be more satisfying for the social experience than for any bonding with nature they induce. Nonetheless, it would be hard to imagine

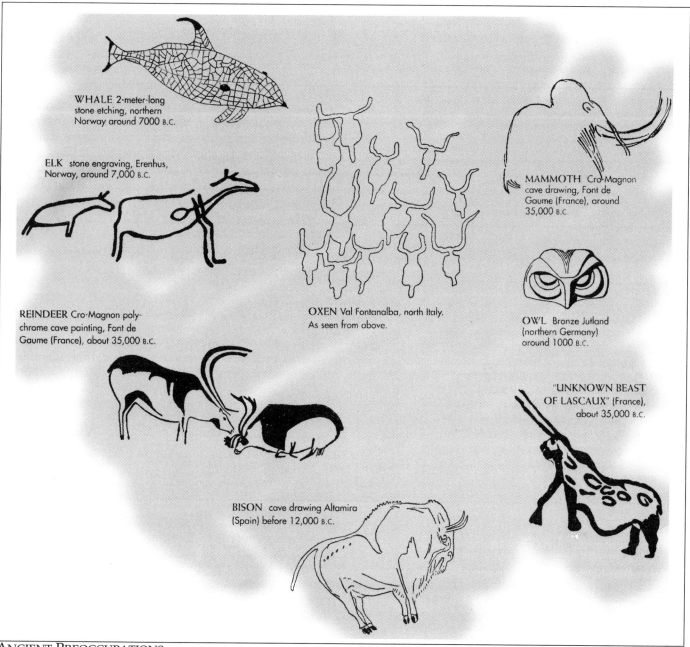

WHALE 2-meter-long stone etching, northern Norway around 7000 B.C.

ELK stone engraving, Erenhus, Norway, around 7,000 B.C.

REINDEER Cro-Magnon poly-chrome cave painting, Font de Gaume (France), about 35,000 B.C.

OXEN Val Fontanalba, north Italy. As seen from above.

BISON cave drawing Altamira (Spain) before 12,000 B.C.

MAMMOTH Cro-Magnon cave drawing, Font de Gaume (France), around 35,000 B.C.

OWL Bronze Jutland (northern Germany) around 1000 B.C.

"UNKNOWN BEAST OF LASCAUX" (France), about 35,000 B.C.

ANCIENT PREOCCUPATIONS TRACINGS OF OLD CAVE PAINTINGS AND OTHER ARTIFACTS SUGGEST THAT HUMANS HAVE LONG BEEN FASCINATED WITH OTHER SPECIES--AND NOT SIMPLY BECAUSE THEY ARE DANGEROUS OR EDIBLE.

any activity having more potential to heighten people's interest in—and their willingness to seriously protect—the other life of the planet than that of closely observing some of that life. It could be of considerable interest to know what really motivates most wildlife watchers, and to what degree that motivation can be harnessed to larger ends.

Certainly, one factor in this movement has been the sharply increased media coverage of environmental issues over the past three decades—particularly the escalating documentation of global biodiversity loss. Articles in WORLD WATCH over the past few years, for example, have documented sharp declines in thousands of species of birds, fish, reptiles, amphibians, marine mammals, and primates. As a species becomes rarer, it acquires a greater curiosity value. When only a few hundred members of a species remain, those last members may ironically attract thousands of humans who paid little attention when the species was common: witness the crowds that gather to observe captive pandas, gorillas, or California condors.

Threatened species lists—and the rescue movements they engender—have also produced a plethora of TV documentaries and coffee-table books celebrating the wonder those embattled creatures inspire. The movements have been aided by advances in camera technology, allowing photographers and filmmakers to record aspects of the private lives of animals that had never been seen before the last decade or so. And for people who want to see the real thing, the advent of compact field guides has greatly eased the way. Beginning in 1934, Roger Tory Peterson, using arrow-marked and simplified paintings, revolutionized field identification with his North American and European bird field guides. Since then, the market has burgeoned and diversified; you can walk into almost any major bookstore and find the *Field Guide to the Palms of the Americas*, or a *Field Guide to the Orchids of Costa Rica and Panama*.

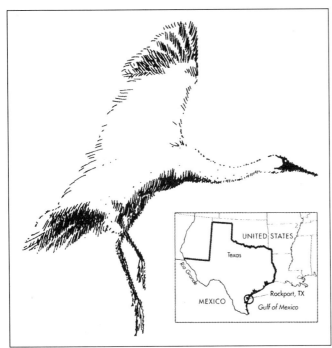

WHOOPING CRANE ONE FLOCK BRINGS $1.2 MILLION A YEAR FROM WATCHERS.

Beyond the rising concern about threatened species and biodiversity loss, some scientists believe humans may harbor an innate affinity for other species—what the evolutionary biologist E.O. Wilson calls "the urge to affiliate with other forms of life," or "biophilia." Over time, what was a physical connection with its origins in hunting and gathering also became one of culture, religion, and spirit. "We are a biological species [that] will find little ultimate meaning apart from the remainder of life," writes Wilson.

This link has always been apparent—as reflected in the great prevalence of animals in the myths, religions, and art of human cultures since prehistoric times. Cro-Magnon cave painters, for example, drew pictures of the creatures they hunted and observed more than 30,000 years ago. Many of the same or similar animals dominate children's story books today. (In 1983, Yale environmental scholar Stephen Kellert found that more than 90 percent of the characters featured in pre-school reading and counting books were animals.) There is the myth of Romulus and Remus, being raised by wolves and then founding Rome. There is the bald eagle becoming a symbol of the United States. Or, there is the turtle recurring as a native American symbol of the world itself. Animals have always been a part of human culture, but what may have happened in the past few decades is that the world's dominant cultures—increasingly overwhelmed by technological and industrial development—have become traumatically separated from what once sustained them.

In any case, we know that across economic, ethnic, and regional lines, people appear to be universally moved by nature—whether in the form of a flying bird, a rolling surf, or the blaze of fall foliage. For many, it takes only a bit of inspiration, perhaps a nature walk led by a school teacher or naturalist, to draw them in. Once hooked, some birding and butterfly-watching enthusiasts research and stalk their quarry almost as though conducting ritualized, non-lethal hunts. Whatever the initial inspiration, most wildlife watchers are driven to learn more about the animals or plants they love.

Citizen Science

This growing interest in watching has brought more than just a surge in hotel reservations and binocular sales. News about impending extinctions, for example, arouses not only curiosity but—often—deep concern. The concern may lead to participation in habitat-saving or species-saving activities, and to a more informed and organized kind of watching—creating a feedback loop that brings still more intensity to the watch. One result has been the rise of a kind of citizen science, in which thousands of watchers collect sightings not just as a pastime, but as a form of data collection for a highly consequential branch of science.

Citizen science now plays a critical role in gathering long-term data on vulnerable wildlife and habitats. For example, the North American Butterfly Association, a nonprofit, 3,500-member conservation organization, holds an annual Fourth of July count, a continent-wide effort in which volunteers identify and tally butterflies living near their homes. The resulting database will provide important insights into butterfly distribution and abundance. Among other findings, the past seven years of observation have enabled researchers to plot major summer concentration areas of the monarch, a widespread migratory butterfly that worries conservationists because much of the population winters in only a few localized and dwindling forests of central Mexico.

Since 1996, the U.S. Geological Survey (USGS) has done the same for toads and frogs with its North American Amphibian Monitoring Program, or NAAMP—an effort spurred by growing global concern over widespread amphibian declines. Volunteers track frog and toad calls during the spring and summer breeding seasons. "It's as close as you're going to get to an idea of what's happening all over the whole patchwork of U.S. and Canadian landscapes," says NAAMP coordinator Linda Weir, who works with states and provinces to set up affiliated monitoring efforts. So far, groups in 29 states have joined, using more than a thousand volunteers to cover about 1,000 designated roadside routes. A similar program recently started in Australia, while regular monitoring programs have been ongoing in Great Britain and a few other European countries. In Britain, for example, the Common Birds Census, conducted since 1970 and carried out primarily by volunteers, has helped create an index on national bird populations. The census has tracked dramatic drops in populations of once-common farmland birds that are thought to be declining due to the destruction of farm hedgerows and increases in pesticide use, and due to harvesting practices that destroy nests and habitat during the birds' breeding seasons. Between 1970 and 1998, for example, the Common Birds Census found an 82 percent decline in grey partridge numbers, a 55 percent decline in song thrush populations, and a 52 percent drop in skylark populations.

One of the oldest citizen science programs is the Christmas Bird Count, which is sponsored by the National Audubon Society and is now in its 100th year. More than 50,000 volunteer birders participated in December 1999 and January 2000, canvassing the wintering grounds of various North American birds. Another well established project is the U.S. Geological Survey's Breeding Bird Survey (BBS), which began in 1966. Volunteers annually cover 3,000 roadside routes, tallying singing and breeding birds across the United States and Canada. Similar surveys are conducted locally in various U.S. states and in Spain, Great Britain, and Australia.

All of this counting of other species could ultimately help drive further shifts toward sustainability in human life itself, by providing inputs to basic policies governing land use, habitat protection, and the like. The Common Birds Census has been instrumental, for example, in persuading the government of the United Kingdom to press for such changes in national agricultural practices as setting harvest times that don't coincide with prime nesting times.

The Backyard Eden

Most of the media's attention to wildlife watching has focused on ecotourism, largely because that's where much of the money is. With tourism booming all over the world, reporters and investors are eager to know what, exactly, people are looking for in their travels. If they'd just as soon choose a ticket for a whooping crane-watching boat as for a theme park ride, that's a significant piece of information. But in fact, the lion's share of nature lovers practice their avocation at home. Of the 63 million wildlife watchers counted in the 1996 National Survey of Fishing, Hunting, and Wildlife-associated Recreation, 44 million reported observing wildlife, mostly birds and mammals, around their own backyards. "Nearly one-third of the adult population of North America dispenses

about a billion pounds of birdseed each year, as well as tons of suet and gourmet seed cakes," writes Stephen W. Kress in *Audubon* magazine. Backyard wildlife enthusiasts spend billions of dollars each year on food, feeders, and binoculars. In Britain, the Royal Society for the Preservation of Birds estimates that two out of three people put out food for garden birds during winter months. Many backyard wildlife watchers also landscape their yards with plants that provide food in more natural forms, such as holly berries or cherries.

STAG IN COPPER, GOLD, AND SILVER, LATE THIRD MILLENNIUM B.C., ALACA HUYUK ROYAL TOMBS, TURKEY

Citizen science has come to the backyards, as well as to public parks and refuges. Project FeederWatch, a joint effort among the Cornell Laboratory of Ornithology, the National Audubon Society, Bird Studies Canada, and the Canadian Nature Federation, has recruited 14,000 backyard watchers to submit data on their homes' winged visitors. The project tracks long-term distribution patterns such as the expanding range of the recently introduced Eurasian collared dove; short-term irruptions of species outside their normal ranges; and the spread of some easily identified diseases, such as conjunctivitis in finches.

Backyards may also engender a greater sense of responsibility because they are small, and stewardship for a small area may make people more conscious of any threats to the area than they'd be in a large park or wilderness. In Great Britain, for example, many gardens are

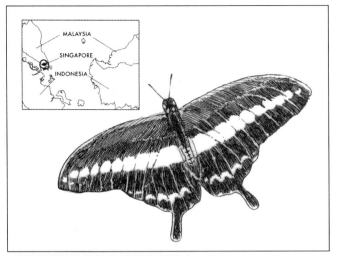

BANDED SWALLOWTAIL BUTTERFLY IN SINGAPORE, SOMETHING TO WATCH BESIDES THE STOCK PRICES.

quite tiny, and birders don't have to look far to see beyond their own property lines. A natural outgrowth of their backyard watching has been the fostering of community projects such as the RSPB's defense of the cirl bunting, a colorful (rust, yellow, and black) European songbird whose last British stronghold is in South Devon. The villages of Bishopsteignton and Stokeinfeignhead, after learning that they were among the cirl bunting's last refuges, each adopted it as their village bird. The upshot is that local farmers have become conscious of its plight and have changed some of their old practices to save it. "There are no real areas of wilderness left in the U.K. and there is not a bit of land that isn't managed for something," says Sue Ellis, a spokesperson for English Nature, an organization that promotes the conservation of England's wildlife and habitats. "Wildlife is on people's doorsteps, so when something happens, people are very, very aware."

Is This a Good Thing, or Not?

While backyard wildlife watchers get less media attention than ecotourists, many of them—in addition to expanding their attentions to community projects—will eventually *become* ecotourists. In 1998, world tourism as a whole generated an estimated $441 billion, according to the Spain-based World Tourism Organization. In her book *Ecotourism and Sustainable Development: Who Owns Paradise?*, Martha Honey calls ecotourism "... the most rapidly growing and most dynamic sector of the tourism market." It is big enough to sway critical decisions in biologically rich but cash-poor countries.

A notable example is Costa Rica, which was little known as a tourist destination two or three decades ago. But as word of the country's verdant rainforests, white-water rivers, and gaudy tropical birds spread, the number of visitors arriving each year from Europe, Japan, and North America jumped from 200,000 to 1 million. Many of them came to see the country's outstanding national parks and reserves, which comprise about a quarter of its total area. Ecotourism has become one of Costa Rica's largest sources of foreign exchange.

In Kenya and Tanzania, too, safari-style ecotourism has become a key revenue source. In 1995, the Kenya Wildlife Service estimated that tourism, 80 percent of which was wildlife watching, was bringing in one-third of the country's foreign exchange. Other countries for which nature-based tourism provides needed foreign exchange include South Africa, Botswana, Belize, Zambia, Ecuador, and Indonesia. In the United States, a group called the Tourism Works for America Council estimated that in 1996, National Park Service lands brought $14.2 billion to local communities and supported almost 300,000 tourism-related jobs.

This isn't always unmitigated good news for local communities and ecosystems, however. The benefits of ecotourism—economic, ecological, and educational—can be offset by all sorts of disbenefits. People eager to see charismatic species can trample less conspicuous ones; building hotels for the visitors cuts large pieces out of the very ecosystems they are coming to see; and the nature trails and jeep roads they use typically lead to increasing fragmentation of what remains. Jet planes and Land Cruisers emit greenhouse gases, the eventual effects of which may weaken ecosystems still further. What ecotourists assume to be harmless observation can turn out to be painful intrusion.

I unwittingly contributed to such an intrusion on Kenya's Masai Mara Reserve, when I joined a safari trip to Kenya in 1995. While traversing the reserve, our van driver suddenly pulled off the road and crashed through the grassland, flushing nesting birds before rumbling within a few feet of a male lion that was lying next to a freshly killed young giraffe. The lion, clearly disturbed by our presence, strained to drag his meal to cover as other safari vehicles rolled through the tail grass towards us. Some of us felt horrible, but I could see that others—their telephoto lenses swiveling—relished being so close.

At the time of my visit, Kenyan tour operators were not forbidden to leave the roads in national reserves. Over the past few years, partly as a result of visitors' outcries, the practice has been banned. But other problems, including water pollution caused by sewage leaching from hotel toilets into nearby wetlands and the widespread gathering of firewood for hotel "safari" campfires and wood-burning furnaces, remain.

As sites grow more popular, other park and reserve managers are beginning to control visitors' movements more carefully. For instance, many parks, including Santa Ana National Wildlife Refuge, now ban vehicular traffic much of the time, with the result that interactions between walkers, bikers, and wildlife are no longer so disruptive. Visitors to Alaska's Denali National Park must leave their cars behind and may only enter the park's wilderness in scheduled buses. The same is true in such

DETAIL OF INTERNAL PANEL FROM A SILVER CALDRON, MIDDLE LA TENE CULTURE (EUROPE), CIRCA FIRST OR SECOND CENTURY B.C.

parks as Mudumalai Wildlife Sanctuary in India, where visitors ride in on jeeps, buses, or on elephant back to observe wild Asian elephants, tigers, gaur (a kind of undomesticated cattle), and other large, sensitive, and potentially dangerous animals.

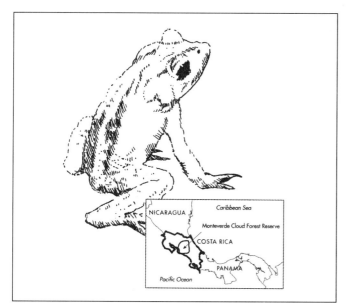

GOLDEN TOAD PEOPLE NEVER REALIZED WHAT THEY WERE STEPPING ON.

Unfortunately, it's often difficult for a group of travelers to know if their tour operator is green or just going *for* the green. "We need to [have] some kind of reviewable ethical standard," says Megan Epler Wood, president of the Vermont-based Ecotourism Society. The Ecotourism Society defines ecotourism as "responsible travel to natural areas that conserves the environment and sustains the well-being of local people." But no global standard or certification process currently exists for tour operations, although Costa Rica and Australia now have strict ecotourism grading standards and efforts are under way to establish them in Kenya.

"We see ecotourism as part of an integrated conservation strategy," says Greta Ryan, manager of ecotourism enterprise development and support at Washington, D.C.-based Conservation International (CI). The nonprofit CI works in 23 countries, with ecotourism now constituting an important component of its work in 17 of them. The key, says Ryan, is to make sure the benefits stay within the community. When this happens, ecotourism tends to reinforce CI's overall program in three ways. By generating local income, it encourages communities to welcome other conservation projects. By alleviating poverty, it reduces the rates of poaching and deforestation. And by making natural assets the centerpieces of the economy, it heightens environmental awareness among both the local people and their visitors.

Often, nature tourists enjoy the sights unaware that their money is being siphoned away from local communities by big-city or out-of-country tour operators who have little ultimate interest in conserving wildlife. "In Kenya, those communities that do not realize a benefit are less likely to consider wildlife positively, and are more likely to want to remove the wildlife from the land," says Neel Inamdar, a director of Eco-Resorts, an ecotourism-oriented travel company, and a board member of the

Kenya-based African Center for Conservation. "Basically, a lot of the parks are not viable by themselves—we have to consider the people on the periphery of the parks. After all, wildlife moves from one place to another." Recent initiatives in areas adjacent to African parks include pilot programs in which villagers or farmers run cooperative wildlife reserves on their lands, guiding and hosting paying visitors instead of converting the areas into agricultural land or livestock range.

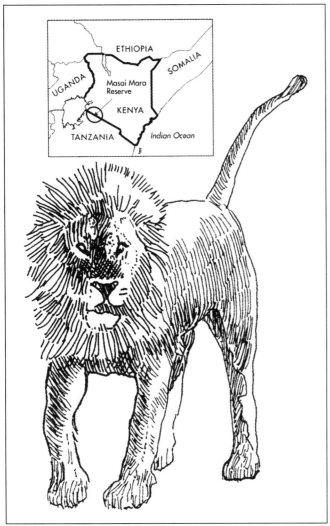

LION ALIVE AND IN THE WILD, IT'S WORTH AN ESTIMATED $575,000.

cepted hypothesis suggests that climate change caused the die-off, another suggests that tourists may have unwittingly contributed to the toad's disappearance by bringing in pathogens on their shoe soles.

On balance, if carefully managed, nature tourism offers large benefits to the environment. Wildlife watchers, a relatively affluent and well educated lot on the whole, are usually willing to pay for their watching—and their economic clout favors protection of the places where they like to do it. A 1995 survey by the Travel Industry Association of America found that 83 percent of U.S. travelers are inclined to support "green" travel companies and are willing to spend, on average, 6.2 percent more for travel services and products provided by environmentally responsible travel suppliers.

BULL COPPER, GOLD, AND SILVER, LATE THIRD MILLENNIUM B.C., ALACA HUYUK ROYAL TOMBS, TURKEY

Some fragile areas, or those within reserves, may simply be unsuited for *any* nature tourism. For instance, there are the rainforest pools once home to golden toads within Costa Rica's Monteverde Cloud Forest Reserve. In the September/October 1990 issue of WORLD WATCH, I quoted Ray Ashton, director of an international consulting firm called Water and Air Research, as saying, "People are tripping over golden toads so they can go see quetzals." Today, the exotic red and green streamer-tailed birds still breed in the reserve, but the golden toad may be extinct. In fact, no one has seen a golden toad since shortly before my article went to press in 1990. While one widely ac-

In the United States, national parks and wildlife refuges now charge entrance fees, and park fees in other countries are being raised. Private landowners, too, are finding they can charge visitors a fee. South Texas ranches that were off-limits a decade or two ago now court birders as a side business. In Cape May, New Jersey, farmers Les and Diane Rea have managed to supplement the income from their 80-acre lima bean farm, which they feared losing a few years ago because of rising costs and pressure from developers, by maintaining habitat for one of the East Coast's most popular birding areas. In 1999,

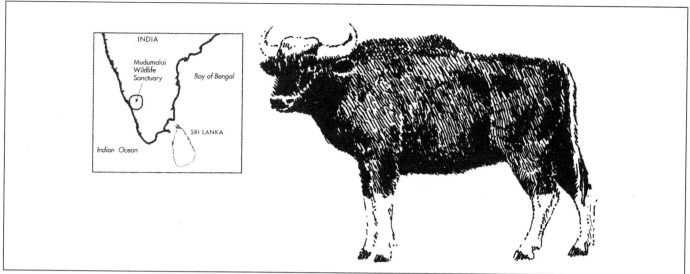

GAUR Yes, it's a wild animal. Compared to a cow, it's huge.

CONNECTICUT WARBLER It helped protect a bean farm from the bull-dozers, and the bean farm returned the favor.

the Cape May Bird Observatory struck an agreement with the Reas to lease birding rights for the property, paying for the lease with funds collected from permit-buying visitors. Now, each spring and fall they welcome scores of paying visitors. The farm draws a dazzling array of winged migrants, including the elusive Connecticut warbler and about three dozen other warbler species.

"Certainly the precedent existed," says Pete Dunne, vice president for natural history information at the New Jersey Audubon Society, who helped the Reas get the program up and running. "Hunters have been doing this for years. We didn't see any reason not to extend it to birding as a way of showing that birders certainly are willing to pay to support their hobby—and as a way of combatting development pressures. Most farmers want to hold onto their farms," says Dunne. "In this case, birders are just another cash crop. You don't have to water them, fertilize them, or till them, and on top of that, they will go to your farm stands and buy vegetables."

Birder, writer, ecotourist, and former WORLD WATCH associate editor Howard Youth lives in Rockville, Maryland.

FOR MORE INFORMATION

North American Butterfly Association (NABA) www.naba.org; 4 Delaware Rd., Morristown, NJ 07960, tel: (973) 285-0907

North American Amphibian Monitoring Program (NAAMP) www.mp1-pwrc.usgs.gov/amphibs.html; 12100 Beech Forest Rd., Gabrielson Bldg. 242, Laurel, MD 20708

Royal Society for the Protection of Birds (RSPB) www.rspb.org.uk; The Lodge, Sandy, Bedfordshire SG19 2DL, United Kingdom, tel: 01767-680551

Breeding Bird Survey (BBS) www.mbr.nbs.gov/bbs

Christmas Bird Count (CBC) www.birdsource.org/cbc/index.html; National Audubon Society, 700 Broadway, New York, NY 10003, tel: (212) 353-0347

Project FeederWatch http://birdsource.cornell.edu/pfw; Cornell Lab of Ornithology, P.O. Box 11, Ithaca, NY 14851-0011, tel: (607) 254-2473.

From *World Watch,* May/June 2000, pp. 12-23. © 2000 by the Worldwatch Institute (www.worldwatch.org). Reprinted by permission.

UNIT 5

Resources: Land, Water, and Air

Unit Selections

Key Points to Consider

- Explain the analogy of the "commons" to describe global resources available to everyone. Is the concept a helpful one in assessing the future of resource management?

- Why is the number of farmers in the world decreasing? What kinds of social, economic, and cultural impacts will be produced by decreasing family farmers and increasing amounts of land farmed by corporate farmers?

- Why are the world's freshwater resources in such short supply? Describe the nature of the accelerating demand for water as a resource and discuss the relationship between that demand and water management.

- Why and how has overfishing contributed to a decline in the food supply from the oceans? Has oceanic pollution contributed to this decline? Explain.

- What are some of the uncertainties about the future impact of global temperature increase on human social and economic systems? What reasons are there to develop extensive monitoring systems to identify the causes and effects of global warming?

 **Links: www.dushkin.com/online/**
These sites are annotated in the World Wide Web pages.

Global Climate Change
    http://www.puc.state.oh.us/consumer/gcc/index.html
National Oceanic and Atmospheric Administration (NOAA)
    http://www.noaa.gov
National Operational Hydrologic Remote Sensing Center (NOHRSC)
    http://www.nohrsc.nws.gov
Virtual Seminar in Global Political Economy/Global Cities & Social Movements
    http://csf.colorado.edu/gpe/gpe95b/resources.html
Websurfers Biweekly Earth Science Review
    http://home.rmi.net/~michaelg/index.html

The worldwide situations regarding reduction of biodiversity, scarce energy resources, and environmental pollution have received the greatest amount of attention among members of the environmentalist community. But there are a number of other resource issues that demonstrate the interrelated nature of all human activities and the environments in which they occur. One such issue is that of declining agricultural land. In the developing world, excessive rural populations have forced the overuse of lands and sparked a shift into marginal areas, and the total availability of new farmland is decreasing at an alarming rate of 2 percent per year. In the developed world, intensive mechanized agriculture has resulted in such a loss of topsoil that some agricultural experts are predicting a decline in food production. Other natural resources, such as minerals and timber, are declining in quantity and quality as well; in some cases they are no longer usable at present levels of technology. The overuse of groundwater reserves has resulted in potential shortages beside which the energy crisis pales in significance. And the very productivity of Earth's environmental systems—their ability to support human and other life—is being threatened by processes that derive at least in part from energy overuse and inefficiency and from pollution. Many environmentalists believe that both the public and private sectors, including individuals, are continuing to act in a totally irresponsible manner with regard to the natural resources upon which we all depend.

Uppermost in the minds of many who think of the environment in terms of an integrated whole, as evidenced by many of the selections in this unit, is the concept of the threshold or critical limit of human interference with natural systems of land, water, and air. This concept suggests that the environmental systems we occupy have been pushed to the brink of tolerance in terms of stability and that destabilization of environmental systems has consequences that can only be hinted at, rather than predicted. Although the broader issue of system change and instability, along with the lesser issues such as the quantity of agricultural land, the quality of iron ore deposits, the sustained yield of forests, or the availability of fresh water seem to be quite diverse, they are all closely tied to a single concept—that of resource marginality.

Many of these ideas are brought together in the lead article of this unit. Research scholars Joanna Burger and Michael Gochfeld revisit one of the most important of all environmental essays in "The Tragedy of the Commons: 30 Years Later." In the original essay, Garrett Hardin invoked the argument of common resources in support of his argument that as human population increased, the global environment would eventually begin to suffer. Burger and Gochfeld note that Hardin's article was seminal in defining the problem of environmental impact in a way that resource managers could deal with and, as a consequence, management of the global commons (fisheries, forests, wildlife, the atmosphere) has received more attention. Whether the commons have received enough attention is for the future to determine.

In "Where Have All the Farmers Gone?" Brian Halweil of the Worldwatch Institute discusses the globalization of industry and trade that is creating a uniform approach to all forms of economic management, including management of agricultural resources.

Halweil notes that increasing agribusiness and decreasing numbers of family farmers represent a loss of both biological and cultural diversity. In the third and fourth articles in the section, the topic of management of water resources is dealt with, albeit on two very different levels: that of groundwater and the boundless ocean. In each article, the concept of marginality is relevant. In the next selection, "Making Every Drop Count," Peter H. Gleick, one the world's foremost authorities on freshwater problems, discusses the problem of the world's dwindling freshwater resources as a result of enhanced technologies that consume more and more of this vital resource. Gleick's concern is whether we will have enough water for even the most basic of human needs if consumption continues at its present rate and if water managers stick to the strategies of the past. The following selection in this section also deals with limits to wise use, in this instance, use of the resources of the world's oceans. In "Oceans Are on the Critical List," researcher Anne Platt McGinn contends that the primary threats to the health of the world's oceanic ecosystems are human-induced and include overharvesting of fish, pollution of both coastal and deep-water zones, introduction of alien species and the consequent threat to oceanic biodiversity, and climate change, which also poses threats to biodiversity. Efforts to protect the oceans, McGinn claims, lag far behind what is needed.

In the fifth article in this unit, the most critical and controversial of the problems that characterize the global atmosphere is dealt with: the continuing accumulation of "greenhouse gases" and the consequent trend toward a warmer Earth—or what has been termed "global warming." In a special report from Time, the global warming issue is explored by a team of investigative reporters. This report, "Feeling the Heat: Life in the Greenhouse," discusses the social, political, economic, and scientific aspects of the global warming debate and concludes that, in spite of scientific and nonscientific rhetoric and argument about global warming, the fact that the globe is becoming warmer is indisputable. It is equally indisputable that human activities are playing some role in the process. While a small segment of the scientific community still does not see human influence in short-term climate trends, the majority believe in a greenhouse effect enhanced by human activity. Interestingly, some of the strongest proponents of action to curb human-induced global warming are some of the biggest energy companies.

In the final article of this unit, "The Human Impact on Climate," two of the world's pre-eminent scientists studying climate change, Thomas Karl and Kevin Trenberth, note that while few scientists disagree on the fact of climate change, many disagree on the question of the exact proposition of the change that should be attributed to natural or human-induced causes.

There are two possible solutions to all these problems posed by the use of increasingly marginal and scarce resources and by the continuing pollution of the global atmosphere. One is to halt the basic cause of the problems—increasing population and consumption. The other is to provide incentives and techniques for the conservation and management of existing resources and for the discovery of alternative resources to eliminate the demand for more marginal resources and the use of heavily polluting ones.

The Tragedy of the Commons: 30 Years Later

by Joanna Burger and Michael Gochfeld

How do we manage resources that seem to belong to everyone? Fish swimming in lakes, game mammals wandering the open plains, and birds migrating overhead belong to everyone and yet are protected by no one. For the sturgeon and bison this lack of protection spelled disaster, for the passenger pigeon, extinction. Today, protecting such common-pool resources has become a challenge, not only on the local scale but on national and global ones as well.

Thirty years ago this December, ecologist Garrett Hardin invoked the analogy of a "commons" in support of his thesis that as human populations increased, there would be increasing pressure on finite resources at both the local and particularly the global levels, with the inevitable result of overexploitation and ruin. He termed this phenomenon the "tragedy of the commons."[1] More specifically, this phrase means that an increase in human population creates an increased strain on limited resources, which jeopardizes sustainability. Hardin argued that common resources could be exploited by anyone who could assert their rights to do so. He painted a bleak picture, emphasizing that the solutions were social rather than technical, and called for privatization or exclusion and for rigorous and even coercive regulation of human population.[2] Recently, he reaffirmed this position.[3]

This article looks at both the positive and negative management of common resources and the legal and ecological progress that has been made since Hardin's original article was written. (See the box, "Hardin's Tragedy of the Commons Thesis.")

Birth of a Discipline

Hardin's original paper was widely cited and stimulated many examples showing that increasing populations did lead to overexploitation, habitat degradation, and species extinctions.[4] Even those ecologists who found Hardin's reliance on coercion distasteful emphasized the consequences of the imbalance between population and resources.[5] Hardin's paper also stimulated many social scientists to alter their perspectives in relation to commons issues, with the result that many examples of both successful and unsuccessful maintenance of common resources have now been published.

The concept of commons is a useful model for understanding environmental management and sustainability. While Hardin believed that ruin was inevitable without coercive population control—an option at odds with our cherished democratic beliefs—recent works by a range of interdisciplinary scientists have identified systems and institutions that do not inevitably lead to overexploitation but that in some cases result in the sustainable use of selected resources, at least on local scales.[6]

While 30 years of research has shown that Hardin's initial thesis emphasizing inevitability and ruin was perhaps too bleak on the local scale, it has been enormously helpful in generating thought-provoking analyses across a wide range of disciplines. His work was widely cited, first by natural scientists and later by social scientists, yet unlike most scientific papers the rate of citation is increasing even 30 years later (see Figure 1). Perhaps its most useful role has been in illustrating the importance of integrating social and political theory with biological data. The traditional theory regarding resource users as unbridled appropriators is being replaced by the recognition that users can communicate and cooperate when it is in their interest to do so and when the resources at their disposal and the sociopolitical context permits it.[7]

What are Commons and Common-Pool Resources?

Common-pool resources (sometimes designated "common property") such as land, fish, and water can be identified and quantified, while the commons is a broader concept that includes the context in which common-pool resources exist and the property system embracing them. Indeed, the switch from discussing the commons to analyzing common-pool resources and property rights illustrates the disagreement many biological and social scientists have with Hardin's original thesis.

In the broad sense, a commons includes the resources held in common by a group of people, all of whom have access and who derive benefit with increasing access. Access may be equal or unequal, and control may be democratic or not. There is some disagreement as to what constitutes a common-pool resource. The term is often restricted to land, grass, wildlife, fish, forests, and water. The concept can also be applied to non-natural resources such as national treasuries, medical care, and the Internet,[8] but the focus of this article will be on the more traditional commons issues of fisheries, recreational areas, public land, and air quality (although atmosphere has been a highly disputed commons with unique qualities to be discussed later). Once these resources could be held in common by small tribes or villages, communities could limit both access to the resources and the amount extracted. Limitation often involved aggression against would-be usurpers.

In many places, this system still exists. In the dry desert lands of northern Namibia and southern Angola, tribal councils control large blocks of land and tribe members are free to build their houses and farm wherever they choose. The councils can mediate disputes, limit intruders, and impose sanctions. While the primary resource held in common is

land for farming and grazing, another very important resource in these arid lands is water for people and livestock. Thus, even where overall population density is low, land is not equally desirable and people congregate near the rivers and marshes, potentially leading to overexploitation of these lands and depletion and fouling of water.

Categories of Commons and Property Rights

Current reexamination of the applicability of common-pool resource management is fitting because the use of many resources has become truly globalized, requiring new and more global solutions. International attention has now focused on various aspects of sustainability. Global economies, multinational corporations, international trade agreements, and international commissions have created an institutional framework in which resource sustainability is one prevailing theme among a virtual cacophony of others.[9] As Elinor Ostrom, professor of polit-

ical theory and policy analysis at Indiana University, points out, it is unclear whether existing international cooperative efforts are adequate to protect essential resources.[10] Rates of population growth and resource consumption vary among regions. The gap in who has access to resources is not narrowing, and there is rapid emergence of new technologies that allow even more efficient exploitation of resources. At the same time, improved communication has heightened expectations of a higher standard of living, even in remote regions of developing nations. The rate of these changes is also accelerating.

The following examples of commons challenges are drawn from fisheries, public land use, and air quality. Each of these represent similar themes but different scales and solutions. Ostrom identifies four properties of these resources that facilitate cooperative management: the resource has not already been depleted beyond hope of recovery; there are reliable indicators of resource condition; the resource is sufficiently predictable; and the distribution of the resource

---

HARDIN'S TRAGEDY OF THE COMMONS THESIS

Hardin based his thesis of the tragedy of the commons on earlier studies written during the late 18th and early 19th centuries. In 1798, Thomas Robert Malthus wrote that human population could grow exponentially, unmatched by resource growth.[1] Charles Darwin's theory of evolution predicted that the characteristics of people who produced more children than others would increase over time. These observations are even more true today given medical care and social systems to protect the children of those who cannot support them. For most of human history, the world seemed like an infinite space with unlimited resources (forests, oceans, wildlife) available for the taking because in nearly every part of the globe there were sufficient resources for the existing, low-density populations. In the past century, however, human population has increased almost everywhere, demonstrating that demand can more than match even very abundant resources.[2]

In 1968, Hardin predicted that with increasing population the eventual fate of all common resources was overexploitation and degradation.[3] His credo, "Freedom in a Commons brings ruin to all," became a universal cry. Others made the same point, although with less flare and consequently less effect. Hardin's concerns focused people's attention on the relationship between individual behavior and resource sustainability.[4] The underlying tenet of his thesis, however, was that populations were increasing beyond the ability of the Earth's resources to support them at a sufficiently high standard of living.

Hardin used William Foster Lloyd's example of herdsmen sharing village lands to graze cattle.[5] Each herdsman derives full benefit from each cow he adds to his herd, while the de-

pletion of grass attributable to that cow is shared among all users. Thus, at each decision point, Hardin argued, each herdsman would choose to add a cow rather than maintain status quo. This leads to each herdsman increasing his herd without limit and to ultimate and inevitable ruin for all. Hardin made several assumptions, including that the world and its resources are finite, human populations will continue to increase, and every person will want to use an increasing share of the resources. Hardin's solution was to have government controls to limit access to the commons or to privatize common-pool resources and, above all, to limit population, even through coercion. Recently Hardin has reaffirmed his predictions, noting that expanding cities must control traffic and parking, nations seek to limit air pollution, and the freedom of the seas is being constrained.[6]

---

1. T. R. Malthus, *Population: The First Essay* (Ann Arbor, Mich.; University of Michigan Press, 1959).

2. J. Cohen, *How Many People Can the Earth Support?* (New York: W. W. Norton, 1995).

3. G. Hardin, "The Tragedy of the Commons," *Science*, 13 December 1968, 1,243–48.

4. G. Hardin and J. Baden, eds., *Managing the Commons* (New York; W. H. Freeman, 1977).

5. W. F. Lloyd, "Population," "Value," "Poor-laws," and "Rent," in *Reprints of Economic Classics* (New York: Kelley, 1968).

6. G. Hardin, "Extensions of 'The Tragedy of the Commons'," *Science*, 1 May 1998, 682–83.

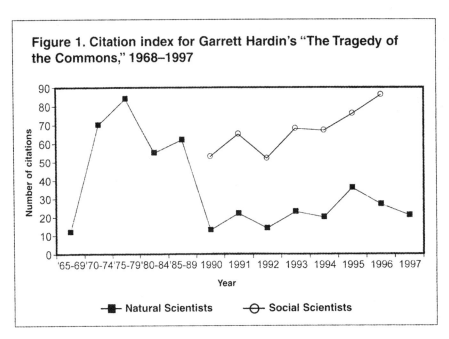

Figure 1. Citation index for Garrett Hardin's "The Tragedy of the Commons," 1968–1997

is sufficiently localized to be studied and controlled by the political entity.

There are also four general categories of property rights: open or uncontrolled access, communal, state, and private[11] (see Table 1). Access refers to who controls access or who has access to the resources under what conditions or for what time period (while subtractability refers to the ability of one user to subtract from the welfare of the others).[12] These categories are not discrete but intergrade, and some common-pool resources can be managed under more than one category.[13] For some fisheries, such as shellfish, the government enforces regulations on seasons, size limits, and overall take, but the local shell-fishermen may claim traditional rights or ownership of particular clam beds that they seed with young shellfish, waiting for them to mature to a marketable size. Infringement of these beds can often lead to violence, as has happened in Maine when interlopers tried to fish for lobster in a territory claimed by someone else.[14]

Different modes of property rights may compete—for example federal versus local government or privatization versus community control. Even where a community or state maintains ownership, restricted access for exploitation of certain resources may be granted by concession. There may be a dissociation between resources: One may own land and trees privately whereas wildlife is communal property, or individuals may own rights to certain trees on communal land.

Basically, there is the question of how access can be controlled or managed, and who wins and who loses. Access can be managed by agreed-upon rights and rules, which are uniformly adhered to or enforced.[15] For example, in a small fishing community without outsiders, fishermen can agree to fish only in certain zones or only at certain times, catching a prescribed amount of fish. As long as everyone follows the rules, and the community governs wisely, the fishermen's extraction would not exceed the carrying capacity or regeneration rate of the fish stocks, the resource would not be depleted, and the situation would be considered sustainable. The failure of someone to follow the agreed-upon rules necessitates sanctions, which must also be agreed upon by the users or commoners.

One difficulty in protecting common-pool resources is that there is often an incongruence between the distribution of the resource and property regimes. Fisheries provide many examples of such disparities; many commercial fish are migratory, making property rights, even those as broad as the 200-mile exclusive economic zones, effective for only part of the year.

Local to Global: The Commons Comes of Age

The greatest changes that have come about since Hardin proposed the tragedy of the commons have been an increase in human populations worldwide, shrink-

ing resources, and the globalization of economies. The focus on commons management as a discrete area of social and economic challenge overlaps broadly with the focus on sustainability of resources. Indeed, those concerned with the sustainability of resources should examine whether a commons represents an appropriate avenue for developing sustainable management, and if so, at what spatial scale.

Temporal scales are important as well, for there are intergenerational aspects that need consideration.[16] The wise or unwise use of resources today directly affects the health and well-being of the next generation. We are borrowing from the next generation at a time when our resources are not only decreasing but population, and thus resource needs, is increasing.

Understanding the use of common-pool resources has been greatly enhanced by two developments: a new economic theory of cooperation that suggests cooperation could lead to the wise and sustained use of common-pool resources; and very detailed empirical work on commons issues that were well founded in theory.[17] Both were essential for the field to move forward. The traditional model of examining commons assumed that each person acted only in his or her own immediate best interest. More recently, economic theory suggests that by cooperating with others (even if there is an initial cost), common users can not only protect the resource but keep it sustainable as well. By examining case studies, researchers found that cooperation can lead to sustainable use and economic viability at least on small regional and short temporal scales.

There are many examples of both failed and successful attempts to manage common resources on local levels. Bonnie McCay, professor of human ecology at Rutgers University, and Fikret Berkes, professor at the Natural Resources Institute at the University of Manitoba, have been instrumental in providing examples where fishermen, hunters, and foresters have established norms, rules, and institutions to successfully extract resources without overexploitation.[18] They argue persuasively that although the tragedy of the commons has been accorded the status of a scientific law, much more detailed study of common-property resource management is needed. There are good examples of self-regulation, including Maine lobstermen and New Jersey fisherman who maintain yields and thus stablize or even raise prices.

Three major categories of environmental problems that are useful in understanding methods of dealing with common-pool resources are fisheries, public lands, and air pollution. They are interesting because they illustrate the two main abuses of the commons: resource depletion and capacity depletion (due to pollution). Fish are a traditional, depletable common-pool resource, public land can fall under either category, and air pollution is a more global issue that affects the world at large. Traditional approaches to air pollution are being complemented by viewing the atmosphere's capacity to absorb pollutants as a commons issue. Although at first air seems markedly different from renewable resources such as fish, firewood, and lumber, viewing the atmosphere in terms of limited capacity means that everyone who introduces pollutants, even though he or she does not consume air, subtracts from its use by others.

Fisheries and Water-Related Issues

Fisheries offer classic examples of commons issues that can be local (a stream or lake), regional (North Atlantic), or global, depending upon the fishery. Fisheries also provide examples of the best and worst management schemes concerning common resources. They are

instructive because there are examples of local and institutional control that have effectively protected a resource and a livelihood in a sustainable manner, as well as examples of massive overexploitation that threaten not only the commercial but the biological viability of some species.[19]

Although the major fisheries challenges involve oceanic species and a system of uncontrolled, open access on the high seas, shrimp farming along the tropical coastlines is one important example of a land-margin commons issue where a combination of private, state, and communal property rights prevail. Increasing market demand for shrimp provides an incentive for the conversion of otherwise valuable mangrove habitats—where many fish species spawn—to shrimp farms, often at the expense rather than enrichment of fishing communities. The shrimp farms are often owned by corporations with the capital for transforming the habitat rather than being managed by local cooperatives. These farms are economically viable for only a few years, after which they are often abandoned.

In the United States, inshore marine resources are managed by state governments as a trust for all citizens. However, even within this government-regulated system fishermen can cooperate locally to preserve a common-pool resource. Lobsters, for example, are a common-pool resource that can be easily overexploited if everyone has a right to harvest and if there are no limits on the number of lobsters each fisherman takes. Additionally, if fishermen can come in from outside the region the problem increases. In Maine, the state government does not limit the number of licenses, but the lobstermen practice exclusion through a system of traditional fishing rights. Acceptance into the lobster fishing community is essential before someone can fish, and thereafter one can extract lobsters only in the territory held by that community. The end result has been a sustainable harvest and higher catches of larger, commercially valuable lobsters by fishermen in the exclusion communities.[20]

McCay and colleagues have also shown that in a trawl fishery in the New York Bight, the fishermen belong to a cooperative that has maintained relatively high prices and sustainability by limiting entry into the local fishery and establishing catch quotas among members. In this

case, self-regulation is both flexible and effective.[21] The trawl fishery illustrates another critical point about commons resources: It is not only important to have sufficient fish to catch and fish populations that are stable but the price must be maintained or the industry will not be viable for the users.

In both of these examples, success has been achieved by local management of a local fishery. The fishermen were effective in excluding outsiders and in limiting the fishing rights of insiders so that fish stocks were sustainable, prices were maintained, and the fishery was viable. Management of fisheries on a regional or global scale is far more problematic because of the difficulty of exclusion or of monitoring catches.

At the opposite end of the spatial spectrum are cases where fisheries management has been ineffective, with declines in the fish stocks so serious that both the fishery and the fish are threatened with extinction.[22] Many of the examples where fish stocks have declined precipitously involve marine fish with wide geographical ranges. Swordfish and bluefin tuna are classic illustrations of Hardin's thesis. They are a common-pool resource that have suffered overexploitation because of the difficulty of exclusion and the pressure from fishermen of many countries who, stimulated by high market value, push for the maximum catch possible.

Carl Safina, conservation writer and director of the Living Oceans Program of the National Audubon Society, has highlighted the plight of bluefin tuna, one of the largest, fastest, and most wide-ranging fish in the ocean.[23] Its west Atlantic breeding population has declined by 90 percent since 1975 (see Figure 2).[24] The International Commission for the Conservation of Atlantic Tunas, responsible for stewardship of these tuna, is made up of members from 20 countries, many of whom are major tuna users. Although the commission's scientific advisory committees repeatedly presented them with data showing drastic declines, the commission continued to allow catches that exceeded the maximum sustainable yield. In this case, the problem has multiple facets: Some countries that catch tuna are not members of the commission and are thus not regulated; some countries that belong to the commission fish under flags from noncommission countries so they elude the regulations; the fishery is pelagic and global, making en-

Table 1. Types of property rights regimes

Type	Description	Example
Open Access	Absence of any well-defined property rights; completely open access to resources that are free to everyone	Recreational fishing in open ocean. Bison and passenger pigeon overharvested leading to decline and even extinction
Common Property	Resource held by community of users who may apportion or regulate access by members and may exclude non-members	Small fishing village that regulates fishing rights among users
State Property	The resource is held by government, which may regulate or exploit the resource or grant public access; government can enforce, sanction, or subsidize the use by some people	Public lands such as national forests or parks where grazing, lumber, or recreational rights are granted by government
Private Property	Individual owns property and has the right to exclude others from use as well as sell or rent the property rights	Private ownership of a woods where owner can sell or rent the land, cut or sell the trees

forcement of regulation difficult if not impossible; and last-ditch efforts to place the species on Convention on International Trade in Endangered Species of Wild Flora and Fauna (CITES) lists, which would limit fishing, have so far failed due to pressure from user nations. Even aggressive efforts by the conservation community have been unable to prevent the continued destruction of bluefin tuna.

On a national scale, the United States successfully excluded foreign fishing fleets from its exclusive economic zone and dominated these waters. Even where outsiders have been excluded, however, the commons problem remains because without institutional controls, insiders are free to exploit resources. As Safina pointed out, excluding outsiders did not prevent U.S. fishermen from overexploiting fish resources despite the establishment of agencies and commissions nominally charged with protecting these resources. The problem partly lies with the membership on such commissions; many members represent fishermen who want to get their share (or more) rather than people who are charged with protecting the fish stocks regardless of the economic pressures.

Although successful management of common-pool resources for sustainability is desirable, there are other approaches that do not incorporate sustainability. Some marine fisheries exemplify an alternative approach involving overcapitalization of fleets, rapid sequential elimination of one common-pool fishery resource after another, and shifting to new resources.[25] This allows the industry to perpetuate itself in the short term with little attention to sustainability of a specific resource. When local resources are exhausted, fishermen must exploit more distant sources or sell their fleets and make other investments.

Moreover, understanding of traditional common-pool resources in fisheries has been expanded to include other coastal resources. Two examples illustrate this point: the serious reduction in horseshoe crabs and shorebirds; and personal watercraft users versus fishermen and other water users.

Horseshoe Crabs and Shorebirds

Since 1990, a directed fishery for horseshoe crabs has developed along the East Coast of North America to fulfill the demand for bait for eels, conch, and other fish. This had led to overexploitation and the reduction in the number of horseshoe crabs spawning in many regions. While this problem may once have been considered a fisheries issue only, it is compounded by the fact that several species of migratory shorebirds are threatened by the massive reduction in horseshoe crab eggs, their major food source on Delaware Bay and other stopover places during their northward migration.[26] Although the animals themselves are transitory, the phenomenon occurs annually and predictably on the same beaches.

Apart from the fishermen, several local communities depend on the tourist income generated by the attraction of huge concentrations of migratory shorebirds and breeding horseshoe crabs. Having large populations of both species available for viewing is a commons resource. The fishermen's direct extraction of crabs and the indirect extraction of birds reduces the resource attractive to the tourists, decreasing their pleasure and ultimately their visits. Although less conspicuous from an ecotourism viewpoint, other species, such as green sea turtles, also depend on horseshoe crabs for food, while a medical industry relies on horseshoe crabs for the production of an important laboratory reagent. Thus, a traditional commons fisheries problem has now emerged as a multispecies conservation problem involving other vertebrates as well as economic issues that affect not only the conservation community but fishing, industry, and tourism. A further complication is that the demand for the eels (for which ground up female horseshoe crab is the only bait) emanates from around the globe. Japan, having depleted its own eel populations and those of nearby Asian countries, now offers extremely high prices for American eels, rendering both eel trapping and horseshoe crabbing economically lucrative. Rapid extraction and rapid financial remuneration is the apparent priority rather than sustainability of the resource. While the Atlantic Coastal Marine Fisheries Commission is responsible for maintaining sustainable horseshoe crab populations, the protection of the shorebirds falls under the U.S. Fish and Wildlife Service, which has a conflicting prerogative.

Personal Watercraft versus Fishermen

Recreational use of public land and waterways, one of the commons resources that Hardin mentioned in his classic paper, has received little attention or rigorous analysis. In this example, the massive increase in the use of personal watercraft (often called "jet skis") threatens a number of common-pool resources: the safe nesting of estuarine birds and other animals, the quality of aquatic vegetation so essential to the production of fish and shellfish, the peace and quiet of residents in shore communities, the physical safety of others using aquatic environments, and the undisturbed fishing of both recreational and commercial fishermen.[27] Fast, noisy, and numerous, these craft speed through habitats inaccessible to boats and are not yet regulated in most areas. However, the U.S. National Park Service is in the process of restricting or eliminating their use.

In many estuaries, commercial fishermen are already reporting decreases in catch because of the physical disturbance caused by personal watercraft, while other users report a serious reduction in aesthetic values such as "peace and quiet." This problem is not limited to coastal environments but threatens inland waterways as well. At issue is the freedom of personal watercraft users to take over aquatic environments where their open access subtracts ecological, aesthetic, and commercial benefits long sought by others.[28] Regulation of their use is in its infancy. Ultimately it may be the fatalities they cause, rather than aesthetic or economic impacts on the commons, that leads to further regulation and exclusion.

Public Land

One commons resource currently under discussion in the United States is the huge tracts of public land used for nuclear weapons production by the U.S. Department of Energy during the Cold War.[29] These are now being considered for transfer back to regional, local, or even private ownership, with the inherent problems of determining access and subtractability. For 50 years the federal government excluded all other users from these lands, which in the future could become commons for recreational, industrial, or agricultural use. Which option will be chosen remains undetermined and is likely to vary from site to site.

The Department of Energy's Savannah River Site in South Carolina is a good example. The site is composed of 800 square kilometers of land alongside the river. It includes habitat for a number of endangered species, such as the red-cockaded woodpecker, the wood stork, and the bald eagle, as well as some of the only remaining pristine Carolina Bay habitats. It also offers excellent hunting, fishing, and forestry opportunities. Only a small portion of the area contains industrial facilities and converting these to alternative industrial applications could be accomplished without detracting from the recreational and other uses of the site. Deciding how these lands will be used is a commons issue because the use by one group of people (expanded industrial development, agriculture, or forestry) could detract from the use of others. There are many users with conflicting ideas and stakes in how these public lands should be used, and the question of winners and losers is not only one of human values but of ecological values as well.

Air Quality as a Common-Pool Concern

Traditional approaches to the commons have usually not dealt with air quality or air pollution. Although the atmosphere is a common-pool resource, it is in its use for waste disposal—where unequal access is very difficult to control—that the resource suffers degradation. Studies of global atmospheric transport reveal that air pollutants travel around the globe, to be deposited thousands of miles away from the source. (See "Atmosphere as a Global Commons" in the March 1998 issue of *Environment* for more on transboundary air pollution problems.)

Air pollution from power plants has long been of concern to downwind receptors. While the downwind states and provinces in northeastern North America were encouraging more stringent air pollution standards to control emissions of acid gases and toxic air pollutants, a serious countervailing force arose in the form of energy deregulation. By requiring states to allow the importation of electric power from any producer, the production of cheaper electricity from more polluting plants might actually be increased through demand from users within the downwind states (who will both provide the incentive for and suffer the consequences of increased energy production). Current legislation in the 12 states that have already deregulated electricity includes a variety of incentives for producers (both in-state and out-of-state) to reduce emissions, including disclosure portfolios that would allow consumers to know the emission characteristics of their vendors. In this example, there is no community of producers or consumers of electricity, but there is a clear community of users of air quality, who may have little prerogative for controlling the quality of their air. The states, which would normally be responsible for protecting their residents' health, are clearly not sufficient, and even the regional or multistate consortiums that have formed may be inadequate to protect this common environment. Moreover, the prerogative of the responsible federal agency, the U.S. Environmental Protection Agency, is in jeopardy unless the final deregulation legislation empowers that agency to improve air pollution standards nationwide (even if the lowered cost of electricity is compromised).

Common-Pool Resources and Conflict Resolution

Our understanding of both common-pool resources and the institutions governing their use comes from a number of case studies of resource management in a variety of cultures. Some of the most enlightening case studies deal with the use of public lands, fisheries, agriculture and irrigation systems, groundwater, and contamination of the air. Conflicts inevitably arise and are resolved differently under different property access systems. By examining what systems have worked as well as which ones have allowed or even accelerated resource depletion and habitat degradation, it is possible to begin to understand the rights, rules, and institutions that govern the wise and sustainable use of common-pool resources. This is the legacy of Hardin's initial article, and the responsibility falls on a wide range of disciplines to accomplish it.

The management of common-pool resources is in various stages of development. Recreational and agricultural lands

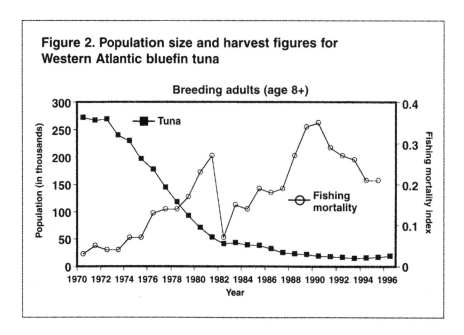

Figure 2. Population size and harvest figures for Western Atlantic bluefin tuna

and forests remain as commons in some regions of the world but are privately or governmentally owned in others. Other common-pool resources, such as clean air and water, are clearly regional or global concerns requiring cooperation among widely dispersed people and governments. In many cases, existing national governments are not presently able to manage them effectively at the national, much less at the global, scale.

The oceans may be in transition from being nationally managed to being regionally or globally controlled, as reflected in the increasing reliance on international treaties to establish exclusive economic zones and international commissions to set quotas, close certain fisheries, and maintain catch statistics. Likewise, there are attempts through international conventions to protect major regional airsheds and even the global atmosphere and ultimately global climate. The Montreal protocol offers an example of a partially successful attempt to retard ozone depletion on a global scale by limiting the use of chlorofluorocarbons.

Social Policy Meets Ecology

Our understanding of common-pool resources is entering a new era of more global influence over resource use and pollution abatement coupled with local institutions managing the resources within their own domains. Ostrom argues that international treaty practices are in a position to take commons management

actions on a global scale. But this will require hard decisions and long-term considerations on the part of user nations. Such decisions are often difficult to make in light of short-term domestic economic constraints influenced by the multinational nature of corporations wielding power and the potential for blackmail by user nations.[30]

The United Nations is the logical forum for developing commons approaches, but its potential is yet to be realized and it seems to be dismissed or ignored in most discussions. Nonetheless, under its aegis a number of attempts are being made. These include the Kyoto protocol on climate change and the Convention on Long-Range Transboundary Air Pollution, which together call for at least a 50 percent reduction in metal emissions and cover basic obligations, cooperative research, reporting, monitoring, compliance, and dispute resolution.[31]

It has become increasingly clear that there are great differences among individuals' access and use of resources, both locally and globally. We will not find one set of rights, rules, and regulations that will fit all common-pool resources. It is also apparent that the social institutions of an ethnically homogeneous, interrelated tribe cooperatively managing a local fishery or plot of land are not a complete model for cooperation among diverse nations managing a global resource.

Social scientists have established a framework for evaluating characteristics

of the user. Commons management is more likely to succeed when the users depend on the resource and share a common understanding of it; when there are grounds for trust; when the users can form an autonomous controlling body; and when they have prior experience with successful management.[32]

Increasing our understanding of how to make the Earth sustainable will require more detailed knowledge about the biology of resources, the social and cultural systems that depend upon these resources, and the economic pressures that govern them. Discussions of sustainability must be based on an understanding of common-pool resource management. The reliance of Western civilization on technological solutions will not solve many of the issues raised by common-pool resources.[33] This is particularly true given the globalization of resource extraction, where the parties that benefit and those that incur the costs are separated geographically and economically and where the benefits are derived by one generation but the costs are incurred by future generations.

Conclusions

Hardin's thesis was seminal in defining the problem of the management of commons and common-pool resources. It met with immediate support by many resource managers who noted that a variety of species had declined dramatically because of overexploitation. Thereafter, social scientists began to note exceptions to his tragedy scenarios, arguing that his thesis was oversimplified and providing examples where institutions allowed people to manage resources sustainably.

Theoretical and empirical studies of common-pool resources have centered on two areas: depletable resources (such as fish, trees, and grasslands) and the depletable waste capacity of resources (such as air and water). While concepts of the commons are equally applicable to both types and remarkable headway has been made in the study of depletable resources, understanding the use of air and water as waste sinks still lags behind. This difference may reflect the more global nature of air and water resources, where the benefits are accrued in one place and the costs are borne by users many thousands of miles away.

The major research thrusts for the future will be in understanding the management of common-pool resources on different temporal and spatial scales, with a view to applying the lessons learned and expanding their applicability. Many of the examples of wise management of common-pool resources have been managed effectively, such as the successful recovery of striped bass along the Atlantic Coast, which required a combination of governmental intervention and user cooperation. The management of global commons resources, particularly fisheries, forests, and wildlife has received considerable attention over the last 10 years and will be an important global issue for many years to come. Although we are making some headway with international treaties to manage global resources (the Montreal protocol and CITES are good examples), other attempts have been disastrous (for example, bluefin tuna). Yet institutions must be developed to deal with these resources or the species are doomed, along with their fisheries.

The difficulty of managing global resources is partly one of attempting to create global treaties and other institutions where there is no global government and few global sanctions. Privatization or long-term governmental stewardship offer alternatives to communal management at the local and regional level, but international conservation will require cooperation across nations where principles of the commons can be invoked to take advantage of common self-interest

in protecting a resource. The management of common-pool resources seems to function best where there are sanctions that everyone agrees to and that can be enforced and where the benefits of management are widely recognized.

Notes

1. G. Hardin, "The Tragedy of the Commons," *Science*, 13 December 1968, 1,243–48.
2. Ibid.
3. G. Hardin, "Extensions of 'The Tragedy of the Commons'," Science, 1 May 1998, 682–83.
4. P. Ehrlich and J. P. Holdren, "Impact of Population Growth," *Science*, March 1971, 1,212–17.
5. B. Commoner, *The Closing Circle* (New York: A. Knopf, 1971); and G. C. Daily and P. R. Ehrlich, "Population, Sustainability, and Earth's Carrying Capacity," *BioScience* 42 (1992): 761–71.
6. D. Feeny, F. Berkes, B. J. McCay, and J. M. Acheson, "The Tragedy of the Commons: Twenty-Two Years Later," *Human Ecology* 18 (1990): 1–19; and F. Berkes, ed., *Common Property Resources: Ecology and Community-Based Sustainable Development* (London; Belhaven Press, 1989).
7. E. Ostrom, "Self-Governance of Common-Pool Resources," in P. Newman, ed., *The New Palgrave Dictionary of Economics and the Law* (London: MacMillan, in press).
8. C. Hess, "Untangling the Web: The Internet as a Commons" (unpublished manuscript presented at the workshop Reinventing the Commons, Transnational Institute, Bonn, Germany, 4–5 November 1995).
9. M. McGinnis and E. Ostrom, "Design Principles for Local and Global Commons," in O. R. Young et al., eds., *International Political Economy and International Institutions* 11 (Cheltenham, U.K.: Edward Elgar Publications, 1996).
10. Ostrom, note 7 above.
11. S. Hanna and M. Monasinghe, eds., *Property Rights and the Environment: Social and Ecological Issues* (Washington, D.C.: The Beijer Institute of Ecological Economics and the World Bank, 1995); and National

Research Council, *Proceedings of the Conference on Common Property Resource Management* (Washington, D.C.: National Academy Press, 1986).
12. F. Berkes, D. Feeny, B. J. McCay, and J. M. Acheson, "The Benefits of the Commons," *Nature*, July 1989, 91–93.
13. Feeny et al., note 6 above.
14. J. M. Acheson, *The Lobster Gangs of Maine* (Hanover, N.H.: University Press of New England, 1988).
15. E. Ostrom, *Governing the Commons: The Evolution of Institutions for Collective Action* (New York: Cambridge University Press, 1990).
16. R. B. Norgaard, "Intergenerational Commons, Globalization, Economism, and Unsustainable Development," *Advances in Human Ecology* 4 (1995): 141–71.
17. Ostrom, note 7 above.
18. B. J. McCay, "Muddling through the Clam Beds: Cooperative Management of New Jersey's Hard Clam Spawner Sanctuaries," *Journal of Shellfish Research* 7 (1988): 327–40; and B. J. McCay and J. M. Acheson, eds., *The Question of the Commons: The Culture and Ecology of Communal Resources* (Tucson, Ariz.: University of Arizona Press, 1987).
19. C. Safina, *Song for the Blue Ocean* (New York: Henry Holt & Co., 1997).
20. Ostrom, note 15 above.
21. Norgaard, note 16 above.
22. C. Safina, "Where Have All the Fishes Gone?," *Issues in Science and Technology* 10 (1994): 37–43.
23. C. Safina, "Bluefin Tuna in the West Atlantic: Negligent Management and the Making of an Endangered Species," *Conservation Biology* 7 (1993): 229–34.
24. International Commission for the Conservation of Atlantic Tunas, *Report of the Standing Committee on Research and Statistics* (Genoa, Italy, 1996).
25. Safina, note 19 above.
26. J. Burger, *A. Naturalist along the Jersey Shore* (New Brunswick, N.J.: Rutgers University Press, 1996).
27. J. Burger, "Effects of Motorboats and Personal Watercraft on Flight Behavior over a Colony of Common Terns," *Condor* 105 (1998): 528–34.
28. J. Burger, "Attitudes about Recreation, Environmental Problems, and Estuarine Health along the New Jer-

sey Shore, U.S.A." *Environmental Management* 22 (1998): 889–96.

29. Commission on Risk Assessment and Risk Management, *Report of the Commission on Risk Assessment and Risk Management* (Washington, D.C.: U.S. Congress, 1997); Department of Energy, *Charting the Course: the Future Use Report*, DOE/EM-0283 (Washington, D.C., 1996); and J. Burger, J. Sanchez, J. W. Gibbons, and M. Gochfeld, "Risk Perception, Federal Spending, and the Savannah River Site: Attitudes of Hunters and Fishermen," *Risk Analysis* 17 (1997): 313–20.

30. Safina, note 22 above.

31. H. E. Ott, "The Kyoto Protocol; Unfinished Business," *Environment*, July/August 1998, 17; Economic and Social Council, "Convention on the Long-Range Transboundary Air Pollution of Heavy Metals" (Åarhus, Denmark: United Nations, 1998).

32. Ostrom, note 7 above.

33. Hardin, note 1 above.

Joanna Burger is a distinguished research professor of biology at Rutgers University. Michael Gochfeld is clinical professor of environmental and community medicine at the University of Medicine and Dentistry of New Jersey's Robert Wood Johnson Medical School. The authors can be reached at the Environmental and Occupational Health Sciences Institute, Piscataway, NJ 08854 (Burger's telephone: (732) 445-4318, e-mail: burger@biology.rutgers.edu; Gochfeld's telephone: (732) 445-0123, ext. 627, e-mail: gochfeld@eohsi.rutgers.edu).

The authors would like to thank several people for comments on the manuscript, including C. Safina, B. McCay, C. Powers, and B. Goldstein. Alana Darnell extracted data from citation abstracts and Robert Ramos prepared the illustrations. Some of the research discussed herein was funded by the New Jersey Department of Environmental Protection, the U.S. Environmental Protection Agency, the Trust for Public Lands, the Department of Energy (cooperative agreement with the Consortium for Risk Evaluation with Stakeholder Participation, DE-FC01-95EW55084) and the National Institute for Environmental Health Sciences (ES05022).

From *Environment*, December 1998, pp. 4–13, 26–27. Reprinted with permission of the Helen Dwight Reid Educational Foundation. Published by Heldref Publications, 1319 Eighteenth St., NW, Washington, DC 20036-1802. © 1998.

Where Have All the Farmers Gone?

The globalization of industry and trade is bringing more and more uniformity to the management of the world's land, and a spreading threat to the diversity of crops, ecosystems, and cultures. As Big Ag takes over, farmers who have a stake in their land—and who often are the most knowledgeable stewards of the land—are being forced into servitude or driven out.

by Brian Halweil

Since 1992, the U.S. Army Corps of Engineers has been developing plans to expand the network of locks and dams along the Mississippi River. The Mississippi is the primary conduit for shipping American soybeans into global commerce—about 35,000 tons a day. The Corps' plan would mean hauling in up to 1.2 million metric tons of concrete to lengthen ten of the locks from 180 meters to 360 meters each, as well as to bolster several major wing dams which narrow the river to keep the soybean barges moving and the sediment from settling. This construction would supplement the existing dredges which are already sucking 85 million cubic meters of sand and mud from the river's bank and bottom each year. Several different levels of "upgrade" for the river have been considered, but the most ambitious of them would purportedly reduce the cost of shipping soybeans by 4 to 8 cents per bushel. Some independent analysts think this is a pipe dream.

Around the same time the Mississippi plan was announced, the five governments of South America's La Plata Basin—Bolivia, Brazil, Paraguay, Argentina, and Uruguay—announced plans to dredge 13 million cubic meters of sand, mud, and rock from 233 sites along the Paraguay-Paraná River. That would be enough to fill a convoy of dump trucks 10,000 miles long. Here, the plan is to straighten natural river meanders in at least seven places, build dozens of locks, and construct a major port in the heart of the Pantanal—the world's largest wetland. The Paraguay-Paraná flows through the center of Brazil's burgeoning soybean heartland—second only to the

United States in production and exports. According to statements from the Brazilian State of Mato Grasso, this "Hidrovía" (water highway) will give a further boost to the region's soybean export capacity.

Lobbyists for both these projects argue that expanding the barge capacity of these rivers is necessary in order to improve competitiveness, grab world market share, and rescue farmers (either U.S. or Brazilian, depending on whom the lobbyists are addressing) from their worst financial crisis since the Great Depression. Chris Brescia, president of the Midwest River Coalition 2000, an alliance of commodity shippers that forms the primary lobbying force for the Mississippi plan, says, "The sooner we provide the waterway infrastructure, the sooner our family farmers will benefit." Some of his fellow lobbyists have even argued that these projects are essential to feeding the world (since the barges can then more easily speed the soybeans to the world's hungry masses) and to saving the environment (since the hungry masses will not have to clear rainforest to scratch out their own subsistence).

Probably very few people have had an opportunity to hear both pitches and compare them. But anyone who has may find something amiss with the argument that U.S. farmers will become more competitive versus their Brazilian counterparts, at the same time that Brazilian farmers will, for the same reasons, become more competitive with their U.S. counterparts. A more likely outcome is that farmers of these two nations will be pitted against each other in a costly race to maximize production, resulting in short-cut practices that essentially strip-mine their soil and throw long-term investments in the land to the wind. Farmers in Iowa will have stronger incentives to plow up land along stream banks, triggering faster erosion of topsoil. Their brethren in Brazil will find themselves needing to cut deeper into the savanna, also accelerating erosion. That will increase the flow of soybeans, all right—both north and south. But it will also further depress prices, so that even as the farmers are shipping more, they're getting less income per ton shipped. And in any case, increasing volume can't help the farmers survive in the long run, because sooner or later they will be swallowed by larger, corporate, farms that can make up for the smaller per-ton margins by producing even larger volumes.

So, how can the supporters of these river projects, who profess to be acting in the farmer's best interests, not notice the illogic of this form of competition? One explanation is that from the advocates' (as opposed to the farmers') standpoint, this competition isn't illogical at all—because the lobbyists aren't really representing farmers. They're working for the commodity processing, shipping, and trading firms who want the price of soybeans to fall, because these are the firms that buy the crops from the farmers. In fact, it is the same three agribusiness conglomerates—Archer Daniels Midland (ADM), Cargill, and Bunge—that are the top soybean processors and traders along both rivers.

Welcome to the global economy. The more brutally the U.S. and Brazilian farmers can batter each-other's prices (and standards of living) down, the greater the margin of profit these three giants gain. Meanwhile, another handful of companies controls the markets for genetically modified seeds, fertilizers, and herbicides used by the farmers—charging oligopolistically high prices both north and south of the equator.

In assessing what this proposed digging-up and reconfiguring of two of the world's great river basins really means, keep in mind that these projects will not be the activities of private businesses operating inside their own private property. These are proposed public works, to be undertaken at huge public expense. The motive is neither the plight of the family farmer nor any moral obligation to feed the world, but the opportunity to exploit poorly informed public sentiments about farmers' plights or hungry masses as a means of usurping public policies to benefit private interests. What gets thoroughly Big Muddied, in this usurping process, is that in addition to subjecting farmers to a gladiator-like attrition, these projects will likely bring a cascade of damaging economic, social, and ecological impacts to the very river basins being so expensively remodeled.

What's likely to happen if the lock and dam system along the Mississippi is expanded as proposed? The most obvious effect will be increased barge traffic, which will accelerate a less obvious cascade of events that has been underway for some time, according to Mike Davis of the Minnesota Department of Natural Resources. Much of the Mississippi River ecosystem involves aquatic rooted plants, like bullrush, arrowhead, and wild celery. Increased barge traffic will kick up more sediment, obscuring sunlight and reducing the depth to which plants can survive. Already, since the 1970s, the number of aquatic plant species found in some of the river has been cut from 23 to about half that, with just a handful thriving under the cloudier conditions. "Areas of the river have reached an ecological turning point," warns Davis. "This decline in plant diversity has triggered a drop in the invertebrate communities that live on these plants, as well as a drop in the fish, mollusk, and bird communities that depend on the diversity of insects and plants." On May 18, 2000, the U.S. Fish and Wildlife Service released a study saying that the Corps of Engineers project would threaten the 300 species of migratory birds and 12 species of fish in the Mississippi watershed, and could ultimately push some into extinction. "The least tern, the pallid sturgeon, and other species that evolved with the ebbs and flows, sandbars and depths, of the river are progressively eliminated or forced away as the diversity of the river's natural habitats is removed to maximize the barge habitat," says Davis.

The outlook for the Hidrovía project is similar. Mark Robbins, an ornithologist at the Natural History Museum at the University of Kansas, calls it "a key step in creating a Florida Everglades-like scenario of destruction in the Pantanal, and an American Great Plains-like scenario in the Cerrado in southern Brazil." The Paraguay-Paraná feeds the Pantanal wetlands, one of the most diverse habitats on the planet, with its populations of woodstorks, snailkites, limpkins, jabirus, and more than 650 other species of birds, as well as more than 400 species of fish and hundreds of other less-studied plants, mussels, and marshland organisms. As the river is dredged and the banks are built up to funnel the surrounding wetlands water into the navigation path, bird nesting habitat and fish spawning grounds will be eliminated, damaging the indigenous and other traditional societies that depend on these resources. Increased barge traffic will suppress river species here just as it will on the Mississippi. Meanwhile, herbicide-intensive soybean monocultures—on farms so enormous that they dwarf even the biggest operations in the U.S. Midwest—are rapidly replacing diverse grasslands in the fragile Cerrado. The heavy plowing and periodic absence of ground cover associated with such farming erodes 100 million tons of soil per year. Robbins notes that "compared to the Mississippi, this southern river system and surrounding grassland is several orders of magnitude more diverse and has suffered considerably less, so there is much more at stake."

Supporters of such massive disruption argue that it is justified because it is the most "efficient" way to do business. The perceived efficiency of such farming might be compared to the perceived efficiency of an energy system based on coal. Burning coal looks very efficient if you ignore its long-term impact on air quality and climate stability. Similarly, large farms look more efficient than small farms if you don't count some of their largest costs—the loss of the genetic diversity that underpins agriculture, the pollution caused by agro-chemicals, and the dislocation of rural cultures. The simultaneous demise of small, independent farmers and rise of multinational food giants is troubling not just for those who empathize with dislocated farmers, but for anyone who eats.

An Endangered Species

Nowadays most of us in the industrialized countries don't farm, so we may no longer really understand that way of life. I was born in the apple orchard and dairy country of Dutchess County, New York, but since age five have spent most of my life in New York City—while most of the farms back in Dutchess County have given way to spreading subdivisions. It's also hard for those of us who get our food from supermarket shelves or drive-thru windows to know how dependent we are on the viability of rural communities.

Whether in the industrial world, where farm communities are growing older and emptier, or in developing nations where population growth is pushing the number of farmers continually higher and each generation is inheriting smaller family plots, it is becoming harder and harder to make a living as a farmer. A combination of falling incomes, rising debt, and worsening rural poverty is forcing more people to either abandon farming as their primary activity or to leave the countryside altogether— a bewildering juncture, considering that farmers produce perhaps the only good that the human race cannot do without.

Since 1950, the number of people employed in agriculture has plummeted in all industrial nations, in some regions by more than 80 percent. Look at the numbers, and you might think farmers are being singled out by some kind of virus:

- In Japan, more than half of all farmers are over 65 years old; in the United States, farmers over 65 outnumber those under 35 by three to one. (Upon retirement or death, many will pass the farm on to children who live in the city and have no interest in farming themselves.)
- In New Zealand, officials estimate that up to 6,000 dairy farms will disappear during the next 10 to 15 years—dropping the total number by nearly 40 percent.
- In Poland, 1.8 million farms could disappear as the country is absorbed into the European Union—dropping the total number by 90 percent.
- In Sweden, the number of farms going out of business in the next decade is expected to reach about 50 percent.

- In the Philippines, Oxfam estimates that over the next few years the number of farm households in the corn–producing region of Mindanao could fall by some 500,000—a 50 percent loss.
- In the United States, where the vast majority of people were farmers at the time of the American Revolution, fewer people are now full-time farmers (less than 1 percent of the population) than are full-time prisoners.
- In the U.S. states of Nebraska and Iowa, between a fifth and a third of farmers are expected to be out of business within two years.

Of course, the declining numbers of farmers in industrial nations does not imply a decline in the importance of the farming sector. The world still has to eat (and 80 million more mouths to feed each year than the year before), so smaller numbers of farmers mean larger farms and greater concentration of ownership. Despite a precipitous plunge in the number of people employed in farming in North America, Europe, and East Asia, half the world's people still make their living from the land. In sub-Saharan Africa and South Asia, more than 70 percent do. In these regions, agriculture accounts, on average, for half of total economic activity.

Some might argue that the decline of farmers is harmless, even a blessing, particularly for less developed nations that have not yet experienced the modernization that moves peasants out of backwater rural areas into the more advanced economies of the cities. For most of the past two centuries, the shift toward fewer farmers has generally been assumed to be a kind of progress. The substitution of high-powered diesel tractors for slow-moving women and men with hoes, or of large mechanized industrial farms for clusters of small "old fashioned" farms, is typically seen as the way to a more abundant and affordable food supply. Our urban-centered society has even come to view rural life, especially in the form of small family-owned businesses, as backwards or boring, fit only for people who wear overalls and go to bed early—far from the sophistication and dynamism of the city.

Urban life does offer a wide array of opportunities, attractions, and hopes—some of them falsely created by urban-oriented commercial media—that many farm families decide to pursue willingly. But city life often turns out to be a disappointment, as displaced farmers find themselves lodged in crowded slums, where unemployment and ill-health are the norm and where they are worse off than they were back home. Much evidence suggests that farmers aren't so much being lured to the city as they are being driven off their farms by a variety of structural changes in the way the global food chain operates. Bob Long, a rancher in McPherson County, Nebraska, stated in a recent *New York Times* article that passing the farm onto his son would be nothing less than "child abuse."

As long as cities are under the pressure of population growth (a situation expected to continue at least for the next three or four decades), there will always be pressure for a large share of humanity to subsist in the countryside. Even in highly urbanized North America and Europe, roughly 25 percent of the population—275 million people—still reside in rural areas. Meanwhile, for the 3 billion Africans, Asians, and Latin Americans who remain in the countryside—and who will be there for the foreseeable future—the marginalization of farmers has set up a vicious cycle of low educational achievement, rising infant mortality, and deepening mental distress.

Hired Hands on Their Own Land

In the 18th and 19th centuries, farmers weren't so trapped. Most weren't wealthy, but they generally enjoyed stable incomes and strong community ties. Diversified farms yielded a range of raw and processed goods that the farmer could typically sell in a local market. Production costs tended to be much lower than now, as many of the needed inputs were home-grown: the farmer planted seed that he or she had saved from the previous year, the farm's cows or pigs provided fertilizer, and the diversity of crops—usually a large range of grains, tubers, vegetables, herbs, flowers, and fruits for home use as well as for sale—effectively functioned as pest control.

Things have changed, especially in the past half-century, according to Iowa State agricultural economist Mike Duffy. "The end of World War II was a watershed period," he says. "The widespread introduction of chemical fertilizers and synthetic pesticides, produced as part of the war effort, set in motion dramatic changes in how we farm—and a dramatic decline in the number of farmers." In the post-war period, along with increasing mechanization, there was an increasing tendency to "outsource" pieces of the work that the farmers had previously done themselves—from producing their own fertilizer to cleaning and packaging their harvest. That outsourcing, which may have seemed like a welcome convenience at the time, eventually boomeranged: at first it enabled the farmer to increase output, and thus profits, but when all the other farmers were doing it too, crop prices began to fall.

Before long, the processing and packaging businesses were adding more "value" to the purchased product than the farmer, and it was those businesses that became the

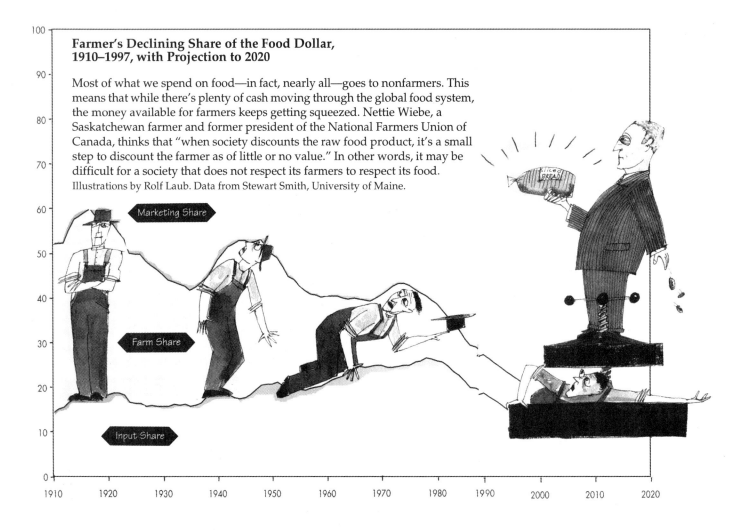

Farmer's Declining Share of the Food Dollar, 1910–1997, with Projection to 2020

Most of what we spend on food—in fact, nearly all—goes to nonfarmers. This means that while there's plenty of cash moving through the global food system, the money available for farmers keeps getting squeezed. Nettie Wiebe, a Saskatchewan farmer and former president of the National Farmers Union of Canada, thinks that "when society discounts the raw food product, it's a small step to discount the farmer as of little or no value." In other words, it may be difficult for a society that does not respect its farmers to respect its food.

Illustrations by Rolf Laub. Data from Stewart Smith, University of Maine.

Marketing Share

Farm Share

Input Share

ConAgra: *Vertical Integration, Horizontal Concentration, Global Omnipresence*

Three conglomerates (ConAgra/DuPont, Cargill/Monsanto, and Novartis/ADM) dominate virtually every link in the North American (and increasingly, the global) food chain. Here's a simplified diagram of one conglomerate.

KEY: Vertical integration of production links, from seed to supermarket  Concentration within a link

INPUTS
Distribution of farm
chemicals, machinery,
fertilizer, and seed

3 companies dominate North American farm machinery sector
6 companies control 63% of global pesticide market
4 companies control 69% of North American seed corn market
3 companies control 71% of Canadian nitrogen fertilizer capacity
ConAgra distributes all of these inputs, and is in a joint venture with DuPont to distribute DuPont's transgenic high-oil corn seed.

FARMS

The farm sector is rapidly consolidating in the industrial world, as farms "get big or get out." Many go under contract with **ConAgra** and other conglomerates; others just go under. In the past 50 years, the number of farmers has declined by 86% in Germany, 85% in France, 85% in Japan, 64% in the U.S., 59% in South Korea, and 59% in the U.K.

GRAIN COLLECTION

A proposed merger of Cargill and Continental Grain will control half of the global grain trade; **ConAgra** has about one-quarter.

GRAIN MILLING

ConAgra and 3 other companies account for 62% of the North American market.

PRODUCTION OF
BEEF, PORK, TURKEY,
CHICKEN, AND SEAFOOD

ConAgra ranks 3rd in cattle feeding and 5th in broiler production.

ConAgra Poultry, Tyson Foods, Perdue, and 3 other companies control 60% of U.S. chicken production

PROCESSING OF
BEEF, PORK, TURKEY,
CHICKEN, AND SEAFOOD

IBP, **ConAgra,** Cargill, and Farmland control 80% of U.S. beef packing

Smithfield, **ConAgra**, and 3 other companies control 75% of U.S. pork packing

SUPERMARKETS

ConAgra divisions own Wesson oil, Butterball turkeys, Swift Premium meats, Peter Pan peanut butter, Healthy Choice diet foods, Hunt's tomato sauce, and about 75 other major brands.

dominant players in the food industry. Instead of farmers outsourcing to contractors, it became a matter of large food processors buying raw materials from farmers, on the processors' terms. Today, most of the money is in the work the farmer no longer does—or even controls. In the United States, the share of the consumer's food dollar that trickles back to the farmer has plunged from nearly 40 cents in 1910 to just above 7 cents in 1997, while the shares going to input (machinery, agrochemicals, and seeds) and marketing (processing, shipping, brokerage, advertising, and retailing) firms have continued to expand. (See graph "Farmer's Declining Share of the Food Dollar") The typical U.S. wheat farmer, for instance, gets just 6 cents of the dollar spent on a loaf of bread—so when you buy that loaf, you're paying about as much for the wrapper as for the wheat.

Ironically, then, as U.S. farms became more mechanized and more "productive," a self-destructive feedback loop was set in motion: over-supply and declining crop prices cut into farmers' profits, fueling a demand for more technology aimed at making up for shrinking margins by increasing volume still more. Output increased dramatically, but expenses (for tractors, combines, fertilizer, and seed) also ballooned—while the commodity prices stagnated or declined. Even as they were looking more and more modernized, the farmers were becoming less and less the masters of their own domain.

On the typical Iowa farm, the farmer's profit margin has dropped from 35 percent in 1950 to 9 percent today. In order to generate the same income, this farm would need to be roughly four times as large today as in 1950—or the farmer would need to get a night job. And that's precisely what we've seen in most industrialized nations: fewer farmers on bigger tracts of land producing a greater share of the total food supply. The farmer with declining margins buys out his neighbor and expands or risks being cannibalized himself.

There is an alternative to this huge scaling up, which is to buck the trend and bring some of the input-supplying and post-harvest processing—and the related profits— back onto the farm. But more self-sufficient farming would be highly unpopular with the industries that now make lucrative profits from inputs and processing. And since these industries have much more political clout than the farmers do, there is little support for rescuing farmers from their increasingly servile condition—and the idea has been largely forgotten. Farmers continue to get the message that the only way to succeed is to get big.

The traditional explanation for this constant pressure to "get big or get out" has been that it improves the efficiency of the food system—bigger farms replace smaller farms, because the bigger farms operate at lower costs. In some respects, this is quite true. Scaling up may allow a farmer to spread a tractor's cost over greater acreage, for example. Greater size also means greater leverage in purchasing inputs or negotiating loan rates—increasingly important as satellite-guided combines and other equip-

ment make farming more and more capital-intensive. But these economies of scale typically level off. Data for a wide range of crops produced in the United States show that the lowest production costs are generally achieved on farms that are much smaller than the typical farm now is. But large farms can tolerate lower margins, so while they may not *produce* at lower cost, they can afford to *sell* their crops at lower cost, if forced to do so—as indeed they are by the food processors who buy from them. In short, to the extent that a giant farm has a financial benefit over a small one, it's a benefit that goes only to the processor—not to the farmer, the farm community, or the environment.

This shift of the food dollar away from farmers is compounded by intense concentration in every link of the food chain—from seeds and herbicides to farm finance and retailing. In Canada, for example, just three companies control over 70 percent of fertilizer sales, five banks provide the vast majority of agricultural credit, two companies control over 70 percent of beef packing, and five companies dominate food retailing. The merger of Philip Morris and Nabisco will create an empire that collects nearly 10 cents of every dollar a U.S. consumer spends on food, according to a company spokesperson. Such high concentration can be deadly for the bottom line, allowing agribusiness firms to extract higher prices for the products farmers buy from them, while offering lower prices for the crop they buy from the farmers.

An even more worrisome form of concentration, according to Bill Heffernan, a rural sociologist at the University of Missouri, is the emergence of several clusters of firms that—through mergers, takeovers, and alliances with other links in the food chain—now possess "a seamless and fully vertically integrated control of the food system from gene to supermarket shelf." (See diagram "ConAgra") Consider the recent partnership between Monsanto and Cargill, which controls seeds, fertilizers, pesticides, farm finance, grain collection, grain processing, livestock feed processing, livestock production, and slaughtering, as well as some well-known processed food brands. From the standpoint of a company like Cargill, such alliances yield tremendous control over costs and can therefore be extremely profitable.

But suppose you're the farmer. Want to buy seed to grow corn? If Cargill is the only buyer of corn in a hundred mile radius, and Cargill is only buying a particular Monsanto corn variety for its mills or elevators or feedlots, then if you don't plant Monsanto's seed you won't have a market for your corn. Need a loan to buy the seed? Go to Cargill-owned Bank of Ellsworth, but be sure to let them know which seed you'll be buying. Also mention that you'll be buying Cargill's Saskferco brand fertilizer. OK, but once the corn is grown, you don't like the idea of having to sell to Cargill at the prices it dictates? Well, maybe you'll feed the corn to your pigs, then, and sell them to the highest bidder. No problem—Cargill's Excel Corporation buys pigs, too. OK, you're moving to the city, and renouncing the farm life! No more home-made grits for breakfast, you're buying corn flakes. Well, good news: Cargill Foods supplies corn flour to the top cereal makers. You'll notice, though, that all the big brands of corn flakes seem to have pretty much the same hefty price per ounce. After all, they're all made by the agricultural oligopoly.

As these vertical food conglomerates consolidate, Heffernan warns, "there is little room left in the global food system for independent farmers"—the farmers being increasingly left with "take it or leave it" contracts from the remaining conglomerates. In the last two decades, for example, the share of American agricultural output produced under contract has more than tripled, from 10 percent to 35 percent—and this doesn't include the contracts that farmers must sign to plant genetically engineered seed. Such centralized control of the food system, in which farmers are in effect reduced to hired hands on their own land, reminds Heffernan of the Soviet-style state farms, but with the Big Brother role now being played by agribusiness executives. It is also reminiscent of the "company store" which once dominated small American mining or factory towns, except that if you move out of town now, the store is still with you. The company store has gone global.

With the conglomerates who own the food dollar also owning the political clout, it's no surprise that agricultural policies—including subsidies, tax breaks, and environmental legislation at both the national and international levels—do not generally favor the farms. For example, the conglomerates command growing influence over both private and public agricultural research priorities, which might explain why the U.S. Department of Agriculture (USDA), an agency ostensibly beholden to farmers, would help to develop the seed-sterilizing Terminator technology—a biotechnology that offers farmers only greater dependence on seed companies. In some cases the influence is indirect, as manifested in government funding decisions, while in others it is more blatant. When Novartis provided $25 million to fund a research partnership with the plant biology department of the University of California at Berkeley, one of the conditions was that Novartis has the first right of refusal for any patentable inventions. Under those circumstances, of course, the UC officials—mindful of where their funding comes from—have strong incentives to give more attention to technologies like the Terminator seed, which shifts profit away from the farmer, than to technologies that directly benefit the farmer or the public at large.

Even policies that are touted to be in the best interest of farmers, like liberalized trade in agricultural products, are increasingly shaped by non-farmers. Food traders, processors, and distributors, for example, were some of the principal architects of recent revisions to the General Agreement on Trade and Tariffs (GATT)—the World Trade Organization's predecessor—that paved the way for greater trade flows in agricultural commodities. Before these revisions, many countries had mechanisms for assuring that their farmers wouldn't be driven out of their own domestic markets by predatory global traders. The traders, however, were able to do away with those protections.

The ability of agribusiness to slide around the planet, buying at the lowest possible price and selling at the highest, has tended to tighten the squeeze already put in place by economic marginalization, throwing every farmer on the planet into direct competition with every other farmer. A recent UN Food and Agriculture Organization assessment of the experience of 16 developing nations in implementing the latest phase of the GATT concluded that "a common reported concern was with a general trend towards the concentration of farms," a process that tends to further marginalize small producers and exacerbate rural poverty and unemployment. The sad irony, according to Thomas Reardon, of Michigan State University, is that while small farmers in all reaches of the world are increasingly affected by cheap, heavily subsidized imports of foods from outside of their traditional rural markets, they are nonetheless often excluded from opportunities to participate in food exports themselves. To keep down transaction costs and to keep processing standardized, exporters and other downstream players prefer to buy from a few large producers, as opposed to many small producers.

As the global food system becomes increasingly dominated by a handful of vertically integrated, international corporations, the servitude of the farmer points to a broader society-wide servitude that OPEC-like food cartels could impose, through their control over food prices and food quality. Agricultural economists have already noted that the widening gap between retail food prices and farm prices in the 1990s was due almost exclusively to exploitation of market power, and not to extra services provided by processors and retailers. It's questionable whether we should pay as much for a bread wrapper as we do for the nutrients it contains. But beyond this, there's a more fundamental question. Farmers are professionals, with extensive knowledge of their local soils, weather, native plants, sources of fertilizer or mulch, native pollinators, ecology, and community. If we are to have a world where the land is no longer managed by such professionals, but is instead managed by distant corporate bureaucracies interested in extracting maximum output at minimum cost, what kind of food will we have, and at what price?

Agrarian Services

No question, large industrial farms can produce lots of food. Indeed, they're designed to maximize quantity. But when the farmer becomes little more than the lowest-cost producer of raw materials, more than his own welfare will suffer. Though the farm sector has lost power and profit, it is still the one link in the agrifood chain accounting for the largest share of agriculture's public goods—including half the world's jobs, many of its most vital communities, and many of its most diverse landscapes. And in providing many of these goods, small farms clearly have the advantage.

Local economic and social stability: Over half a century ago, William Goldschmidt, an anthropologist working at the USDA, tried to assess how farm structure and size affect the health of rural communities. In California's San Joaquin Valley, a region then considered to be at the cutting edge of agricultural industrialization, he identified two small towns that were alike in all basic economic and geographic dimensions, including value of agricultural production, except in farm size. Comparing the two, he found an inverse correlation between the sizes of the farms and the well-being of the communities they were a part of.

The small-farm community, Dinuba, supported about 20 percent more people, and at a considerably higher level of living—including lower poverty rates, lower levels of economic and social class distinctions, and a lower crime rate—than the large-farm community of Arvin. The majority of Dinuba's residents were independent entrepreneurs, whereas fewer than 20 percent of Arvin's residents were—most of the others being agricultural laborers. Dinuba had twice as many business establishments as Arvin, and did 61 percent more retail business. It had more schools, parks, newspapers, civic organizations, and churches, as well as better physical infrastructure—paved streets, sidewalks, garbage disposal, sewage disposal and other public services. Dinuba also had more institutions for democratic decision making, and a much broader participation by its citizens. Political scientists have long recognized that a broad base of independent entrepreneurs and property owners is one of the keys to a healthy democracy.

The distinctions between Dinuba and Arvin suggest that industrial agriculture may be limited in what it can do for a community. Fewer (and less meaningful) jobs, less local spending, and a hemorrhagic flow of profits to absentee landowners and distant suppliers means that industrial farms can actually be a net drain on the local economy. That hypothesis has been corroborated by Dick Levins, an agricultural economist at the University of Minnesota. Levins studied the economic receipts from Swift County, Iowa, a typical Midwestern corn and soybean community, and found that although total farm sales are near an all-time high, farm income there has been dismally low—and that many of those who were once the financial stalwarts of the community are now deeply in debt. "Most of the U.S. Corn Belt, like Swift County, is a colony, owned and operated by people who don't live there and for the benefit of those who don't live there," says Levin. In fact, most of the land in Swift County is rented, much of it from absentee landlords.

This new calculus of farming may be eliminating the traditional role of small farms in anchoring rural economies—the kind of tradition, for example, that we saw in the emphasis given to the support of small farms by Japan, South Korea, and Taiwan following World War II. That emphasis, which brought radical land reforms and targeted investment in rural areas, is widely cited as having been a major stimulus to the dramatic economic boom those countries enjoyed.

Not surprisingly, when the economic prospects of small farms decline, the social fabric of rural communities begins to tear. In the United States, farming families are more than twice as likely as others to live in poverty. They have less education and lower rates of medical protection, along with higher rates of infant mortality, alcoholism, child abuse, spousal abuse, and mental stress. Across Europe, a similar pattern is evident. And in sub-Saharan Africa, sociologist Deborah Bryceson of the Netherlands-based African Studies Centre has studied the dislocation of small farmers and found that "as de-agrarianization proceeds, signs of social dysfunction associated with urban areas [including petty crime and breakdowns of family ties] are surfacing in villages."

People without meaningful work often become frustrated, but farmers may be a special case. "More so than other occupations, farming represents a way of life and defines who you are," says Mike Rosemann, a psychologist who runs a farmer counseling network in Iowa. "Losing the family farm, or the prospect of losing the family farm, can generate tremendous guilt and anxiety, as if one has failed to protect the heritage that his ancestors worked to hold onto." One measure of the despair has been a worldwide surge in the number of farmers committing suicide. In 1998, over 300 cotton farmers in Andhra Pradesh, India, took their lives by swallowing pesticides that they had gone into debt to purchase but that had nonetheless failed to save their crops. In Britain, farm workers are two-and-a-half times more likely to commit suicide than the rest of the population. In the United States, official statistics say farmers are now five times as likely to commit suicide as to die from farm accidents, which have been traditionally the most frequent cause of unnatural death for them. The true number may be even higher, as suicide hotlines report that they often receive calls from farmers who want to know which sorts of accidents (Falling into the blades of a combine? Getting shot while hunting?) are least likely to be investigated by insurance companies that don't pay claims for suicides.

Whether from despair or from anger, farmers seem increasingly ready to rise up, sometimes violently, against government, wealthy landholders, or agribusiness giants. In recent years we've witnessed the Zapatista revolution in Chiapas, the seizing of white-owned farms by landless blacks in Zimbabwe, and the attacks of European farmers on warehouses storing genetically engineered seed. In the book *Harvest of Rage*, journalist Joel Dyer links the 1995 Oklahoma City bombing that killed nearly 200 people—

In the Developing World, an Even Deeper Farm Crisis

"One would have to multiply the threats facing family farmers in the United States or Europe five, ten, or twenty times to get a sense of the handicaps of peasant farmers in less developed nations," says Deborah Bryceson, a senior research fellow at the African Studies Centre in the Netherlands. Those handicaps include insufficient access to credit and financing, lack of roads and other infrastructure in rural areas, insecure land tenure, and land shortages where population is dense.

Three forces stand out as particularly challenging to these peasant farmers:

Structural adjustment requirements, imposed on indebted nations by international lending institutions, have led to privatization of "public commodity procurement boards" that were responsible for providing public protections for rural economies. "The newly privatized entities are under no obligation to service marginal rural areas," says Rafael Mariano, chairman of a Filipino farmers' union. Under the new rules, state protections against such practices as dumping of cheap imported goods (with which local farmers can't compete) were abandoned at the same time that state provision of health care, education, and other social services was being reduced.

Trade liberalization policies associated with structural adjustment have reduced the ability of nations to protect their agricultural economies even if they want to. For example, the World Trade Organization's Agreement on Agriculture will forbid domestic price support mechanisms and tariffs on imported goods—some of the primary means by which a country can shield its own farmers from overproduction and foreign competition.

The growing emphasis on agricultural grades and standards—the standardizing of crops and products so they can be processed and marketed more "efficiently"—has tended to favor large producers, and to marginalize smaller ones. Food manufacturers and supermarkets have emerged as the dominant entities in the global agri-food chain, and with their focus on brand consistency, ingredient uniformity, and high volume, smaller producers often are unable to deliver—or aren't even invited to bid.

Despite these daunting conditions, many peasant farmers tend to hold on long after it has become clear that they can't compete. One reason, says Peter Rosset of the Institute for Food and Development Policy, is that "even when it gets really bad, they will cling to agriculture because of the fact that it at least offers some degree of food security—that you can feed yourself." But with the pressures now mounting, particularly as export crop production swallows more land, even that fallback is lost.

---

as well as the rise of radical right and antigovernment militias in the U.S. heartland—to a spreading despair and anger stemming from the ongoing farm crisis. Thomas Homer-Dixon, director of the Project on Environment, Population, and Security at the University of Toronto, regards farmer dislocation, and the resulting rural unemployment and poverty, as one of the major security threats for the coming decades. Such dislocation is responsible for roughly half of the growth of urban populations across the Third World, and such growth often occurs in volatile shantytowns that are already straining to meet the basic needs of their residents. "What was an extremely traumatic transition for Europe and North America from a rural society to an urban one is now proceeding at two to three times that speed in developing nations," says Homer-Dixon. And, these nations have considerably less industrialization to absorb the labor. Such an accelerated transition poses enormous adjustment challenges for India and China, where perhaps a billion and a half people still make their living from the land.

Ecological stability: In the Andean highlands, a single farm may include as many as 30 to 40 distinct varieties of potato (along with numerous other native plants), each having slightly different optimal soil, water, light, and temperature regimes, which the farmer—given enough time—can manage. (In comparison, in the United States, just four closely related varieties account for about 99 percent of all the potatoes produced.) But, according to Karl Zimmerer, a University of Wisconsin sociologist, declining farm incomes in the Andes force more and more growers into migrant labor forces for part of the year, with serious effects on farm ecology. As time becomes constrained, the farmer manages the system more homogenously—cutting back on the number of traditional varieties (a small home garden of favorite culinary varieties may be the last refuge of diversity), and scaling up production of a few commercial varieties. Much of the traditional crop diversity is lost.

Complex farm systems require a highly sophisticated and intimate knowledge of the land—something small-scale, full-time farmers are more able to provide. Two or three different crops that have different root depths, for example, can often be planted on the same piece of land, or crops requiring different drainage can be planted in close proximity on a tract that has variegated topography. But these kinds of cultivation can't be done with heavy tractors moving at high speed. Highly site-specific and management-intensive cultivation demands ingenuity and awareness of local ecology, and can't be achieved by heavy equipment and heavy applications of agrochemicals. That isn't to say that being small is always sufficient to ensure ecologically sound food production, because economic adversity can drive small farms, as well as big ones, to compromise sustainable food production by transmogrifying the craft of land stewardship into the crude labor of commodity production. But a large-scale, highly mechanized farm is simply not equipped to preserve landscape complexity. Instead, its normal modus is to use blunt management tools, like crops that have been genetically engineered to churn out insecticides, which obviate the need to scout the field to see if spraying is necessary at all.

In the U.S. Midwest, as farm size has increased, cropping systems have gotten more simplified. Since 1972, the number of counties with more than 55 percent of their acreage planted in corn and soybeans has nearly tripled, from 97 to 267. As farms scaled up, the great simplicity of managing the corn-soybean rotation—an 800 acre farm, for instance, may require no more than a couple of weeks planting in the spring and a few weeks harvesting in the fall—became its big selling point. The various arms of the agricultural economy in the region, from extension services to grain elevators to seed suppliers, began to solidify around this corn-soybean rotation, reinforcing the farmers' movement away from other crops. Fewer and fewer farmers kept livestock, as beef and hog production became "economical" only in other parts of the country where it was becoming more concentrated. Giving up livestock meant eliminating clover, pasture mixtures, and a key source of fertilizer in the Midwest, while creating tremendous manure concentrations in other places.

But the corn and soybean rotation—one monoculture followed by another—is extremely inefficient or "leaky" in its use of applied fertilizer, since low levels of biodiversity tend to leave a range of vacant niches in the field, including different root depths and different nutrient preferences. Moreover, the Midwest's shift to monoculture has subjected the country to a double hit of nitrogen pollution, since not only does the removal and concentration of livestock tend to dump inordinate amounts of feces in the places (such as Utah and North Carolina) where the livestock operations are now located, but the monocultures that remain in the Midwest have much poorer nitrogen retention than they would if their cropping were more complex. (The addition of just a winter rye crop to the corn-soy rotation has been shown to reduce nitrogen runoff by nearly 50 percent.) And maybe this disaster-in-the-making should really be regarded as a triple hit, because in addition to contaminating Midwestern water supplies, the runoff ends up in the Gulf of Mexico, where the nitrogen feeds massive algae blooms. When the algae die, they are decomposed by bacteria, whose respiration depletes the water's oxygen—suffocating fish, shellfish,

and all other life that doesn't escape. This process periodically leaves 20,000 square kilometers of water off the coast of Louisiana biologically dead. Thus the act of simplifying the ecology of a field in Iowa can contribute to severe pollution in Utah, North Carolina, Louisiana, *and* Iowa.

The world's agricultural biodiversity—the ultimate insurance policy against climate variations, pest outbreaks, and other unforeseen threats to food security—depends largely on the millions of small farmers who use this diversity in their local growing environments. But the marginalization of farmers who have developed or inherited complex farming systems over generations means more than just the loss of specific crop varieties and the knowledge of how they best grow. "We forever lose the best available knowledge and experience of place, including what to do with marginal lands not suited for industrial production," says Steve Gleissman, an agroecologist at the University of California at Santa Cruz. The 12 million hogs produced by Smithfield Foods Inc., the largest hog producer and processor in the world and a pioneer in vertical integration, are nearly genetically identical and raised under identical conditions—regardless of whether they are in a Smithfield feedlot in Virginia or Mexico.

As farmers become increasingly integrated into the agribusiness food chain, they have fewer and fewer controls over the totality of the production process—shifting more and more to the role of "technology applicators," as opposed to managers making informed and independent decisions. Recent USDA surveys of contract poultry farmers in the United States found that in seeking outside advice on their operations, these farmers now turn first to bankers and then to the corporations that hold their contracts. If the contracting corporation is also the same company that is selling the farm its seed and fertilizer, as is often the case, there's a strong likelihood that the company's procedures will be followed. That corporation, as a global enterprise with no compelling local ties, is also less likely to be concerned about the pollution and resource degradation created by those procedures, at least compared with a farmer who is rooted in that community. Grower contracts generally disavow any environmental liability.

And then there is the ecological fallout unique to large-scale, industrial agriculture. Colossal confined animal feeding operations (CAFOs)—those "other places" where livestock are concentrated when they are no longer present on Midwestern soy/corn farms—constitute perhaps the most egregious example of agriculture that has, like a garbage barge in a goldfish pond, overwhelmed the scale at which an ecosystem can cope. CAFOs are increasingly the norm in livestock production, because, like crop monocultures, they allow the production of huge populations of animals which can be slaughtered and marketed at rock-bottom costs. But the disconnection between the livestock and the land used to produce their feed means that such CAFOs generate gargantuan amounts of waste, which the surrounding soil cannot possibly absorb. (One farm in Utah will raise over five million hogs in a year, producing as much waste each day as the city of Los Angeles.) The waste is generally stored in large lagoons, which are prone to leak and even spill over during heavy storms. From North Carolina to South Korea, the overwhelming stench of these lagoons—a combination of hydrogen sulfide, ammonia, and methane gas that smells like rotten eggs—renders miles of surrounding land uninhabitable.

A different form of ecological disruption results from the conditions under which these animals are raised. Because massive numbers of closely confined livestock are highly susceptible to infection, and because a steady diet of antibiotics can modestly boost animal growth, overuse of antibiotics has become the norm in industrial animal production. In recent months, both the Centers for Disease Control and Prevention in the United States and the World Health Organization have identified such industrial feeding operations as principal causes of the growing antibiotic resistance in food-borne bacteria like *salmonella* and *campylobacter*. And as decisionmaking in the food chain grows ever more concentrated—confined behind fewer corporate doors—there may be other food safety issues that you won't even hear about, particularly in the burgeoning field of genetically modified organisms (GMOs). In reaction to growing public concern over GMOs, a coalition that ingenuously calls itself the "Alliance for Better Foods"—actually made up of large food retailers, food processors, biotech companies and corporate-financed farm organizations—has launched a $50 million public "educational" campaign, in addition to giving over $676,000 to U.S. lawmakers and political parties in 1999, to head off the mandatory labeling of such foods.

Perhaps most surprising, to people who have only casually followed the debate about small-farm values versus factory-farm "efficiency," is the fact that a wide body of evidence shows that small farms are actually more productive than large ones—by as much as 200 to 1,000 percent greater output per unit of area. How does this jive with the often-mentioned productivity advantages of large-scale mechanized operations? The answer is simply that those big-farm advantages are always calculated on the basis of how much of *one crop* the land will yield per acre. The greater productivity of a smaller, more complex farm, however, is calculated on the basis of how much food *overall* is produced per acre. The smaller farm can grow several crops utilizing different root depths, plant heights, or nutrients, on the same piece of land simultaneously. It is this "polyculture" that offers the small farm's productivity advantage.

To illustrate the difference between these two kinds of measurement, consider a large Midwestern corn farm. That farm may produce more corn per acre than a small farm in which the corn is grown as part of a polyculture that also includes beans, squash, potato, and "weeds" that serve as fodder. But in overall output, the polycrop—under close supervision by a knowledgeable farmer—produces much more food overall, whether you measure in weight, volume, bushels, calories, or dollars.

The inverse relationship between farm size and output can be attributed to the more efficient use of land, water, and other agricultural resources that small operations afford, including the efficiencies of intercropping various plants in the same field, planting multiple times during the year, targeting irrigation, and integrating crops and livestock. So in terms of converting inputs into outputs, society would be better off with small-scale farmers. And as population continues to grow in many nations, and the agricultural resources per person continue to shrink, a small farm structure for agriculture may be central to meeting future food needs.

Rebuilding Foodsheds

Look at the range of pressures squeezing farmers, and it's not hard to understand the growing desperation. The situation has become explosive, and if stabilizing the erosion of farm culture and ecology is now critical not just to farmers but to everyone who eats, there's still a very challenging question as to what strategy can work. The agribusiness giants are deeply entrenched now, and scattered protests could have as little effect on them as a mosquito bite on a tractor. The prospects for farmers gaining political strength on their own seem dim, as their numbers—at least in the industrial countries—continue to shrink.

A much greater hope for change may lie in a joining of forces between farmers and the much larger numbers of other segments of society that now see the dangers, to their own particular interests, of continued restructuring of the countryside. There are a couple of prominent models for such coalitions, in the constituencies that have joined forces to fight the Mississippi River Barge Capacity and Hidrovía Barge Capacity projects being pushed forward in the name of global soybean productivity.

The American group has brought together at least the following riverbedfellows:

- National environmental groups, including the Sierra Club and National Audubon Society, which are alarmed at the prospect of a public commons being damaged for the profit of a small commercial interest group;
- Farmers and farmer advocacy organizations, concerned about the inordinate power being wielded by the agribusiness oligopoly;
- Taxpayer groups outraged at the prospect of a corporate welfare payout that will drain more than $1 billion from public coffers;
- Hunters and fishermen worried about the loss of habitat;

- Biologists, ecologists, and birders concerned about the numerous threatened species of birds, fish, amphibians, and plants;
- Local-empowerment groups concerned about the impacts of economic globalization on communities;
- Agricultural economists concerned that the project will further entrench farmers in a dependence on the export of low-cost, bulk commodities, thereby missing valuable opportunities to keep money in the community through local milling, canning, baking, and processing.

A parallel coalition of environmental groups and farmer advocates has formed in the Southern hemisphere to resist the Hidrovía expansion. There too, the river campaign is part of a larger campaign to challenge the hegemony of industrial agriculture. For example, a coalition has formed around the Landless Workers Movement, a grassroots organization in Brazil that helps landless laborers to organize occupations of idle land belonging to wealthy landlords. This coalition includes 57 farm advocacy organizations based in 23 nations. It has also brought together environmental groups in Latin America concerned about the related ventures of logging and cattle

ranching favored by large landlords; the mayors of rural towns who appreciate the boost that farmers can give to local economies; and organizations working on social welfare in Brazil's cities, who see land occupation as an alternative to shantytowns.

The Mississippi and Hidrovía projects, huge as they are, still constitute only two of the hundreds of agro-industrial developments being challenged around the world. But the coalitions that have formed around them represent the kind of focused response that seems most likely to slow the juggernaut, in part because the solutions these coalitions propose are not vague or quixotic expressions of idealism, but are site-specific and practical. In the case of the alliance forming around the Mississippi River project, the coalition's work has included questioning the assumptions of the Corps of Engineers analysis, lobbying for stronger antitrust examination of agribusiness monopolies, and calling for modification of existing U.S. farm subsidies, which go disproportionately to large farmers. Environmental groups are working to re-establish a balance between use of the Mississippi as a barge mover and as an intact watershed. Sympathetic agricultural extensionists are promoting alternatives to the standard corn-soybean rotation, including certified organic crop production, which can simultaneously bring down

Past and Future: Connecting the Dots

Given the direction and speed of prevailing trends, how far can the decline in farmers go? The lead editorial in the September 13, 1999 issue of *Feedstuffs*, an agribusiness trade journal, notes that "Based on the best estimates of analysts, economists and other sources interviewed for this publication, American agriculture must now quickly consolidate all farmers and livestock producers into about 50 production systems… each with its own brands," in order to maintain competitiveness. Ostensibly, other nations will have to do the same in order to keep up.

To put that in perspective, consider that in traditional agriculture, each farm is an independent production system. In this map of Ireland's farms circa 1930, each dot represents 100 farms, so the country as a whole had many thousands of independent production systems. But if the *Feedstuffs* prognosis were to come to pass, this map would be reduced to a single dot. And even an identically keyed map of the much larger United States would show the country's agriculture reduced to just one dot.

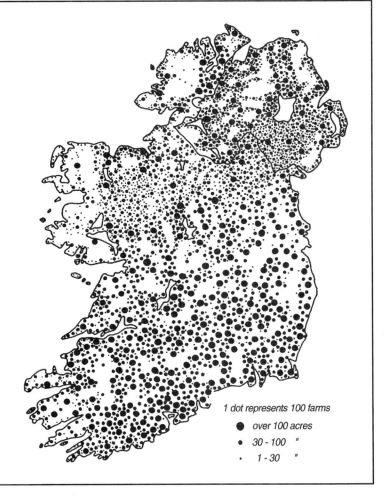

1 dot represents 100 farms

● *over 100 acres*

• *30 - 100 "*

· *1 - 30 "*

input costs and garner a premium for the final product, and reduce nitrogen pollution.

The United States and Brazil may have made costly mistakes in giving agribusiness such power to reshape the rivers and land to its own use. But the strategy of interlinked coalitions may be mobilizing in time to save much of the world's agricultural health before it is too late. Dave Brubaker, head of the Spira/GRACE Project on Industrial Animal Production at the Johns Hopkins University School of Public Health, sees these diverse coalitions as "the beginning of a revolution in the way we look at the food system, tying in food production with social welfare, human health, and the environment." Brubaker's project brings together public health officials focused on antibiotic overuse and water contamination resulting from hog waste; farmers and local communities who oppose the spread of new factory farms or want to close down existing ones; and a phalanx of natural allies with related campaigns, including animal rights activists, labor unions, religious groups, consumer rights activists, and environmental groups.

"As the circle of interested parties is drawn wider, the alliance ultimately shortens the distance between farmer and consumer," observes Mark Ritchie, president of the Institute for Agriculture and Trade Policy, a research and advocacy group often at the center of these partnerships. This closer proximity may prove critical to the ultimate sustainability of our food supply, since socially and ecologically sound buying habits are not just the passive *result* of changes in the way food is produced, but can actually be the most powerful *drivers of* these changes. The explosion of farmers' markets, community-supported agriculture, and other direct buying arrangements between farmers and consumers points to the growing numbers of nonfarmers who have already shifted their role in the food chain from that of choosing from the tens of thousands of food brands offered by a few dozen companies to bypassing such brands altogether. And, since many of the additives and processing steps that take up the bulk of the food dollar are simply the inevitable consequence of the ever-increasing time commercial food now spends in global transit and storage, this shortening of distance between grower and consumer will not only benefit the culture and ecology of farm communities. It will also give us access to much fresher, more flavorful, and more nutritious food. Luckily, as any food marketer can tell you, these characteristics aren't a hard sell.

Brian Halweil is a staff researcher at the Worldwatch Institute.

From *World Watch,* September/October 2000, pp. 12-28. © 2000 by the Worldwatch Institute (www.worldwatch.org). Reprinted by permission.

Safeguarding Our Water

Making Every Drop Count

We drink it, we generate electricity with it, we soak our crops with it. And we're stretching our supplies to the breaking point. Will we have enough clean water to satisfy all the world's needs?

by Peter H. Gleick

The history of human civilization is entwined with the history of the ways we have learned to manipulate water resources. The earliest agricultural communities emerged where crops could be cultivated with dependable rainfall and perennial rivers. Simple irrigation canals permitted greater crop production and longer growing seasons in dry areas. Five thousand years ago settlements in the Indus Valley were built with pipes for water supply and ditches for wastewater. Athens and Pompeii, like most Greco-Roman towns of their time, maintained elaborate systems for water supply and drainage.

As towns gradually expanded, water was brought from increasingly remote sources, leading to sophisticated engineering efforts, such as dams and aqueducts. At the height of the Roman Empire, nine major systems, with an innovative layout of pipes and well-built sewers, supplied the occupants of Rome with as much water per person as is provided in many parts of the industrial world today. During the industrial revolution and population explosion of the 19th and 20th centuries, the demand for tens of thousands of monumental engineering projects designed to control floods, protect clean water supplies, and provide water for irrigation and hydropower brought great benefits to hundreds of millions of people. Thanks to improved sewer systems, water-related diseases such as cholera and typhoid, once endemic throughout the world, have largely been conquered in the more industrial nations. Vast cities, incapable of surviving on their local resources, have bloomed in the desert with water brought from hundreds and even thousands of miles away. Food production has kept pace with soaring populations mainly because of the expansion of artificial irrigation systems that make possible the growth of 40 percent of the world's food. Nearly one fifth of all the electricity generated worldwide is produced by turbines spun by the power of falling water.

> **The water lost from Mexico City's leaky supply system is enough to meet the needs of a city the size of Rome**

Yet there is a dark side to this picture: despite our progress, half of the world's population still suffers with water services inferior to those available to the ancient Greeks and Romans. As the latest United Nations report on access to water reiterated in November of last year, more than one billion people lack access to clean drinking water; some two and a half billion do not have adequate sanitation services. Preventable water-related diseases kill an estimated 10,000 to 20,000 children every day, and the latest evidence suggests that we are falling behind in efforts to solve these problems. Massive cholera outbreaks appeared in the mid-1990s in Latin America, Africa and Asia. Millions of people in Bangladesh and India drink water contaminated with arsenic. And the surging populations throughout the developing world are intensifying the pressures on limited water supplies.

The effects of our water policies extend beyond jeopardizing human health. Tens of millions of people have been forced to move from their homes—often with little warning or compensation—to make way for the reservoirs behind dams. More than 20 percent of all freshwater fish species are now threatened or endangered because dams and water withdrawals have destroyed the free-flowing river ecosystems where they thrive. Certain irrigation practices degrade soil quality and reduce agricultural productivity, heralding a premature end to the green revolution. Groundwater aquifers are being pumped down faster than they are naturally replenished in parts of India, China, the U.S. and elsewhere. And disputes over shared water resources have led to violence and continue to raise local, national and even international tensions [*see box, "Continuing Conflict over Freshwater"*].

At the outset of the new millennium, however, the way resource planners think about water is beginning to change. The focus is slowly shifting back to the provision of basic human and environ-

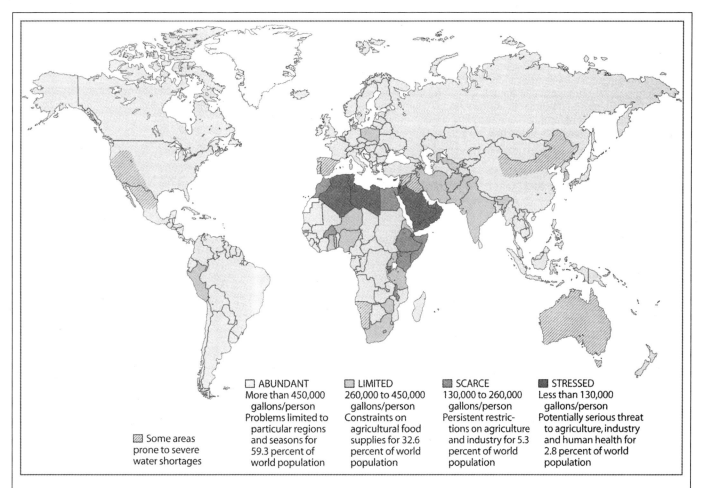

ABUNDANT
More than 450,000 gallons/person Problems limited to particular regions and seasons for 59.3 percent of world population

LIMITED
260,000 to 450,000 gallons/person Constraints on agricultural food supplies for 32.6 percent of world population

SCARCE
130,000 to 260,000 gallons/person Persistent restrictions on agriculture and industry for 5.3 percent of world population

STRESSED
Less than 130,000 gallons/person Potentially serious threat to agriculture, industry and human health for 2.8 percent of world population

Some areas prone to severe water shortages

Where the Water Will Be in 2025

Annual Global Water Withdrawals

Total Withdrawals

Per Capita Withdrawals

The total amount of water withdrawn globally from rivers, underground aquifers and other sources has increased ninefold since 1900 *(chart).* Water use per person has only doubled in that time, however, and it has even declined slightly in recent years. Despite this positive trend, some experts worry that improvements in water-use efficiency will fail to keep pace with projected population growth. Estimated annual water availability per person in 2025 *(map)* reveals that at least 40 percent of the world's 7.2 billion people may face serious problems with agriculture, industry or human health if they must rely solely on natural endowments of freshwater. Severe water shortages could also strike particular regions of water-rich countries, such as the U.S. and China.

People's access to water also depends on factors not reflected here, such as political and economic conditions, changing climate patterns and available technology. —*P.H.G.*

mental needs as the top priority—ensuring "some for all, instead of more for some," as put by Kader Asmal, former minister for water affairs and forestry in South Africa. To accomplish these goals and meet the demands of booming populations, some water experts now call for using existing infrastructure in smarter ways rather than building new facilities, which is increasingly considered the option of last, not first, resort. The challenges we face are to use the water we have more efficiently, to rethink our priorities for water use and to identify alternative supplies of this precious resource.

This shift in philosophy has not been universally accepted, and it comes with strong opposition from some established water organizations. Nevertheless, it may be the only way to address successfully the pressing problems of providing everyone with clean water to drink, adequate water to grow food and a life free from preventable water-related illness. History shows that although access to clean drinking water and sanitation ser-

vices cannot guarantee the survival of a civilization, civilizations most certainly cannot prosper without them.

Damage from Dams

Over the past 100 years, humankind has designed networks of canals, dams and reservoirs so extensive that the resulting redistribution of freshwater from one place to another and from one season to the next accounts for a small but

Continuing Conflict over Freshwater

Myths, legends and written histories reveal repeated controversy over freshwater resources since ancient times. Scrolls from Mesopotamia, for instance, indicate that the states of Umma and Lagash in the Middle East clashed over the control of irrigation canals some 4,500 years ago.

Throughout history, water has been used as a military and political goal, as a weapon of war and even as a military target. But disagreements most often arise from the fact that water resources are not neatly partitioned by the arbitrary political borders set by governments. Today nearly half of the land area of the world lies within international river basins, and the watersheds of 261 major rivers are shared by two or more countries. Overlapping claims to water resources have often provoked disputes, and in recent years local and regional conflicts have escalated over inequitable allocation and use of water resources.

A small sampling of water conflicts that occurred in the 20th century demonstrates that treaties and other international diplomacy can sometimes encourage opposing countries to cooperate—but not always before blood is shed. The risk of future strife cannot be ignored: disputes over water will become more common over the next several decades as competition for this scarce resource intensifies. —*P.H.G.*

U.S. 1924

Local farmers dynamite the Los Angeles aqueduct several times in an attempt to prevent diversions of water from the Owens Valley to Los Angeles.

India and Pakistan 1947 to 1960

Partitioning of British Indian awkwardly divides the waters of the Indus River valley between India and Pakistan. Competition over irrigation supplies incites numerous conflicts between the two nations; in one case, India stems the flow of water into Pakistani irrigation canals. After 12 years of World Bank–led negotiations, a 1960 treaty helps to resolve the discord.

Egypt and Sudan 1958

Egypt sends troops into contested territory between the two nations during sensitive negotiations concerning regional politics and water from the Nile. Signing of a Nile waters treaty in 1959 eases tensions.

Israel, Jordan and Syria 1960s and 1970s

Clashes over allocation, control and diversion of the Yarmouk and Jordan rivers continue to the present day.

South Africa 1990

A pro-apartheid council cuts off water to 50,000 black residents of Wesselton Township after protests against wretched sanitation and living conditions.

Iraq 1991

During the Persian Gulf War, Iraq destroys desalination plants in Kuwait. A United Nations coalition considers using the Ataturk Dam in Turkey to shut off the water flow of the Euphrates River to Iraq.

India 1991 to present

An estimated 50 people die in violence that continues to erupt between the Indian states of Karnataka and Tamil Nadu over the allocation of irrigation water from the Cauvery River, which flows from one state into the other.

Yugoslavia 1999

NATO shuts down water supplies in Belgrade and bombs bridges on the Danube River, disrupting navigation.

A comprehensive chronology of water-related conflicts can be found at www.worldwater.org/conflictIntro.htm

---

measurable change in the wobble of the earth as it spins. The statistics are staggering. Before 1900 only 40 reservoirs had been built with storage volumes greater than 25 billion gallons; today almost 3,000 reservoirs larger than this inundate 120 million acres of land and hold more than 1,500 cubic miles of water—as much as Lake Michigan and Lake Ontario combined. The more than 70,000 dams in the U.S. are capable of capturing and storing half of the annual river flow of the entire country.

In many nations, big dams and reservoirs were originally considered vital for national security, economic prosperity and agricultural survival. Until the late 1970s and early 1980s, few people took into account the environmental consequences of these massive projects. To-

day, however, the results are clear: dams have destroyed the ecosystems in and around countless rivers, lakes and streams. On the Columbia and Snake rivers in the northwestern U.S., 95 percent of the juvenile salmon trying to reach the ocean do not survive passage through the numerous dams and reservoirs that block their way. More than 900 dams on New England and European rivers block Atlantic salmon from their spawning grounds, and their populations have fallen to less than 1 percent of historical levels. Perhaps most infamously, the Aral Sea in central Asia is disappearing because water from the Amu Darya and Syr Darya rivers that once sustained it has been diverted to irrigate cotton. Twenty-four species of fish formerly found only in that sea are currently thought to be extinct.

As environmental awareness has heightened globally, the desire to protect—and even restore—some of these natural resources has grown. The earliest environmental advocacy groups in the U.S. mobilized against dams proposed in places such as Yosemite National Park in California and the Grand Canyon in Arizona. In the 1970s plans in the former Soviet Union to divert the flow of Siberian rivers away from the Arctic stimulated an unprecedented public outcry, helping to halt the projects. In many developing countries, grassroots opposition to the environmental and social costs of big water projects is becoming more and more effective. Villagers and community activists in India have encouraged a public debate over major dams. In China, where open disagreement with government policies is strongly discouraged, protest against the monumental Three Gorges Project has been unusually vocal and persistent.

Why should we raise all water to drinkable standards and then use it to flush toilets?

Until very recently, international financial organizations such as the World Bank, export-import banks and multilateral aid agencies subsidized or paid in full for dams or other water-related civil engineering projects—which often have price tags in the tens of billions of dollars. These organizations are slowly beginning to reduce or eliminate such subsidies, putting more of the financial burden on already strained national economies. Having seen so much ineffective development in the past—and having borne the associated costs (both monetary and otherwise) of that development—many governments are unwilling to pay for new structures to solve water shortages and other problems.

A handful of countries are even taking steps to remove some of the most egregious and damaging dams. For example, in 1998 and 1999 the Maisons-Rouges and Saint-Etienne-du-Vigan dams in the Loire River basin in France were demolished to help restore fisheries in the region. In 1999 the Edwards Dam, which was built in 1837 on the Kennebec River in Maine, was dismantled to open up an 18-mile stretch of the river for fish spawning; within months Atlantic salmon, American shad, river herring, striped bass, shortnose sturgeon, Atlantic Sturgeon, rainbow smelt and American eel had returned to the upper parts of the river. Altogether around 500 old, dangerous or environmentally harmful dams have been removed from U.S. rivers in the past few years.

Fortunately—and unexpectedly—the demand for water is not rising as rapidly as some predicted. As a result, the pressure to build new water infrastructures has diminished over the past two decades. Although population, industrial output and economic productivity have continued to soar in developed nations, the rate at which people withdraw water from aquifers, rivers and

lakes has slowed. And in a few parts of the world, demand has actually fallen.

Demand Is Down—But for How Long?

What explains this remarkable turn of events? Two factors: people have figured out how to use water more efficiently, and communities are rethinking their priorities for water use. Throughout the first three quarters of the 20th century, the quantity of freshwater consumed per person doubled on average; in the U.S., water withdrawals increased 10-fold while the population quadrupled. But since 1980 the amount of water consumed per person has actually decreased, thanks to a range of new technologies that help to conserve water in homes and industry. In 1965, for instance, Japan used approximately 13 million gallons of water to produce $1 million of commercial output; by 1989 this had dropped to 3.5 million gallons (even accounting for inflation)—almost a quadrupling of water productivity. In the U.S., water withdrawals have fallen by more than 20 percent from their peak in 1980.

As the world's population continues to grow, dams, aqueducts and other kinds of infrastructure will still have to be built, particularly in developing countries where basic human needs have not been met. But such projects must be built to higher standards and with more accountability to local people and their environment than in the past. And even in regions where new projects seem warranted, we must find ways to meet demands with fewer resources, minimum ecological disruption and less money.

The fastest and cheapest solution is to expand the productive and efficient use of water. In many countries, 30 percent of more of the domestic water supply never reaches its intended destinations, disappearing from leaky pipes, faulty equipment or poorly maintained distribution systems. The quantity of water that Mexico City's supply system loses is enough to meet the needs of a city the size of Rome, according to recent estimates. Even in more modern systems, losses of 10 to 20 percent are common.

When water does reach consumers, it is often used wastefully. In homes, most water is literally flushed away. Before 1990 most toilets in the U.S. drew about six gallons of water for each flush. In 1992 the U.S. Congress passed a na-

tional standard mandating that all new residential toilets be low-flow models that require only 1.6 gallons per flush—a 70 percent improvement with a single change in technology. It will take time to replace all older toilets with the newer, better ones. A number of cities, however, have found the water conservation made possible by the new technology to be so significant—and the cost of saving that water to be so low—that they have established programs to speed up the transition to low-flow toilets.

Even in the developing world, technologies such as more efficient toilets have a role to play. Because of the difficulty of finding new water sources for Mexico City, city officials launched a water conservation program that involved replacing 350,000 old toilets. The replacements have already saved enough water to supply an additional 250,000 residents. And numerous other options for both industrial and nonindustrial nations are available as well, including better leak detection, less wasteful washing machines, drip irrigation and water-conserving plants in outdoor landscaping.

The amount of water needed for industrial applications depends on two factors: the mix of goods and services demanded by society and the processes chosen to generate them. For instance, producing a ton of steel before World War II required 60 to 100 tons of water. Current technology can make a ton of steel with less than six tons of water. Replacing old technology with new techniques reduces water needs by a factor of 10. Producing a ton of aluminum, however, requires only one and a half tons of water. Replacing the use of steel with aluminum, as has been happening for years in the automobile industry, can further lower water use. And telecommuting from home can save the hundreds of gallons of water required to produce, deliver and sell a gallon of gasoline, even accounting for the water required to manufacture our computers.

The largest single consumer of water is agriculture—and this use is largely inefficient. Water is lost as it is distributed to farmers and applied to crops. Consequently, as much as half of all water diverted for agriculture never yields any food. Thus, even modest improvements in agricultural efficiency could free up huge quantities of water. Growing tomatoes with traditional irrigation systems may require 40 percent more water than growing tomatoes with drip systems.

Even our diets have an effect on our overall water needs. Growing a pound of corn can take between 100 and 250 gallons of water, depending on soil and climate conditions and irrigation methods. But growing the grain to produce a pound of beef can require between 2,000 and 8,500 gallons. We can conserve water not only by altering how we choose to grow our food but also by changing what we choose to eat.

Shifting where people use water can also lead to tremendous gains in efficiency. Supporting 100,000 high-tech California jobs requires some 250 million gallons of water a year; the same amount of water used in the agricultural sector sustains fewer than 10 jobs—a stunning difference. Similar figures apply in many other countries. Ultimately these disparities will lead to more and more pressure to transfer water from agricultural uses to other economic sectors. Unless the agricultural community embraces water conservation efforts, conflicts between farmers and urban water users will worsen.

The idea that a planet with a surface covered mostly by water could be facing a water shortage seems incredible. Yet 97 percent of the world's water is too salty for human consumption or crops, and much of the rest is out of reach in deep groundwater or in glaciers and ice caps. Not surprisingly, researchers have investigated techniques for dipping into the immense supply of water in the oceans. The technology to desalinate brackish water or saltwater is well developed, but it remains expensive and is currently an option only in wealthy but dry areas near the coast. Some regions, such as the Arabian Gulf, are highly dependent on desalination, but the process remains a minor contributor to overall water supplies, providing less than 0.2 percent of global withdrawals.

With the process of converting saltwater to freshwater so expensive, some companies have turned to another possibility: moving clean water in ships or even giant plastic bags from regions with an abundance of the resource to those places around the globe suffering from a lack of water. But this approach, too, may have serious economic and political constraints.

Rather than seeking new distant sources of water, smart planners are beginning to explore using alternative *kinds* of water to meet certain needs. Why should communities raise all water to drinkable standards and then use that expensive resource for flushing toilets or watering lawns? Most water ends up flowing down the drain after a single use, and developed countries spend billions of dollars to collect and treat this wastewater before dumping it into a river or the ocean. Meanwhile, in poorer countries, this water is often simply returned untreated to a river or lake where it may pose a threat to human health or the environment. Recently attention has begun to focus on reclaiming and reusing this water.

Wastewater can be treated to different levels suitable for use in a variety of applications, such as recharging groundwater aquifers, supplying industrial processes, irrigating certain crops or even augmenting potable supplies. In Windhoek, Namibia, for instance, residents have used treated wastewater since 1968 to supplement the city's potable water supply; in drought years, such water has constituted up to 30 percent of Windhoek's drinking water supply. Seventy percent of Israeli municipal wastewater is treated and reused, mainly for agricultural irrigation of nonfood crops. Efforts to capture, treat and reuse more wastewater are also under way in neighboring Jordan. By the mid-1990s residents of California relied on more than 160 billion gallons of reclaimed water annually for irrigating landscapes, golf courses and crops, recharging groundwater aquifers, supplying industrial processes and even flushing toilets.

New approaches to meeting water needs will not be easy to implement: economic and institutional structures still encourage the wasting of water and the destruction of ecosystems. Among the barriers to better water planning and use are inappropriately low water prices, inadequate information on new efficiency technologies, inequitable water allocations, and government subsidies for growing water-intensive crops in arid regions or building dams.

Part of the difficulty, however, also lies in the prevalence of old ideas among water planners. Addressing the world's basic water problems requires fundamental changes in how we think about water, and such changes are coming about slowly. Rather than trying endlessly to find enough water to meet hazy projections of future desires, it is time to find a way to meet our present and future needs with the water that is already available, while preserving the ecological cycles that are so integral to human well-being.

Further Information

THE WORLD'S WATER 1998–1999. Peter H. Gleick. Island Press, 1998.

INTERNATIONAL RIVER BASINS OF THE WORLD. Aaron T. Wolf et al. in *Water Resources Development*, Vol. 15, No. 4, pages 387–427; December 1999.

THE WORLD'S WATER 2000–2001. Peter H. Gleick, Island Press, 2000.

Information on the world's water resources can be found at www.worldwater.org

United Nations Environment Program Global Environment Monitoring System's Freshwater Quality Program can be found at www.cciw.ca/gems/

VISION 21: A SHARED VISION FOR HYGIENE, SANITATION AND WATER SUPPLY. Water Supply and Sanitation Collaborative Council. Available at www.wscc.org/vision21/docs/index.html

PETER H. GLEICK is director of the Pacific Institute for Studies in Development, Environment and Security, a nonprofit policy research think tank based in Oakland, Calif. Gleick cofounded the institute in 1987. He is considered one of the world's leading experts on freshwater problems, including sustainable use of water, water as it relates to climate change, and conflicts over shared water resources.

OCEANS
Are on the Critical List

The primary threats to the planet's seas—overfishing, habitat degradation, pollution, introduction of alien species, and climate change—are largely human-induced.

BY ANNE PLATT MCGINN

O CEANS FUNCTION as a source of food and fuel, a means of trade and commerce, and a base for cities and tourism. Worldwide, people obtain much of their animal protein from fish. Ocean-based deposits meet one-fourth of the world's annual oil and gas needs, and more than half of world trade travels by ship. More important than these economic figures, however, is the fact that humans depend on oceans for life itself. Harboring a greater variety of animal body types (phyla) than terrestrial systems and supplying more than half of the planet's ecological goods and services, the oceans play a commanding role in the Earth's balance of life.

Due to their large physical volume and density, oceans absorb, store, and transport vast quantities of heat, water, and nutrients. The oceans store about 1,000 times more heat than the atmosphere does, for example. Through processes such as evaporation and photosynthesis, marine systems and species help regulate the climate, maintain a livable atmosphere, convert solar energy into food, and break down natural wastes. The value of these "free" services far surpasses that of

ocean-based industries. Coral reefs alone, for instance, are estimated to be worth $375,000,000,000 annually by providing fish, medicines, tourism revenues, and coastal protection for more than 100 countries.

Despite the importance of healthy oceans to our economy and well-being, we have pushed the world's oceans perilously close to—and in some cases past—their natural limits. The warning signs are clear. The share of overexploited marine fish species jumped from almost none in 1950 to 35% in 1996, with an additional 25% nearing full exploitation. More than half of the world's coastlines and 60% of the coral reefs are threatened by human activities, including intensive coastal development, pollution, and overfishing.

In January, 1998, as the United Nations was launching the Year of the Ocean, more than 1,600 marine scientists, fishery biologists, conservationists, and oceanographers from across the globe issued a joint statement entitled "Troubled Waters." They agreed that the most pressing threats to ocean health are human-induced, in-

cluding species overexploitation, habitat degradation, pollution, introduction of alien species, and climate change. The impacts of these five threats are exacerbated by poorly planned commercial activities and coastal population growth.

Yet, many people still consider the oceans as not only inexhaustible, but immune to human interference. Because scientists just recently have begun to piece together how ocean systems work, society has yet to appreciate—much less protect—the wealth of oceans in its entirety. Indeed, current courses of action are rapidly undermining this wealth.

Nearly 1,000,000,000 people, predominantly in Asia, rely on fish for at least 30% of their animal protein supply. Most of these fish come from oceans, but with increasing frequency they are cultured on farms rather than captured in the wild. Aquaculture, based on the traditional Asian practice of raising fish in ponds, constitutes one of the fastest-growing sectors in world food production.

In addition to harvesting food from the sea, people have traditionally relied on oceans as a transportation route. Sea trade currently is dominated by multinational companies that are more influenced by the rise and fall of stock prices than by the tides and trade winds. Modern fishing trawlers, oil tankers, aircraft carriers, and container ships follow a path set by electronic beams, satellites, and computers.

Society derives a substantial portion of energy and fuel from the sea—a trend that was virtually unthinkable a century ago. In an age of falling trade barriers and mounting pressures on land-based resources, new ocean-based industries such as tidal and thermal energy production promise to become even more vital to the workings of the world economy. Having increased sixfold between 1955 and 1995, the volume of international trade is expected to triple again by 2020, according to the U.S. National Oceanographic and Atmospheric Administration, and 90% of it is expected to move by ocean.

In contrast to familiar fishing grounds and sea passageways, the depths of the ocean were long believed to be a vast wasteland that was inhospitable, if not completely devoid of life. Since the first deployment of submersibles in the 1930s and more advanced underwater acoustics and pressure chambers in the 1960s, scientific and commercial exploration has helped illuminate life in the deep sea and the geological history of the ancient ocean. Mining for sand, gravel, coral, and minerals (including sulfur and, most recently, petroleum) has taken place in shallow waters and continental shelves for decades, although offshore mining is severely restricted in some national waters.

Isolated, but highly concentrated, deep-sea deposits of manganese, gold, nickel, and copper, first discovered in the late 1970s, continue to tempt investors. These valuable nodules have proved technologically difficult and expensive to extract, given the extreme pressures and depths of their location. An international compromise on the deep seabed mining provisions of the Law of the Sea in 1994 has opened the way to some mining in international waters, but it appears unlikely to lead to much as long as mineral prices remain low, demand is largely met from the land, and the cost of underwater operations remains prohibitively high.

Perhaps more valuable than the mineral wealth in oceans are still-undiscovered living resources—new forms of life, potential medicines, and genetic material. For example, in 1997, medical researchers stumbled across a compound in dogfish that stops the spread of cancer by cutting off the blood supply to tumors. The promise of life-saving cures from marine species is gradually becoming a commercial reality for bioprospectors and pharmaceutical companies as anti-inflammatory and cancer drugs have been discovered and other leads are being pursued.

Tinkering with the ocean for the sake of shortsighted commercial development, whether for mineral wealth or medicine, warrants close scrutiny, however. Given how little we know—a mere 1.5% of the deep sea has been explored, let alone adequately inventoried—any development could be potentially irreversible in these unique environments. Although seabed mining is subjected to some degree of international oversight, prospecting for living biological resources is completely unregulated.

During the past 100 years, scientists who work both underwater and among marine fossils found high in mountains have shown that the tree of life has its evolutionary roots in the sea. For about 3,200,000,000 years, all life on Earth was marine. A complex and diverse food web slowly evolved from a fortuitous mix there of single-celled algae, bacteria, and several million trips around the sun. Life remained sea-bound until 245,000,000 years ago, when the atmosphere became oxygen-rich.

Thanks to several billion years' worth of trial and error, the oceans today are home to a variety of species that have no descendants on land. Thirty-two out of 33 animal life forms are represented in marine habitats. (Only insects are missing.) Fifteen of these are exclusively marine phyla, including those of comb jellies, peanut worms, and starfish. Five phyla, including that of sponges, live predominantly in salt water. Although, on an individual basis, marine species account for just nine percent of the 1,800,000 species described for the entire planet, there may be as many as 10,000,000 additional species in the sea that as yet have not been classified.

In addition to hosting a vast array of biological diversity, the marine environment performs such vital functions as oxygen production, nutrient recycling, storm protection, and climate regulation—services that often are taken for granted.

Marine biological activity is concentrated along the world's coastlines (where sunlit surface waters receive nutrients and sediments from land-based runoff, river deltas, and rainfall) and in upwelling systems (where cold, nutrient-rich, deep-water currents run up against continental margins). It provides 25% of the planet's primary biological productivity and an estimated 80–90% of the global commercial fish catch. It is estimated that coastal environments account for 38% of the goods and services provided by the Earth's ecosystems, while open oceans contribute an additional 25%. The value of all marine goods and services is estimated at 21 trillion dollars annually, 70% more than terrestrial systems come to.

Oceans are vital to both the chemical and biological balance of life. The same mechanism that created the present atmosphere—photosynthesis—continues to feed the marine food chain. Phytoplankton—tiny microscopic plants—take carbon dioxide (CO_2) from the atmosphere and convert it into oxygen and simple sugars, a form of carbon that can be consumed by marine animals. Other types of phytoplankton process nitrogen and sulfur, thereby helping the oceans function as a biological pump.

Although most organic carbon is consumed in the marine food web and eventually returned to the atmosphere via respiration, the unused balance rains down to the deep waters that make up the bulk of the ocean, where it is stored temporarily. Over the course of millions of years, these deposits have accumulated to the point that most of the world's organic carbon, approximately 15,000,000 gigatons (a gigaton equals 1,000,000,000 tons), is sequestered in marine sediments, compared with 4,000 gigatons in land-based reserves. On an annual basis, about one-third of the world's carbon emissions—around two gigatons—is taken up by oceans, an amount roughly equal to the uptake by land-based resources. If deforestation continues to diminish the ability of forests to absorb carbon dioxide, oceans are expected to play a more important role in regulating the planet's CO_2 budget in the future as human-induced emissions keep rising.

Perhaps no other example so vividly illustrates the connections between the oceans and the atmosphere than El Niño. Named after the Christ child because it usually appears in December, the El Niño Southern Oscillation takes place when trade winds and ocean surface currents in the eastern and central Pacific Ocean reverse direction. Scientists do not know what triggers the shift, but the aftermath is clear: Warm surface waters essentially pile up in the eastern Pacific and block deep, cold waters from upwelling, while a low pressure system hovers over South America, collecting heat and moisture that would otherwise be distributed at sea. This produces severe weather conditions in many parts of the world—increased precipitation, heavy flooding, drought, fire, and deep freezes—which, in turn, have enormous economic impact. During the 1997–98 El Niño, for instance, Argentina lost more than $3,000,000,000 in agricultural products due to these ocean-climate reactions, and Peru reported a 90% drop in anchovy harvests compared with the previous year.

A sea of problems

As noted earlier, the primary threats to oceans are largely human-induced and synergistic. Fishing, for example, has drastically altered the marine food web and underwater habitat areas. Meanwhile, the ocean's front line of defense—the coastal zone—is crumbling from years of degradation and fragmentation, while its waters have been treated as a waste receptacle for generations. The combination of overexploitation, the loss of buffer areas, and a rising tide of pollution has suffocated marine life and the livelihoods based on it in some areas. Upsetting the marine ecosystem in these ways has, in turn, given the upper hand to invasive species and changes in climate.

Overfishing poses a serious biological threat to ocean health. The resulting reductions in the genetic diversity of the spawning populations make it more difficult for the species to adapt to future environmental changes. The orange roughy, for instance, may have been fished down to the point where future recovery is impossible. Moreover, declines in one species can alter predator-prey relations and leave ecosystems vulnerable to invasive species. The overharvesting of triggerfish and pufferfish for souvenirs on coral reefs in the Caribbean has sapped the health of the entire reef chain. As these fish declined, populations of their prey—sea urchins—exploded, damaging the coral by grazing on the protective layers of algae and hurting the local reef-diving industry.

These trends have enormous social consequences as well. The welfare of more than 200,000,000 people around the world who depend on fishing for their income and food security is severely threatened. As the fish disappear, so do the coastal communities that depend on fishing for their way of life. Subsistence and small-scale fishers, who catch nearly half of the world's fish, suffer the greatest losses as they cannot afford to compete with large-scale vessels or changing technology. Furthermore, the health of more than 1,000,000,000 poor consumers who depend on minimal quantities of fish to constitute their diets is at risk as an ever-growing share of fish—83% by value—continues to be exported to industrial countries each year.

Despite a steadily growing human appetite for fish, large quantities are wasted each year because the fish are undersized or a nonmarketable sex or species, or because a fisher does not have a permit to catch them and must therefore throw them out. The United Nations' Food and Agricultural Organization estimates that discards of fish alone—not counting marine mammals, seabirds, and turtles—total 20,000,000 tons, equivalent to one-fourth of the annual marine catch. Many of these fish do not survive the process of getting entangled in gear, being brought on board, and then tossed back to sea.

Another threat to habitat areas stems from trawling, with nets and chains dragged across vast areas of mud, rocks, gravel, and sand, essentially sweeping everything in the vicinity. By recent estimates, all the ocean's continental shelves are trawled by fishers at least once every two years, with some areas hit several times a season. Considered a major cause of habitat degradation, trawling disturbs bottom-dwelling communities as well as localized species diversity and food supplies.

The conditions that make coastal areas so productive for fish—proximity to nutrient flows and tidal mixing and their place at the crossroads between land and water—also make them vulnerable to human assault. Today, nearly 40% of the world lives within 60 miles of a coastline. As more people move to coastal areas and further stress the seams between land and sea, coastal ecosystems are losing ground.

Human activities on land cause a large portion of offshore contamination. An estimated 44% of marine pollution comes from land-based pathways, flowing down rivers into tidal estuaries, where it bleeds out to sea; an additional 33% is airborne pollution that is carried by winds and deposited far off shore. From nutrient-rich sediments, fertilizers, and human waste to toxic heavy metals and synthetic chemicals, the outfall from human society ends up circulating in the fluid and turbulent seas.

Excessive nutrient loading has left some coastal systems looking visibly sick. Seen from an airplane, the surface waters of Manila Bay in the Philippines resemble green soup due to dense carpets of algae. Nitrogen and phosphorus are necessary for life and, in limited quantities, can help boost plant productivity, but too much of a good thing can be bad. Excessive nutrients build up and create conditions that are conducive to outbreaks of dense algae blooms, also known as "red tides." The blooms block sunlight, absorb dissolved oxygen, and disrupt food-web dynamics. Large portions of the Gulf of Mexico are now considered a biological "dead zone" due to algal blooms.

The frequency and severity of red tides has increased in the past couple of decades. Some experts link the recent outbreaks to increasing loads of nitrogen and phosphorus from nutrient-rich wastewater and agricultural runoff in poorly flushed waters.

Organochlorines, a fairly recent addition to the marine environment, are proving to have pernicious effects. Synthetic organic compounds such as chlordane, DDT, and PCBs are used for everything from electrical wiring to pesticides. Indeed, one reason they are so difficult to control is that they are ubiquitous. The organic form of tin (tributyltin), for example, is used in most of the world's marine paints to keep barnacles, seaweed, and other organisms from clinging to ships. Once the paint is dissolved in the water, it accumulates in mollusks, scallops, and rock crabs, which are consumed by fish and marine mammals.

As part of a larger group of chemicals known collectively as persistent organic pollutants (POPs), these compounds are difficult to control because they do not degrade easily. Highly volatile in warm temperatures, POPs tend to circulate toward colder environments where the conditions are more stable, such as the Arctic Circle. Moreover, they do not dissolve in water, but are lipid-soluble, meaning that they accumulate in the fat tissues of fish that are then consumed by predators at a more concentrated level.

POPs have been implicated in a wide range of animal and human health problems—from suppression of immune systems, leading to higher risk of illness and infection, to disruption of the endocrine system, which is linked to birth defects and infertility. Their continued use in many parts of the world poses a threat to marine life and fish consumers everywhere.

Because marine species are extremely sensitive to fluctuations in temperature, changes in climate and atmospheric conditions pose high risks to them. Recent evidence shows that the thinning ozone layer above Antarctica has allowed more ultraviolet-B radiation to penetrate the waters. This has affected photosynthesis and the growth of phytoplankton and macroalgae. The effects are not limited to the base of the food chain. By striking aquatic species during their most vulnerable stages of life and reducing their food supply at the same time, increases in UV-B could have devastating impacts on world fisheries production.

Because higher temperatures cause water to expand, a warming world may trigger more frequent and damaging storms. Ironically, the coastal barriers, seawalls, jetties, and levees that are designed to protect human settlements from storm surges likely exacerbate the problem of coastal erosion and instability, as they create deeper inshore troughs that boost wave intensity and sustain winds.

Depending on the rate and extent of warming, global sea levels may rise as much as three feet by 2100—up to five times as much as during the last century. Such a rise would flood most of New York City, including the entire subway system and all three major airports. Economic damages and losses could cost the global economy as much as $970,000,000,000 in 2100, according to the Organisation for Economic Co-operation and Development. The human costs would be unimaginable, especially in the low-lying, densely populated river deltas of Bangladesh, China, Egypt, and Nigeria.

These damages could be just the tip of the iceberg. Warmer temperatures would likely accelerate polar ice cap melting and could boost this rising wave by several feet. Just four years after a large portion of Antarctica melted, another large ice sheet fell off into the Southern Sea in February, 1998, rekindling fears that global warming could ignite a massive thaw that would flood coastal areas worldwide. Because oceans play such a vital role in regulating the Earth's climate and maintaining a healthy planet, minor changes in ocean circulation or in its temperature or chemical balance could have repercussions many orders of magnitude larger than the sum of human-induced wounds.

While understanding past climatic fluctuations and predicting future developments are an ongoing challenge for scientists, there is clear and growing evidence of the overuse—indeed abuse—that many marine ecosystems and species are suffering from direct human actions. The situation is probably much worse, for many sources of danger are still unknown or poorly monitored. The need to take preventive and decisive action on behalf of oceans is more important than ever.

Saving the oceans

Scientists' calls for precaution and protective measures are largely ignored by policymakers, who focus on enhancing commerce, trade, and market supply and look to extract as much from the sea as possible, with little regard for the effects on marine species or habitats. Overcoming the interest groups that favor the status quo will require engaging all potential stakeholders and reformulating the governance equation to incorporate the stewardship obligations that come with the privilege of use.

Fortunately for the planet, a new sea ethic is emerging. From tighter dumping regulations to recent international agreements, policymakers have made initial progress toward the goal of cleaning up humans' act. Still, much more is needed in the way of public education to build political support for marine conservation.

To boost ongoing efforts, two key principles are important. First, any dividing up of the waters should be based on equity, fairness, and need as determined by dependence on the resource and the best available scientific knowledge, not simply on economic might and political pressure. In a similar vein, resource users should be responsible for their actions, with decision-making and accountability shared by stakeholders and government officials. Second, given the uncertainty in scientific knowledge and management capabilities, it is necessary to err on the side of caution and take a precautionary approach.

Replanting mangroves and constructing artificial reefs are two concrete steps that help some fish stocks rebound quickly while letting people witness firsthand the results of their labors. Once they see the immediate payoff of their work, they are more likely to stay involved in longer-term protection efforts, such as marine sanctuaries, which involve removing an area from use entirely.

Marine protected areas are an important tool to help marine scientists and resource planners incorporate an ecologically based approach to oceans protection. By limiting accessibility and easing pressures on the resource, these areas allow stocks to rebound and profits to return. Globally, more than 1,300 marine and coastal sites have some form of protection, but most lack effective on-the-ground management.

Meanwhile, efforts to establish marine refuges and parks lag far behind similar efforts on land. The World Heritage Convention, which identifies and protects areas of special significance to mankind, identifies just 31 sites that include either a marine or a coastal component, out of a total of 522. John Waugh, Senior Program Officer of the World Conservation Union-U.S., and others argue that the World Heritage List could be extended to a number of marine hotspots and should include representative areas of the continental shelf, the deep sea, and the open ocean. Setting these and other areas aside as off-limits to commercial development can help advance scientific understanding of marine systems and provide refuge for threatened species.

To address the need for better data, coral reef scientists have enlisted the help of recreational scuba divers. Sport divers who volunteer to collect data are given basic training to identify and survey fish and coral species and conduct rudimentary site assessments. The data then are compiled and put into a global inventory that policymakers use to monitor trends and to target intervention. More efforts like these— that engage the help of concerned individuals and volunteers—could help overcome funding and data deficiencies and build greater public awareness of the problems plaguing the world's oceans.

Promoting sustainable ocean use also means shifting demand away from environmentally damaging products and extraction techniques. To this end, market forces, such as charging consumers more for particular fish and introducing industry codes of conduct, can be helpful. In April, 1996, the World Wide Fund for Nature teamed up with one of the world's largest manufacturers of seafood products, Anglo-Dutch Unilever, to create economic incentives for sustainable fishing. Implemented through an independent Marine Stewardship Council, fisheries products that are harvested in a sustainable manner will qualify for an ecolabel. Similar efforts could help convince industries to curb wasteful practices and generate greater consumer awareness of the need to choose products carefully.

Away from public oversight, companies engaged in shipping, oil and gas extraction, deep-sea mining, bioprospecting, and tidal and thermal energy represent a coalition of special interests whose activities help determine the fate of the oceans. It is crucial to get representatives of these industries engaged in implementing a new ocean charter that supports sustainable use.

Their practices not only affect the health of oceans, they help decide the pace of a transition toward a more sustainable energy economy, which, in turn, affects the balance between climate and oceans.

Making trade data and industry information publicly available is an important way to build industry credibility and ensure some degree of public oversight. While regulations are an important component of environmental protection, pressure from consumers, watchdog groups, and conscientious business leaders can help develop voluntary codes of action and standard industry practices that can move industrial sectors toward cleaner and greener operations. Economic incentives targeted to particular industries, such as low-interest loans for thermal projects, can aid companies in making a quicker transition to sustainable practices.

The fact that oceans are so central to the global economy and to human and planetary health may be the strongest motivation for protective action. Although the range of assaults and threats to ocean health are broad, the benefits that oceans provide are invaluable and shared by all. These huge bodies of water represent an enormous opportunity to forge a new system of cooperative, international governance based on shared resources and common interests. Achieving these far-reaching goals, however, begins with the technically simple, but politically daunting, task of overcoming several thousand years' worth of ingrained behavior. It requires seeing oceans not as an economic frontier for exploitation, but as a scientific frontier for exploration and a biological frontier for careful use.

For generations, oceans have drawn people to their shores for a glimpse of the horizon, a sense of scale, and awe at nature's might. Today, oceans offer careful observers a different kind of awe—a warning that humans' impacts on the Earth are exceeding natural bounds and in danger of disrupting life. Protection efforts already lag far behind what is needed. How humans choose to react will determine the future of the planet. Oceans are not simply one more system under pressure—they are critical to man's survival. As Carl Safina writes in *The Song for the Blue Ocean*, "we need the oceans more than they need us."

Anne Platt McGinn *is a senior researcher, Worldwatch Institute, Washington, D.C.*

From *USA Today Magazine*, January 2000, pp. 32–35. © 2000 by the Society for the Advancement of Education. Reprinted by permission.

SPECIAL REPORT • GLOBAL WARMING

FEELING THE HEAT
LIFE IN THE GREENHOUSE

Except for nuclear war or a collision with an asteroid, no force has more potential to damage our planet's web of life than global warming. It's a "serious" issue, the White House admits, but nonetheless George W. Bush has decided to abandon the 1997 Kyoto treaty to combat climate change—an agreement the U.S. signed but the new President believes is fatally flawed. His dismissal last week of almost nine years of international negotiations sparked protests around the world and a face-to-face disagreement with German Chancellor Gerhard Schröder. Our special report examines the signs of global warming that are already apparent, the possible consequences for our future, what we can do about the threat and why we have failed to take action so far.

By MICHAEL D. LEMONICK

There is no such thing as normal weather. The average daytime high temperature for New York City this week should be 57°F, but on any given day the mercury will almost certainly fall short of that mark or overshoot it, perhaps by a lot. Manhattan thermometers can reach 65° in January every so often and plunge to 50° in July. And seasons are rarely normal. Winter snowfall and summer heat waves beat the average some years and fail to reach it in others. It's tough to pick out overall changes in climate in the face of these natural fluctuations. An unusually warm year, for example, or

even three in a row don't necessarily signal a general trend.

Yet the earth's climate does change. Ice ages have frosted the planet for tens of thousands of years at a stretch, and periods of warmth have pushed the tropics well into what is now the temperate zone. But given the normal year-to-year variations, the only reliable signal that such changes may be in the works is a long-term shift in worldwide temperature.

And that is precisely what's happening. A decade ago, the idea that the planet was warming up as a result of human activity was largely

theoretical. We knew that since the Industrial Revolution began in the 18th century, factories and power plants and automobiles and farms have been loading the atmosphere with heat-trapping gases, including carbon dioxide and methane. But evidence that the climate was actually getting hotter was still murky.

Not anymore. As an authoritative report issued a few weeks ago by the United Nations-sponsored Intergovernmental Panel on Climate Change makes plain, the trend toward a warmer world has unquestionably begun. Worldwide temperatures have climbed more than 1°F over the

MAKING THE CASE THAT OUR CLIMATE IS CHANGING

From melting glaciers to rising oceans, the signs are everywhere. Global warming can't be blamed for any particular heat wave, drought or deluge, but scientists say a hotter world will make such extreme weather more frequent—and deadly.

EXHIBIT A

Thinning Ice

ANTARCTICA, home to these Adélie penguins, is heating up. The annual melt season has increased up to three weeks in 20 years.

MOUNT KILIMANJARO has lost 75% of its ice cap since 1912. The ice on Africa's tallest peak could vanish entirely within 15 years.

LAKE BAIKAL in eastern Siberia now feezes for the winter 1.1 days later than it did a century ago.

VENEZUELAN mountaintops had six glaciers in 1972. Today only two remain.

EXHIBIT B

Hotter Times

TEMPERATURES SIZZLED from Kansas to New England last May.

CROPS WITHERED and Dallas temperatures topped 100°F for 29 days straight in a Texas hot spell that struck during the summer of 1998.

INDIA'S WORST heat shock in 50 years killed more than 2,500 people in May 1998.

CHERRY BLOSSOMS in Washington bloom seven days earlier in the spring than they did in 1970.

EXHIBIT C

Wild Weather

HEAVY RAINS in England and Wales made last fall Britain's wettest three-month period on record.

FIRES due to dry conditions and record-breaking heat consumed 20% of Samos Island, Greece, last July.

FLOODS along the Ohio River in March 1997 caused 30 deaths and at least $500 million in property damage.

HURRICAN FLOYD brought flooding rains and 130-m.p.h. winds through the Atlantic seabord in September 1999, killing 77 people and leaving thousands homeless.

EXHIBIT D

Nature's Pain

PACIFIC SALMON populations fell sharply in 1997 and 1998, when local ocean temperatures rose 6°F.

POLAR BEARS in Hudson Bay are having fewer cubs, possibly as a result of earlier spring ice breakup.

CORAL REEFS suffer from the loss of algae that color and nourish them. The process, called bleaching, is caused by warmer oceans.

DISEASES like dengue fever are expanding their reach northward in the U.S.

BUTTERFLIES are relocating to higher latitudes. The Edith's Checkerspot butterfly of western North America has moved almost 60 miles north in 100 years.

---

past century, and the 1990s were the hottest decade on record. After analyzing data going back at least two decades on everything from air and ocean temperatures to the spread and retreat of wildlife, the IPCC asserts that this slow but steady warming has had an impact on no fewer than 420 physical processes and animal and plant species on all continents.

Glaciers, including the legendary snows of Kilimanjaro, are disappearing from mountaintops around the globe. Coral reefs are dying off as the seas get too warm for comfort. Drought is the norm in parts of Asia and Africa. El Niño events, which trigger devastating weather in the eastern Pacific, are more frequent.

The Arctic permafrost is starting to melt. Lakes and rivers in colder climates are freezing later and thawing earlier each year. Plants and animals are shifting their ranges poleward and to higher altitudes, and migration patterns for animals as diverse as polar bears, butterflies and beluga whales are being disrupted.

Faced with these hard facts, scientists no longer doubt that global warming is happening, and almost nobody questions the fact that humans are at least partly responsible. Nor are the changes over. Already, humans have increased the concentration of carbon dioxide, the most abundant heat-trapping gas in the atmosphere, to 30% above pre-industrial levels—and each year the rate of increase gets faster. The obvious conclusion: temperatures will keep going up.

Unfortunately, they may be rising faster and heading higher than anyone expected. By 2100, says the IPCC, average temperatures will increase between 2.5°F and 10.4°F—more than 50% higher than predictions of just a half-decade ago. That may not seem like much, but consider that it took only a 9°F shift to end the last ice age. Even at the low end, the changes could be problematic enough, with storms getting more frequent and intense, droughts more pronounced, coastal areas ever more severely eroded by rising seas, rainfall scarcer on agricultural land and ecosystems thrown out of balance.

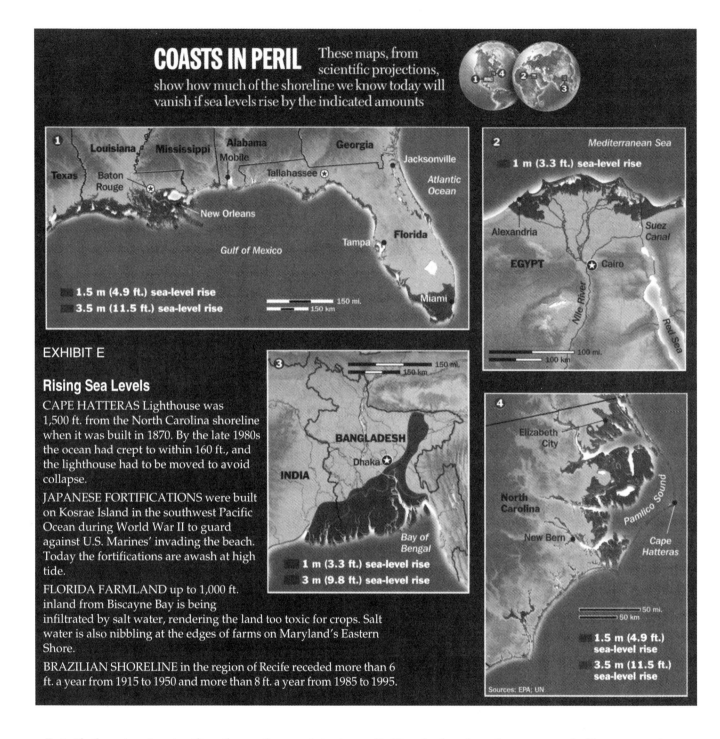

COASTS IN PERIL These maps, from scientific projections, show how much of the shoreline we know today will vanish if sea levels rise by the indicated amounts

EXHIBIT E

Rising Sea Levels

CAPE HATTERAS Lighthouse was 1,500 ft. from the North Carolina shoreline when it was built in 1870. By the late 1980s the ocean had crept to within 160 ft., and the lighthouse had to be moved to avoid collapse.

JAPANESE FORTIFICATIONS were built on Kosrae Island in the southwest Pacific Ocean during World War II to guard against U.S. Marines' invading the beach. Today the fortifications are awash at high tide.

FLORIDA FARMLAND up to 1,000 ft. inland from Biscayne Bay is being infiltrated by salt water, rendering the land too toxic for crops. Salt water is also nibbling at the edges of farms on Maryland's Eastern Shore.

BRAZILIAN SHORELINE in the region of Recife receded more than 6 ft. a year from 1915 to 1950 and more than 8 ft. a year from 1985 to 1995.

But if the rise is significantly larger, the result could be disastrous. With seas rising as much as 3 ft., enormous areas of densely populated land—coastal Florida, much of Louisiana, the Nile Delta, the Maldives, Bangladesh—would become uninhabitable. Entire climatic zones might shift dramatically, making central Canada look more like central Illinois, Georgia more like Guatemala. Agriculture would be thrown into turmoil. Hundreds of millions of people would have to migrate out of unlivable regions.

Public health could suffer. Rising seas would contaminate water supplies with salt. Higher levels of urban ozone, the result of stronger sunlight and warmer temperatures, could worsen respiratory illnesses. More frequent hot spells could lead to a rise in heat-related deaths. Warmer temperatures could widen the range of disease-carrying rodents and bugs, such as mosquitoes and ticks, increasing the incidence of dengue fever, malaria, encephalitis, Lyme disease and other afflictions. Worst of all, this increase in temperatures is happening at a pace that outstrips anything the earth has seen in the past 100 million years. Humans will have a hard enough time adjusting, especially in poorer countries, but for wildlife, the changes could be devastating.

Like any other area of science, the case for human-induced global warming has uncertainties—and like many pro-business lobbyists, President Bush has proclaimed those uncertainties a reason to study the problem further rather than act. But while the evidence is circumstantial, it is powerful, thanks to the IPCC's painstaking research. The U.N.-sponsored group was organized in the late 1980s. Its mission: to sift through climate-related studies from a dozen different fields and integrate them into a coherent picture. "It isn't just the work of a few green people," says Sir John Houghton, one of the early leaders who at the time ran the British Meteorological Office. "The IPCC scientists come from a wide range of backgrounds and countries."

Measuring the warming that has already taken place is relatively simple; the trick is unraveling the causes and projecting what will happen over the next century. To do that, IPCC scientists fed a wide range of scenarios involving varying estimates of population and economic growth, changes in technology and other factors into computers. That process gave them about 35 estimates, ranging from 6 billion to 35 billion tons, of how much excess carbon dioxide will enter the atmosphere.

Then they loaded those estimates into the even larger, more powerful computer programs that attempt to model the planet's climate. Because no one climate model is considered definitive, they used seven different versions, which yielded 235 independent predictions of global temperature increase. That's where the range of 2.5°F to 10.4°F (1.4°C to 5.8°C) comes from.

The computer models were criticized in the past largely because the climate is so complex that the limited hardware and software of even a half-decade ago couldn't do an adequate simulation. Today's climate models, however, are able to take into account the heat-trapping effects not just of CO_2 but also of other greenhouse gases, including methane. They can also factor in natural variations in the sun's energy and the effect of substances like dust from volcanic eruptions and particulate matter spewed from smokestacks.

That is one reason the latest IPCC predictions for temperature increase are higher than they were five years ago. Back in the mid-1990s, climate models didn't include the effects of the El Chichon and Mount Pinatubo volcanic eruptions, which threw enough dust into the air to block out some sunlight and slow down the rate of warming. That effect has dissipated, and the heating should start to accelerate. Moreover, the IPCC noted, many countries have begun to reduce their emissions of sulfur dioxide in order to fight acid rain. But sulfur dioxide particles, too, reflect sunlight; without this shield, temperatures should go up even faster.

The models still aren't perfect. One major flaw, agree critics and champions alike, is that they don't adequately account for clouds. In a warmer world, more water will evaporate from the oceans and presumably form more clouds. If they are billowy cumulus clouds, they will tend to shade the planet and slow down warming; if they are high, feathery cirrus clouds, they will trap even more heat.

Research by M.I.T. atmospheric scientist Richard Lindzen suggests that warming will tend to make cirrus clouds go away. Another critic, John Christy of the University of Alabama in Huntsville, says that while the models reproduce the current climate in a general way, they fail to get right the amount of warming at different levels in the atmosphere. Neither Lindzen nor Christy (both IPCC authors) doubts, however, that humans are influencing the climate. But they question how much—and how high temperatures will go. Both scientists are distressed that only the most extreme scenarios, based on huge population growth and the maximum use of dirty fuels like coal, have made headlines.

It won't take the greatest extremes of warming to make life uncomfortable for large numbers of people. Even slightly higher temperatures in regions that are already drought- or flood-prone would exacerbate those conditions. In temperate zones, warmth and increased CO_2 would make some crops flourish—at first. But beyond 3° of warming, says Bill Easterling, a professor of geography and agronomy at Penn State and a lead author of the IPCC report, "there would be a dramatic turning point. U.S. crop yields would start to decline rapidly." In the tropics, where crops are already at the limit of their temperature range, the decrease would start right away.

Even if temperatures rise only moderately, some scientists fear, the climate would reach a "tipping point"—a point at which even a tiny additional increase would throw the system into violent change. If peat bogs and Arctic permafrost warm enough to start releasing the methane stored within them, for example, that potent greenhouse gas would suddenly accelerate the heat-trapping process.

By contrast, if melting ice caps dilute the salt content of the sea, major ocean currents like the Gulf Stream could slow or even stop, and so would their warming effects on northern regions. More snowfall reflecting more sunlight back into space could actually cause a net cooling. Global warming could, paradoxically, throw the planet into another ice age.

Even if such a tipping point doesn't materialize, the more drastic effects of global warming might be only postponed rather than avoided. The IPCC's calculations end with the year 2100, but the warming won't. World Bank chief scientist, Robert Watson, currently serving as IPCC chair, points out that the CO_2 entering the atmosphere today will be there for a century. Says Watson: "If we stabilize (CO_2 emissions) now, the concentration will continue to go up for hundreds of years. Temperatures will rise over that time."

That could be truly catastrophic. The ongoing disruption of ecosystems and weather patterns would be bad enough. But if temperatures reach the IPCC's worst-case levels and stay there for as long as 1,000 years, says Michael Oppenheimer, chief scientist at Environmental Defense, vast ice sheets in Greenland and Antarctica could melt, raising sea level more than 30 ft. Florida would be history, and every city on the U.S. Eastern seaboard would be inundated.

In the short run, there's not much chance of halting global warming, not even if every nation in the world ratifies the Kyoto Protocol tomorrow. The treaty doesn't require reductions in carbon dioxide emissions until 2008. By that time, a great deal of damage will already have been done. But we can slow things down. If action today can keep the climate from eventually reaching an unstable tipping point or can finally begin to reverse the warming trend a century from now, the effort would hardly be futile. Humanity embarked unknowingly on the dangerous experiment of tinkering with the climate of our planet. Now that we know what we're doing, it would be utterly foolish to continue.

Reported by David Bjerklie,
Robert H. Boyle and
Andrea Dorfman/New York and
Dick Thompson/Washington

The Human Impact on Climate

How much of a disruption do we cause? The much-awaited answer could be ours by 2050, but only if nations of the world commit to long-term climate monitoring now

by Thomas R. Karl and Kevin E. Trenberth

The balance of evidence suggests a discernible human influence on global climate." With these carefully chosen words, the Intergovernmental Panel on Climate Change (jointly supported by the World Meteorological Organization and the United Nations Environmental Program) recognized in 1995 that human beings are far from inconsequential when it comes to the health of the planet. What the panel did not spell out—and what scientists and politicians dispute fiercely—is exactly when, where and how much that influence has and will be felt.

So far the climate changes thought to relate to human endeavors have been relatively modest. But various projections suggest that the degree of change will become dramatic by the middle of the 21st century, exceeding anything seen in nature during the past 10,000 years. Although some regions may benefit for a time, overall the alterations are expected to be disruptive or even severe. If researchers could clarify the extent to which specific activities influence climate, they would be in a much better position to suggest strategies for ameliorating the worst disturbances. Is such quantification possi-

ble? We think it is and that it can be achieved by the year 2050—but only if that goal remains an international priority.

Despite uncertainties about details of climate change, our activities clearly affect the atmosphere in several troubling ways. Burning of fossil fuels in power plants and automobiles ejects particles and gases that alter the composition of the atmosphere. Visible pollution from sulfur-rich fuels includes micron-size particles called aerosols, which often cast a milky haze in the sky. These aerosols temporarily cool the atmosphere because they reflect some of the sun's rays back to space, but they stay in the air for only a few days before rain sweeps them to the planet's surface. Certain invisible gases deliver a more lasting impact. Carbon dioxide remains in the atmosphere for a century or more. Worse yet, such greenhouse gases trap some of the solar radiation that the planet would otherwise radiate back to space, creating a "blanket" that insulates and warms the lower atmosphere.

Indisputably, fossil-fuel emissions alone have increased carbon dioxide concentrations in the atmo-

sphere by about 30 percent since the start of the Industrial Revolution in the late 1700s. Oceans and plants help to offset this flux by scrubbing some of the gas out of the air over time, yet carbon dioxide concentrations continue to grow. The inevitable result of pumping the sky full of greenhouse gases is global warming. Indeed, most scientists agree that the earth's mean temperature has risen at least 0.6 degree Celsius (more than one degree Fahrenheit) over the past 120 years, much of it caused by the burning of fossil fuels.

The global warming that results from the greenhouse effect dries the planet by evaporating moisture from oceans, soils and plants. Additional moisture in the atmosphere provides a swollen reservoir of water that is tapped by all precipitating weather systems, be they tropical storms, thundershowers, snowstorms or frontal systems. This enhanced water cycle brings on more severe droughts in dry areas and leads to strikingly heavy rain or snowfall in wet regions, which heightens the risk of flooding. Such weather patterns have burdened many parts of the world in recent decades.

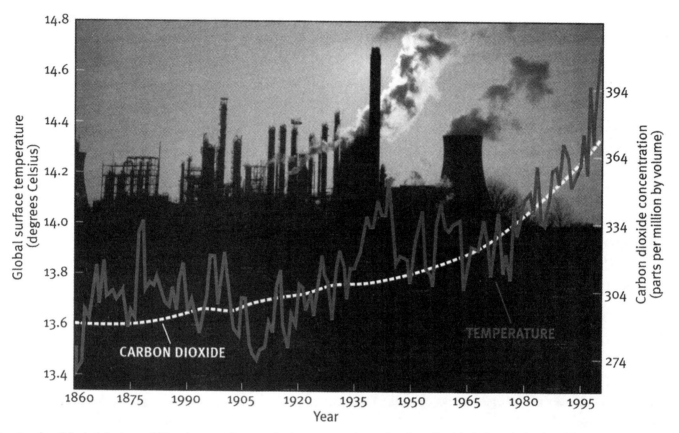

Burning fossil fuels (photograph) has increased atmospheric concentrations of carbon dioxide (white dashes) and has contributed to a rise in global surface temperatures during the past 140 years (gray line).

Chinch Gryniewicz / Corbis; LAURIE GRACE (Graph)

Human activities aside from burning fossil fuels can also wreak havoc on the climate system. For instance, the conversion of forests to farmland eliminates trees that would otherwise absorb carbon from the atmosphere and reduce the greenhouse effect. Fewer trees also mean greater rainfall runoff, thereby increasing the risk of floods.

It is one thing to have a sense of the factors that can bring about climate change. It is another to know how the human activity in any given place will affect the local and global climate. To achieve that aim, those of us who are concerned about the human influence on climate will have to be able to construct more accurate climate models than have ever been designed before. We will therefore require the technological muscle of supercomputers a million times faster than those in use today. We will also have to continue to disentangle the myriad interactions among the oceans, atmosphere and biosphere to know exactly what variables to feed into the computer models.

Most important, we must be able to demonstrate that our models accurately simulate past and present climate change before we can rely on models to predict the future. To do that, we need long-term records. Climate simulation and prediction will come of age only with an ongoing record of changes as they happen.

Computers and Climate Interactions

For scientists who model climate patterns, everything from the waxing and waning of ice ages to the desertification of central Africa plays out inside the models run on supercomputers. Interactions among the components of the climate system—the atmosphere, oceans, land, sea ice, freshwater and biosphere—behave according to physical laws represented by dozens of mathematical equations. Modelers instruct the computers to solve these equations for each box in a three-dimensional grid that covers the globe. Because nature is not constrained by boxes, the chore is not only to incorporate the correct mathematics within each box but also to describe appropriately the transfer of energy and mass into and out of the boxes.

The computers at the world's preeminent climate-modeling facilities can perform between 10 and 50 billion operations per second, but with

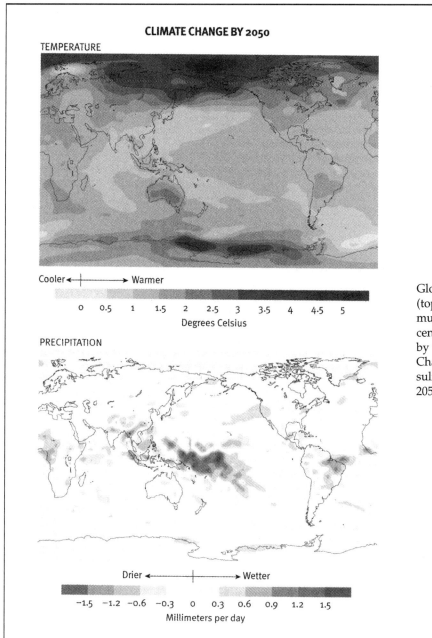

CLIMATE CHANGE BY 2050

TEMPERATURE

Cooler ← | → Warmer

0   0.5   1   1.5   2   2.5   3   3.5   4   4.5   5
Degrees Celsius

PRECIPITATION

Drier ← | → Wetter

−1.5   −1.2   −0.6   −0.3   0   0.3   0.6   0.9   1.2   1.5
Millimeters per day

Global warming of up to five degrees Celsius (top) could enhance precipitation (bottom) in much of the world by the middle of the 21st century. These simulations use 1992 estimates by the Intergovernmental Panel on Climate Change for emissions of greenhouse gases and sulfate aerosols between the years 2000 and 2050.

NOAA/GEOPHYSICAL FLUID DYNAMICS LABORATORY

so many evolving variables, the simulation of a single century can take months. The time it takes to run a simulation, then, limits the resolution (or number of boxes) that can be included within climate models. For typical models designed to mimic the detailed evolution of weather systems, boxes in the three-dimensional grid measure about 250 kilometers (156 miles) square in the horizontal direction and one kilometer in the vertical. Tracking patterns within smaller areas thus proves especially difficult.

Even the most sophisticated of our current global models cannot directly simulate conditions such as cloud cover and the formation of rain. Powerful thunderstorm clouds that can unleash sudden downpours often operate on scales of less than 10 kilometers, and raindrops condense at submillimeter scales. Because each of these events happens in a region smaller than the volume of the smallest grid unit, their characteristics must be inferred by elaborate statistical techniques.

Such small-scale weather phenomena develop randomly. The frequency of these random events can differ extensively from place to place, but most agents that alter climate, such as rising levels of greenhouse gases, affect all areas of the planet much more uniformly. The variability of weather will increasingly mask large-scale climate activity as smaller regions are considered. Lifting that mask thus drains computer time, because it requires running several simulations, each with slightly different starting conditions. The climate features that occur in every simulation constitute the climate "signal," whereas those that are not reproducible are considered weather-related climate "noise."

Conservative estimates indicate that computer-processing speed will have increased by well over a million times by 2050. With that computational power, climate modelers could perform many simulations with different starting conditions and better distinguish climate signals from climate noise. We could also routinely run longer simulations of hundreds of years with less than one-kilometer horizontal resolution and an average of 100-meter vertical resolution over the oceans and atmosphere.

Faster computers help to predict climate change only if the mathematical equations fed into them perfectly describe what happens in nature. For example, if a model atmosphere is simulated to be too cold by four degrees C (not uncommon a decade ago), the simulation will indicate that the atmosphere can hold about 20 percent less water than its actual capacity—a significant error that renders meaningless any subsequent estimates of evaporation and precipitation. Another problem is that we do not yet know how to replicate adequately all the processes that influence climate, such as hiccups in the carbon cycle and modifications in land use. What is more, these changes can initiate feedback cycles that, if ignored, can lead the model astray. Raising temperature, for example, sometimes enhances another variable, such as moisture content of the atmosphere, which in turn amplifies the original perturbation. (In this case, more moisture in the air causes increased warming because water vapor is a powerful greenhouse gas.)

Researchers are only beginning to realize how much some of these positive feedbacks influence the planet's life-giving carbon cycle. The 1991 eruption of Mount Pinatubo in the Philippines, for instance, belched out enough ash and sulfur dioxide to cause a temporary global cooling as those compounds interacted with water droplets in the air to block some of the sun's incoming radiation. This depleted energy can inhibit carbon dioxide uptake in the plants.

Using land in a different way can perturb continental and regional climate systems in ways that are difficult to translate into equations. Clearing forests for farming and ranching brightens the land surface. Croplands are lighter-colored than dark forest and thus reflect more solar radiation, which tends to cool the atmosphere, especially in autumn and summer.

Dearth of Data

Climate simulations can never move out of the realm of good guesses without accurate observations to validate them and to show that the models do indeed reflect reality. In other words, to reduce our uncertainty about the sensitivity of the climate system to human activity, we need to know how the climate has changed in the past. We must be capable of adequately simulating conditions before the Industrial Revolution and especially since that time, when humans have altered irrevocably the composition of the atmosphere.

To understand climate from times prior to the development of weather-tracking satellites and other instruments, we rely on indicators such as air and chemicals trapped in ice cores, the width of tree rings, coral growth, and sediment deposits on the bottoms of oceans and lakes. These snapshots provide us with information that aids in piecing together past conditions. To truly understand the present climate, however, we require more than snapshots of physical, chemical and biological quantities; we also need the equivalent of long-running videotape records of the currently evolving climate. Ongoing measurements of sea ice, snow cover, soil moisture, vegetative cover, and ocean temperature and salinity are just some of the variables involved.

But the present outlook is grim: no U.S. or international institution has the mandate or resources to monitor long-term climate. Scientists currently compile their interpretations of climate change from large networks of satellites and surface sensors such as buoys, ships, observatories, weather stations and airplanes that are being operated for other purposes, such as short-term weather forecasting. As a result, depictions of past climate variability are often equivocal or missing.

The National Oceanic and Atmospheric Administration operates many of these networks, but it does not have the resources to commit to a long-term climate-monitoring program. Even the National Aeronautics and Space Administration's upcoming Earth Observing System, which entails launching several sophisticated satellites to monitor various aspects of global systems, does not include the continuity of a long-term climate observation program in its mission statement.

Whatever the state of climate monitoring may be, another challenge in the next decade will be to ensure that the quantities we do measure actually represent real multidecadal changes in the environment. In other words, what happens if we use a new camera or point it in a different direction? For instance, a satellite typically lasts only four years or so before it is replaced with another in a different orbit. The replacement usually has new instruments and observes the earth at a different time of day. Over a period of years, then, we end up measuring not only climate variability but also the changes introduced by observing the climate in a different way. Unless precautions are taken to quantify the modifications in observing technology and sampling methods before the older technology is replaced, climate records could be rendered useless because it will be impossible to compare the new set of data with its older counterpart.

Future scientists must be able to evaluate their climate simulations with unequivocal data that are properly archived. Unfortunately, the data we have archived from satellites and critical surface sensors are in jeopardy of being lost forever. Long-term surface observations in the U.S. are still being recorded on outdated punched paper tapes or are stored on decaying paper or on old computer hardware. About half the data from our new Doppler radars are lost because the recording system relies on people to deal with the details of data preservation during severe weather events, when warnings and other critical functions are a more immediate concern.

Can We Realize the Vision?

Over the next 50 years we can broadly understand, if we choose to, how human beings are affecting the global, regional and even small-scale aspects of climate. But waiting until then to take action would be foolhardy. Long lifetimes of carbon dioxide and other greenhouse gases in the atmosphere, coupled with the climate's typically slow response to evolving conditions, mean that even if we cut back on harmful human activities today, the planet very likely will still undergo substantial change.

Glaciers melting in the Andes highlands and elsewhere are already confirming the reality of a warming planet. Rising sea level—and drowning coastlines—testify to the projected global warming of perhaps two degrees C or more by the end of the next century. Climate change will in all likelihood capture the most attention when its effects exacerbate other pressures on society. The spread of settlements into coastal regions and low-lying areas vulnerable to flooding is just one of the initial difficulties that we will most likely face. But as long as society can fall back on the uncertainty of human impact on climate, legislative mandates for changing standards of fossil-fuel emissions or forest clear-cutting will be hard fought.

The need to foretell how much we influence our world argues for doing everything we can to develop comprehensive observing and data-archiving systems now. The resulting information could feed models that help make skillful predictions of climate several years in advance. With the right planning we could be in a position to predict, for example, exactly how dams and reservoirs might be better designed to accommodate anticipated floods and to what extent greenhouse gas emissions from new power plants will warm the planet.

Climate change is happening now, and more change is certain. We can act to slow it down, and we can sensibly plan for it, but at present we are doing neither. To anticipate the true shape of future climate, scientists must overcome the obstacles we have outlined above. The need for greater computer power and for a more sophisticated understanding of the nuances of climate interactions should be relatively easy to overcome. The real stumbling block is the long-term commitment to global climate monitoring. How can we get governments to commit resources for decades of surveys, particularly when so many governments change hands with such frequency?

If we really want the power to predict the effects of human activity by 2050—and to begin addressing the disruption of our environment—we must pursue another path. We have a tool to clear such a path: the United Nations Framework Convention on Climate Change, signed by President George Bush in 1992. The convention binds together 179 governments with a commitment to remedy damaging human influence on global climate. The alliance took a step toward stabilizing greenhouse gas emissions by producing the Kyoto Protocol in 1997, but long-term global climate-monitoring systems remain unrealized.

FURTHER INFORMATION

GLOBAL WARMING: IT'S HAPPENING. Kevin E. Trenberth in *natural-SCIENCE*, Vol. 1, Article 9; 1997. Available at naturalscience.com/ns/articles/01-09/ns/_ket.html on the World Wide Web.

ADEQUACY OF CLIMATE OBSERVING SYSTEMS, 1999. Commission on Geosciences, Environment, and Resources. National Academy Press, 1999. Available at www.nap.edu/books/0309063906/html/ on the World Wide Web.

CLIMATE CHANGE AND GREENHOUSE GASES. Tamara S. Ledley et al. in *EOS*, Vol. 80, No. 39, pages 453–458; Sept. 28, 1999. Available at www.agu.org/eos_elec/99148e.html on the World Wide Web.

The United Nations Framework Convention on Climate Change and Kyoto Protocol updates are available at www.unfccc.org/ on the World Wide Web.

THOMAS R. KARL has directed the National Climatic Data Center (NCDC) in Asheville, N.C., since March 1998. The center is part of the National Oceanic and Atmospheric Administration and serves as the world's largest active archive of climate data. Karl, who has worked at the center since 1980, has focused much of his research on climate trends and extreme weather. He also writes reports for the Intergovernmental Panel on Climate Change (IPCC), the official science source for international climate change negotiations.

KEVIN E. TRENBERTH directs the Climate Analysis section at the National Center for Atmospheric Research (NCAR) in Boulder, Colo., where he studies El Niño and climate variability. After several years in the New Zealand Meteorological Service, he became a professor of atmospheric sciences at the University of Illinois in 1977 and moved to NCAR in 1984. Trenberth also co-writes IPCC reports with Karl.

UNIT 6

Pollution: The Hazards of Growth

Unit Selections

Key Points to Consider

• What is the "cycle of biological nitrogen" and how do human activities contribute to it? Are there mechanisms to deal with an overabundance of nitrogen in ecological systems?

• Why is groundwater pollution so difficult to trace and to monitor? What mechanisms might be employed to reduce the contributions of agriculture and industry to the contamination of the world's important freshwater supply?

• Describe the differences between "point source" and "nonpoint source" pollution. What are some of the differences in how these pollution sources can be managed? Is government regulation the only or best answer to the water pollution problem? Explain.

• What are some of the most significant improvements in environmental quality made during the last 30 years in the United States? Do you think the U.S. environment is better or worse than it was 30 years ago? Defend your answer.

 **Links: www.dushkin.com/online/**
These sites are annotated in the World Wide Web pages.

IISDnet
http://iisd1.iisd.ca
Persistant Organic Pollutants (POP)
http://irptc.unep.ch/pops/
School of Labor and Industrial Relations (SLIR): Hot Links
http://www.lir.msu.edu/hotlinks/
Space Research Institute
http://arc.iki.rssi.ru/Welcome.html
Worldwatch Institute
http://www.worldwatch.org

Of all the massive technological changes that have combined to create our modern industrial society, perhaps none has been as significant for the environment as the chemical revolution. The largest single threat to environmental stability is the proliferation of chemical compounds for a nearly infinite variety of purposes, including the universal use of organic chemicals (fossil fuels) as the prime source of the world's energy systems. The problem is not just that thousands of new chemical compounds are being discovered or created each year, but that their long-term environmental effects are often not known until an environmental disaster involving humans or other living organisms occurs. The problem is exacerbated by the time lag that exists between the recognition of potentially harmful chemical contamination and the cleanup activities that are ultimately required.

A critical part of the process of dealing with chemical pollutants is the identification of toxic and hazardous materials, a problem that is intensified by the myriad ways in which a vast number of such materials, natural and man-made, can enter environmental systems. Governmental legislation and controls are important in correcting the damage produced by toxic and hazardous materials such as DDT or PCBs or CFCs, in limiting fossil fuel burning, or in preventing the spread of living organic hazards such as pests and disease-causing agents. Unfortunately, as evidenced by most of the articles in this unit, we are losing the battle against harmful substances regardless of legislation, and chemical pollution of the environment is probably getting worse rather than better.

The first article in this unit deals with one of the newest forms of pollution resulting from the chemical revolution: the increase in nitrogen in soil, water, and air as a consequence of heavy applications of artificial fertilizers that utilize nitrogen; fossil fuel combustion that releases nitrogen in combination with other elements (particularly oxygen); and of other human-related sources. In "Toxic Fertility," Danielle Nierenberg, of the staff of *World Watch* magazine, notes that we have reached the point where, in the annual nitrogen cycle, more nitrogen is produced by human activity than is produced by natural processes. Much of the excess nitrogen produced by fertilizers, fossil fuels, and other sources is "fixed" within ecosystems and, therefore, has the capacity of functioning as an ecological poison. The second article in this section also places an emphasis on organic or biological pollution related to what may be humanity's most important environmental problem: the quality of the global supply of freshwater. In "Groundwater Shock: The Polluting of the World's Major Freshwater Stores," Payal Sampat notes that the vast majority (97 percent) of the world's freshwater supply lies not in the visible surface systems of lakes and streams but in underground aquifers. This precious reserve, used for virtually every purpose from drinking to irrigating crops, is becoming polluted by surface processes related to agricultural, commercial, industrial, domestic, transportation, and other human activities. Sampat notes that while much of the world worries about what is happening in the atmosphere (global warming), what happens below our feet may ultimately be of as much concern.

The section's third selection continues with the theme of water pollution and the interrelationship between polluter and pollutant. In "Water Quality: The Issues" author Mary Cooper notes the tremendous strides that have been made in cleaning up the nation's water supply since the passage of the Clean Water Act of 1972. But she also notes that nearly 40 percent of the inland and coastal waters in the United States are unfit for fishing, swimming, or drinking. A basic part of the problem, she claims, is that the Clean Water legislation dealt only with pollution from point sources: for example, factories, power plants, and wastewater treatment facilities. Much of the water pollution anywhere is attributable to nonpoint sources, such as runoff from crop and animal agriculture, urban areas, roads, and forests. By their nature, nonpoint sources are much more difficult to control, and Cooper advocates the development of nonregulatory approaches to mitigate this form of water pollution.

Finally, the concluding article in the section, "It's a Breath of Fresh Air," offers a breath of optimism. Science reporter David Whitman catalogues the progress that has been made in environmental quality in the United States since President Richard Nixon warned in 1970 that by the year 2000, the United States would be "a country in which we can't drink the water, where we can't breathe the air." Whitman notes that while global issues are far from resolved, the progress made in the United States provides reason for hope. It is worth recalling, says Whitman, "that doom-and-gloom environmental predictions have proved wrong more often than they have proved right."

The pollution problem might appear nearly impossible to solve. Yet as the last article notes, solutions exist: massive cleanup campaigns to remove existing harmful chemicals from the environment and to severely restrict their future use; strict regulation of the production, distribution, use, and disposal of potentially hazardous chemicals; the development of sound biological techniques to replace existing uses of chemicals for such purposes as pest control; the adoption of energy and material resource conservation policies; and more conservative and protective agricultural and construction practices. We now possess the knowledge and the tools to ensure that environmental cleanup is carried through. (It will not be an easy task, and it will be terribly expensive. It will also demand a new way of thinking about humankind's role in the environmental systems upon which all life forms depend.) If we do not complete the task, the support capacity of the environment may be damaged or diminished beyond our capacities to repair it. The consequences would be fatal for all who inhabit this planet.

Toxic Fertility

Over the past half century, the amount of biologically active nitrogen circulating through the world's living things has probably doubled. In unnatural excess, an essential nutrient is becoming a kind of ecological poison.

by Danielle Nierenberg

LAST DECEMBER, talks on the climate treaty reached an impasse. The treaty process is supposed to result in a blueprint for reducing carbon emissions. But when delegates met in the Hague, in the Netherlands, for their sixth official conference of the parties, the agenda focused not so much on cutting fossil fuel use as on the issue of "carbon sinks." Sinks are areas, primarily young forests, that are absorbing more carbon than they are releasing. Since they draw carbon from the atmosphere, sinks offer an attractive accounting option for the United States and some other nations that have high carbon emissions. These nations want to claim a "carbon credit" against their emissions, on the strength of their sinks. How big a credit—if any—should be allowed? In one way or another, that question underlay much of the discussion, and the delegates weren't able to agree on an answer. They failed, in other words, to agree on a way to balance the global carbon budget.

Apart from the immediate reasons for concern over this failure, there is the matter of another unbalanced natural budget. Nitrogen, like carbon, plays a key role in the vast biochemical cycles of life. And increasingly, the nitrogen cycle is being reshaped by human activity—a process that could eventually affect virtually every ecosystem on earth. Our economies are in urgent need of a "nitrogen audit."

Like carbon, nitrogen is a basic ingredient of living things. It's found, for example, in DNA, in proteins, and in chlorophyll, the pigment that drives photosynthesis. Nitrogen shares another key characteristic of carbon: it's very common. It comprises a whopping 78 percent of the atmosphere. But nearly all of this atmospheric nitrogen is elemental dinitrogen, or N_2—it exists in the form of two nitrogen atoms linked together. Elemental nitrogen cannot be metabolized by most living things. Nitrogen becomes biologically active only when it is "fixed"— that is, incorporated into certain other molecules, primarily ammonium (NH_4) and nitrate (NO_3). Fixed nitrogen flows throughout the food web: it is absorbed first by plants, then by plant-eating animals, then by their parasites and predators. Death at each stage of the way releases nitrogen compounds to begin the cycle anew. The fixation process is what makes the nitrogen cycle so different from the carbon cycle. Despite the abundance of elemental nitrogen, fixed nitrogen is frequently what scientists call a "limiting nutrient." Under normal natural conditions, it is often in short supply, so the level of available nitrogen is a key regulator of ecological processes.

The gate-keepers to the biological part of the nitrogen cycle are certain micro-organisms capable of fixing elemental atmospheric nitrogen. Some of these organisms live in soil, often in close association with plants that belong to the bean family. This relationship benefits both parties: the plants get the nitrogen compounds; the microbes get carbohydrates, which the plants produce through photosynthesis. (Sometimes the plants themselves are said to be nitrogen fixing, but this is a kind of terminological shorthand.) Nitrogen fixing occurs in water as well. One of the biggest mysteries of the nitrogen cycle involves marine plankton. These microscopic plants are fixing enormous quantities of nitrogen, but their role in the global cycle has yet to be clearly defined. Finally, in addition to these living portals, there is an inanimate natural process that fixes large quantities of nitrogen: lightning fuses nitrogen and oxygen to create nitrate.

Recent human activity has greatly increased the rate at which nitrogen is being fixed. Since the 1950s, the amount of nitrogen circulating through living things is thought to have doubled. And increasingly, in forests and fields, in rivers and along the coasts, scientists are blaming excess fixed nitrogen for a range of ecological problems—some of them obvious, others very subtle. Any one of these problems can usually be linked to some local or regional cause; the nitrogen balance of a river, for example, might be upset by increased sewage outflow. But when you step back and look at the cycle from a global perspective, three general activities emerge as the primary reasons for the growing fixed nitrogen glut.

Fertilizing the Nitrogen Cycle

Annual releases of fixed nitrogen caused by human activity

Source	Millions of tons
Fertilizer	80
Nitrogen-fixing crops	40
Fossil fuels	20
Biomass burning	40
Wetland drainage	10
Land clearing	20
Total human releases	210
Total natural fixed-nitrogen production *	140

(*) Terrestrial sources only; marine sources have not yet been reliably estimated.

Source: World Resources Institute, "Global Nitrogen Glut" table, available at www.wri.org/wri/wr-98-99/nutrient.htm.

First, coal and oil combustion is releasing a huge, long-buried reservoir of fixed nitrogen by burning the residues of ancient plants, in the form of coal and oil. The fossil fuel economy is disrupting not just the carbon cycle but the nitrogen cycle as well. Second, the progressive destruction of forests and wetlands is releasing the nitrogen contained in these natural areas, just as it releases the carbon. Taken together, these two activities are releasing about 90 million tons of fixed nitrogen annually; that's about 43 percent of the human addition to the nitrogen cycle. (See table above.)

The remainder of the human addition—some 120 million tons—comes from agriculture. Nitrogen-fixing crops

produce about a third of that amount; the rest comes from artificial fertilizer. Fixed nitrogen is the basic component of fertilizer. Through its dependence on artificial fertilizer, modern conventional agriculture has become, in a sense, a form of industrial nitrogen management. This is a relatively recent development in agricultural history. Low-cost techniques for synthesizing ammonia emerged shortly after the Second World War. Cheap ammonia led to mass production of artificial fertilizer and heralded what the ecologist and nitrogen expert David Tilman has called "the 35 most glorious years of agricultural production."

For farmers in the industrialized countries—and increasingly, in the developing countries as well—this limiting nutrient is now available in virtually limitless quantities. As is typical of cheap commodities, a great deal of it is wasted. Fertilizer is often very inefficiently applied; much of it never reaches the crop. It leaches out of the fields and into the streams, or it's converted into a nitrogenous gas like nitrous oxide and escapes into the atmosphere.

Nearly all crops grown in the industrialized countries are now nitrogen-saturated—that is, they're being exposed to more nitrogen than they can use. But fertilizer production continues to grow, on the strength of developing world demand. At current rates of production, fertilizer is adding some 80 million tons of fixed nitrogen to the cycle; by 2020 that burden is expected to reach 134 million tons—just 6 million tons shy of the total input from all natural terrestrial sources combined.

ON THE MORNING OF JUNE 22, 1995, the wall of an artificial "waste lagoon" gave way at a factory farm in North Carolina. Some 95 million liters of putrefying hog urine and feces spilled out of the lagoon, washed across several fields and a road, then poured into the New River. Millions of fish and other aquatic organisms died in what became one of the worst incidents of water pollution in the state's history. Unfortunately, it wasn't an isolated incident—or at least not for long. Very large livestock farms, known as "concentrated animal feeding operations" or CAFOs, had become a major part of the state's agricultural sector, and the CAFOs had begun to hemorrhage waste.

A couple of weeks after the New River spill, 34 million liters of poultry waste flowed down a creek and into the Northeast Cape Fear River. In August of that year, another 3.8 million liters of hog waste ended up in the Cape Fear Estuary, reports the January/February 2000 issue of *American Scientist*. But the worst was yet to come. The hurricanes of 1998 and 1999 brought a series of massive floods to North Carolina's seaboard and untold millions of liters of hog waste were washed out of various CAFO lagoons. The state's coastal ecosystems have yet to recover.

Like sewage, CAFO pollution is extremely high in nitrogen, and most of that nitrogen comes from the artificial fertilizer used to grow the animal feed. You could say that CAFOs are a consequence of chemical fertilizer, because fertilizer has allowed for the "uncoupling" of livestock and crop. When farmers get their fertilizer out of a bag,

they don't need manure. And feed corn can be shipped to CAFOs just as readily as fertilizer can be shipped to corn growers. In each case, the basic input is no longer produced by the landscape in which it is used, so the local ecology no longer effectively limits the intensity of production. The environmental costs of this fractured system

Change in the Terrestrial Nitrogen Cycle

When inert atmospheric nitrogen (N$_2$) is fixed—that is, bonded to oxygen or hydrogen—it becomes an essential plant nutrient. But too much fixed nitrogen can upset basic physiological and ecological functions. Under normal natural conditions, the amount of fixed nitrogen is usually fairly limited. On land, fixation occurs naturally only in certain soil microbes and during lightning strikes, which bond nitrogen and oxygen. (Nitrogen is fixed in the oceans as well, by some types of plankton.)

The left side of this diagram shows the terrestrial nitrogen cycle as it would naturally function. The right side shows some of the ways in which human activity is increasing the amount of fixed nitrogen in the cycle.

1 *The burning of coal and oil is releasing nitrogen that was naturally fixed—but millions of years ago when these fossil fuels were living plants. Some nitrogen is also fixed directly, as a byproduct of combustion.*

2 *and* **3** *Fertilizer production is artificially fixing a large amount of additional nitrogen. Much of this is released into the environment either directly, when the fertilizer is used on crops, or indirectly, in the manure of livestock fed on the fertilized grain.*

4 *The cultivation of bean sand other leguminous crops, which grow in close association with nitrogen-fixing microbes, uses a natural mechanism of nitrogen fixing—but on a scale that is unnaturally large and unnaturally intense, because it involves extensive monocultures.*

Finally, the destruction of forests and wetlands (not shown here), does not add fixed nitrogen to the cycle as a whole, but it releases large amounts of fixed nitrogen from long-term confinement in those ecosystems.

ILLUSTRATION BY MICHAEL ROTHMAN, COURTESY LARS HEDIN, DEPARTMENT OF ECOLOGY AND EVOLUTIONARY BIOLOGY, CORNELL UNIVERSITY.

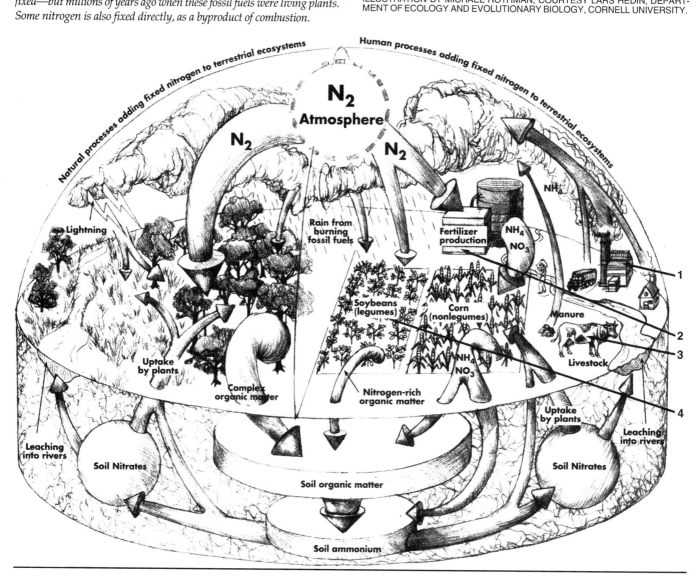

are likely to make it untenable over the long term. But at least for the present, fertilizer is the source of about a third of human dietary protein (from both animals and plants), according to Vaclav Smil, a professor at the University of Manitoba who has written extensively on global biochemical cycles.

The logistics of managing CAFO waste are formidable. Each of the 50,000 or so sows in one of those North Carolina facilities will produce about 20 piglets over the course of a year. A sow and its piglets will excrete some 1.9 tons of waste annually—that's enough manure to fill a pick-up truck. The waste cannot simply accumulate in lagoons, since lagoon space is obviously limited, so CAFOs require huge amounts of "spreadable acreage"—cropland on nearby farms where manure can be spread, sprayed, or injected. But crops can only use so much nitrogen. Adding too much can actually reduce yields; like people, plants can overeat, and excessive nitrogen uptake tends to interfere with a plant's ability to manufacture the various chemicals needed for its metabolism. Too much nitrogen can also throw the soil community out of balance by favoring only those organisms that thrive in high-nitrogen conditions, at the expense of many other organisms.

If you're trying to do a conscientious job of it, finding adequate spreadable acreage is a very difficult task indeed. For each of those sow and piglet units, a CAFO should ideally have access to 1.2 hectares (3 acres) of land. (This ratio is actually determined not by the manure's nitrogen concentration but by the concentration of phosphorus, which is also frequently a limiting nutrient and therefore capable of causing some of the same "over-fertilizing" effects that nitrogen does. Sufficient spreadable acreage for the phosphorus will—at least in the case of pig manure—accommodate the nitrogen too.) A 50,000 animal CAFO would need about 60,000 spreadable hectares. Inevitably, given the size of the CALFOs, that ideal is not attained. Frequently, far too much manure is spread on the fields. Or the manure may be spread at the wrong times in the growing season, when the crops cannot effectively take up the nutrients. Or sometimes the manure is spread on fields of nitrogen-fixing crops like soybeans and alfalfa, which require little or no additional fertilizer. In North America, it's estimated that only about half of livestock waste is now effectively fed into the crop cycle. Much of the remainder ends up as pollution—of the water, the air, and the soil itself.

Take the water pollution first. Nitrate contamination of groundwater can create serious risks for public health. (See "Groundwater Shock," January/ February 2000.) For example, high nitrate levels in wells near feedlot operations have been linked to greater risk of miscarriage. In extreme cases, nitrate contamination can cause methemoglobinemia, or "blue-baby syndrome," a form of infant poisoning in which the blood's ability to transport oxygen is greatly reduced, sometimes to the point of death. Nitrate water pollution is a serious ecological concern as well, even when it doesn't involve millions of liters of hog feces. Per-

haps the most obvious form of ecological disruption involves algal blooms, explosive growths of algae and cyanobacteria (so-called "blue-green algae") that can suffocate many other aquatic organisms. There are more subtle wildlife effects as well; some amphibian declines, for example, appear to be caused by chronic exposure to elevated nitrate. (See "Amphibia Fading," July/August 2000.)

But as anyone who lives near a CAFO can tell you, groundwater contamination is hardly the most noticeable environmental effect. If raw manure is exposed to air, up to 95 percent of the nitrogen in it will escape into the atmosphere as gaseous ammonia (NH_3). In the vicinity of a CAFO, the process results in an olfactory experience that is difficult to forget. But that's not all it does, since the nitrogen doesn't usually stay airborne for long—it's usually deposited within 80 to 160 kilometers of its source. In the words of Merrit Frey, who studies factory farming for the Clean Water Network, a coalition of U.S.-based nonprofits concerned about water quality, the ammonia from CAIFOs is "not just a localized odor nuisance, but a regional environmental problem." Once it falls from the sky, it tends to contribute to the same problems that result from the more direct forms of soil and water pollution.

As livestock production continues to intensify, such problems are likely to become more common. Already, North Carolina's 7 million factory-raised hogs create more waste than do the 6.5 million human residents of the state. Intensive livestock production is the rule in the United States and western Europe, and producers are increasingly interested in setting up similar operations elsewhere. In part, such interest is the result of the growing regulatory attention that established operations are now attracting. In the United States, for example, CAFO waste is the target of new regulations recently proposed by the U.S. EPA, and of proposed amendments to the Clean Water Act. But the export of the CAFO model is also partly a variation on a standard economic theme: investment in developing markets. The demand for meat is growing in the developing world, and the costs of producing it are generally much lower there than in the United States or Europe.

China, for example, is interested in boosting domestic meat production to satisfy growing domestic demand, according to David Brubaker, an expert on factory farming at the Johns Hopkins University School of Public Health in Maryland. Brubaker says that several U.S. agribusinesses are tying to sell the Chinese on CAFO production of hogs, poultry, and cattle. In the Philippines, two such corporations, Tyson Foods and Purina Mills, opened a hog breeding facility near Manila in 1998; the facility can produce 100,000 hogs per year. Richard Levins, an agricultural economist at the University of Minnesota, says that even Canada is becoming a prime location for hog CAFO developers looking for lots of space, few people, and relatively lax environmental regulations.

Taken globally, livestock production has become a major outlet pipe for much of the nitrogen that humanity is injecting into the natural cycle. The planet's population of some 2.5 billion pigs and cattle void more than 80 million tons of waste nitrogen annually. The entire human population, in comparison, produces just over 30 million tons. Manure, once a valuable farm resource, is now being produced in such quantities that it might be better considered toxic waste.

HERE AND THERE, ALONG CHINA'S 18,000-kilometer coastline, fishers and fish farmers have long contended with an unwelcome form of marine bounty. Red tides, the toxic blooms of certain algae species, are a natural phenomenon in these waters during the spring and summer, but the frequency of the blooms is increasing. Scientists suspect that the algae are responding to two factors. Sea surface temperatures in the region are rising—a likely effect of climate change—and more and more nitrogen-rich waste is pouring into China's coastal waters. The annual load of such pollution—in the form of sewage, industrial waste, and farm runoff—now exceeds 8 billion tons.

Warmer, nutrient rich water is ideal habitat for algae, and the resulting red tides have been poisoning not just fish and shellfish, but any people unfortunate enough to consume the contaminated seafood. Nontoxic algal blooms are on the upsurge too. Even though these algae produce no poisons, they use up most of the water's available oxygen as they decompose. The resulting hypoxic "dead zones"—areas where dissolved oxygen concentrations are too low to support most forms of life—can last for months.

Algal blooms are hardly unique to China. Perhaps the most famous of these events is the recurrent bloom in the Gulf of Mexico, off the coast of Texas and Louisiana. Almost every spring, and increasingly at other seasons as well, thick clouds of algae form in these waters. The Gulf dead zone is vast—in 1999 it covered 18,000 square kilometers, about the size of the state of New Jersey—and it does millions of dollars in damage to the region's fisheries every year. Here too, nitrogen is the key factor (although increasing loads of phosphorus and silica are also apparently feeding the algae). Most of the nitrogen appears to be coming from sewage and agricultural runoff. According to a report by the White House Office of Science and Technology Policy, manure alone contributes some 15 percent of the nitrogen that makes its way into the Gulf. That's more than all nonfarm industrial nitrogen sources combined.

But farm runoff and sewage aren't the only components of the nitrogen cycle that are promoting algal blooms. In the Baltic Sea, the primary burden of excess nitrogen comes from fossil fuel emissions. A full third of the nitrogen entering the Sea—by far the largest share of the excess load—consists of nitrogen oxides produced by the combustion of coal and oil in the surrounding countries. The Baltic is a naturally low-nitrogen environment that supports a unique community of organisms adapted to

those circumstances. But as the nitrogen levels have increased, cyanobacteria are responding with large, eerily beautiful blooms. The blooms are shading out the sea's "forests" of bladderwrack, a species of brown seaweed that requires clear, well lit waters. The bladderwrack is prime spawning and nursery habitat for many fish species, now threatened by the seaweed's decline. And as the blooms decompose, they steal oxygen from the water—a form of change to which the Baltic is especially sensitive, since its waters are relatively low-oxygen to begin with. The drop in oxygen is working change in the seafloor community, where bristleworms, which tolerate hypoxia, are replacing the once-dominant mussels. This change, in turn, is likely to restructure the foodweb, since mussels are an important prey item for many fish that won't eat bristleworms.

Algal blooms and dead zones are now a regular feature of coastal life in many other places around the world—off the coast of New England, for instance, off the west coast of India, and off Japan and Korea as well. Most of the world's major coastal ecosystems appear to be suffering some degree [of] algae-induced hypoxia. And toxic blooms are an increasingly conspicuous part of this problem. According to scientists at the U.N. Food and Agriculture Organization (FAO) and the International Oceanographic Commission, the 1980s and 1990s saw a global upsurge in red tides. As these episodes have become more common, so has the number of algae species known to be involved. In the early 1990s, only about 20 species were known to produce toxic blooms; today, at least 85 have been identified.

SOME EFFECTS OF NITROGEN POLLUTION are much more subtle than algal blooms, but arguably even more dangerous. For example, the nitrogen oxides produced through fossil fuel combustion are a major component of the acid rain that is attacking soil and fresh water in many parts of the world. Waters that become increasingly acidic support fewer forms of aquatic life. In a similar fashion, the acidification of soils tends to impoverish the soil community. That's partly because the acid releases aluminum ions from the mineral matrix in which they are usually embedded. Free aluminum is toxic to plants—and to many aquatic organisms if it washes into streams. The acid also causes certain minerals to leach out of the soil. Calcium, magnesium, and potassium are essential plant nutrients and are often in relatively short supply. Where they become rare, plant growth is likely to slow, and the more mineral-hungry species may fade from the scene.

These minerals come into the soil through the weathering of rock. Since weathering is a very slow process, acidification could reduce the productivity of affected soils for centuries—or longer. The extreme case appears to involve cropland, where too much fertilizer may perform the role of acid rain. According to Phillip Barak, a professor of soil chemistry and plant nutrition at the University of Wisconsin, nitrogen-induced mineral leaching in some U.S. cropland has artificially "aged" these soils by the equivalent of 5,000 years.

Over the past couple of decades, the industrialized countries have made considerable progress in reducing emissions of one main ingredient of acid rain: the sulfur dioxide produced by the combustion of sulfur-contaminated coal. The switch to low sulfur coal and the installation of smoke stack "scrubbers" on coal-burning powerplants have greatly reduced this type of pollution. But incremental solutions aren't as useful in reducing the nitrogen compounds released by fossil fuel combustion. The fixed nitrogen in fossil fuels cannot be readily removed—and additional nitrogen is fixed as a byproduct of the combustion itself.

Excess soil nitrogen is sickening forests and fields in other ways as well. Nitrogen pollution may reduce cold hardiness in certain tree species, making them more liable to injury or death during the winter. Too much nitrogen also tends to reduce fine root density, which in turn restricts water and nutrient uptake, making plants more susceptible to drought.

On a community level, nitrogen, loading is a homogenizing force, because it strongly favors fast-growing plant species that can use extra nutrient, at the expense of slower-growing species that cannot. Affected areas may show robust growth but that growth is likely to be quite uniform. Surveys in Great Britain by the Institute for Terrestrial Ecology found that near major sources of nitrogen pollution, such as large poultry farms, the ground-layer flora of nearby woodlands was dominated by dense stands of a few rank, tall grass species. The farther away from the farms the researchers went, the more varied the woodland flora became. This homogenizing tendency is also apparent in the heathlands of Northern Europe, particularly in the Netherlands, which has the greatest concentration of livestock on the planet. Formerly a diverse assemblage of shrubs and herbs, these heathlands are increasingly dominated by invasive grasses and trees. Excess nitrogen is indeed fertilizing more and more of the world's wild communities. But it is promoting the growth of a few opportunistic species, at the expense of a more diverse whole.

OVER THE PAST FEW YEARS, Indian researchers have been detecting substantial quantities of nitrous oxide rising out of the Arabian Sea, off India's west coast. The Sea's emissions are now thought to contribute up to 21 percent of total output of the gas from the world's oceans. In the upper atmosphere, nitrous oxide tends to deplete the stratospheric ozone layer, which shields the Earth against harmful ultraviolet radiation. Nitrous oxide is also a potent greenhouse gas. On a molecule-per-molecule basis, it's 200 times more effective at retaining heat than is CO_2. Luckily, it's far less common than CO_2, but it still accounts for 2 to 3 percent of overall greenhouse warming.

Why is the Arabian Sea exhaling so much of this gas? Rajiv Naqvi, a researcher at India's National Institute of Oceanography sees a link with intensifying agriculture. Over the past four decades or so, fertilizer use has grown substantially on India's western coastal plain, as it has in most of the country's croplands. (According to FAO statistics, Indian fertilizer consumption nearly doubled from 1965 to 1998, rising to over 11 million tons.) As in many of the world's coastal waters, the resulting nitrogen-laden runoff is feeding cyanobacteria and other plankton. Cyanobacteria generally produce some nitrous oxide as they grow, but their activity in these waters is affected by another powerful factor: India's June to September monsoons. Since freshwater is lighter than salt water, the hard monsoon rainfall tends to blanket the ocean surface, reducing the aeration of the saltier water beneath. That drops the oxygen level, and the lack of oxygen in turn causes the cyanobacteria to metabolize nitrogen in a way that releases larger amounts of nitrous oxide. This effect appears to have been exacerbated in recent years because of unusually heavy monsoon rains, a possible result of climate change.

Ice core data indicate that the atmospheric concentration of nitrous oxide was quite stable until about a century ago, when it began to increase. The current rate of increase is estimated at 0.2 to 0.3 percent per year. Atmospheric concentrations are now about 10 percent higher than they were at the beginning of the century.

In the abstract, this is what seems to be happening in the Arabian Sea: an increase in fixed nitrogen, possibly combined with climate change in the form of intensifying monsoons, is putting more pressure on the climate system. There is a subtle connection here between the nitrogen and carbon cycles, but this is just one of many such links. Pamela Matson, Director of Stanford University's Earth Systems Program, puts it simply: "In order to understand carbon and Earth's other cycles, we have to understand how nitrogen is changing." How will the changing nitrogen cycle affect the changing carbon cycle?

Given the climate treaty negotiations, there is a nearly irresistible political impulse to look for links that might be increasing carbon absorption from the atmosphere. Perhaps the nitrogen glut will increase the carbon sinks. At first glance, that looks like a reasonable expectation. After all, nitrogen is often a limiting nutrient; more of it should mean more plant growth, which should increase CO_2 absorption. But it doesn't seem to be that simple.

In the oceans, plankton growth out beyond the coastal waters is often limited not by nitrogen availability but by the availability of iron, another essential nutrient. So extra nitrogen doesn't automatically translate into extra CO_2 absorption. (It's true that some scientists advocate seeding the oceans with iron to increase CO_2 absorption, but given all the unforeseen problems we've already created with excess nitrogen and carbon, it would hardly be wise policy to interfere with yet another cycle.) And even in the case of the coastal algal blooms, which may increase

CO_2 absorption, at least periodically, it's not clear whether that extra CO_2 will remain in the waters over the long term, or end up back in the atmosphere relatively quickly.

On land, the issue is primarily a matter of forest growth, since forests generally store more carbon than other types of terrestrial ecosystems. It's true that excess nitrogen may cause forests to grow more rapidly over the short term. But over the long term, the prospects for such forests are fairly dismal, given the acidification, aluminum poisoning, and other forms of physiological and ecological disruption that nitrogen loading tends to cause. And a declining forest is more likely to be releasing carbon than absorbing it. Despite the connections between the nitrogen cycle and the carbon cycle, there is no good reason to assume that disruption of the former will partly "cancel out" disruptions in the latter.

IN TERMS OF ITS TECHNICAL DETAIL, stabilizing the nitrogen cycle is likely to be just as demanding a task as is stabilizing the carbon cycle. But perhaps the most obvious aspect of this problem is its familiarity: the types of change that would make the most difference are already standard items on the environmental agenda. Three basic reforms appear to be necessary if we are to achieve major reductions of our fixed nitrogen emissions. We will need to convert the dominant mode of agricultural production from its current, "high input" paradigm to one that emphasizes organic production. (See "Where Have All the Farmers Gone?" September/October 2000.) We will need to convert our fossil fuel-based energy economy to one based on sunlight, wind, geothermal, and other forms of renewable energy. And finally, we will need to slow and eventually reverse the destruction of the planet's remaining natural areas, especially its remaining forests.

These are enormous goals, of course, but each of them shares some characteristics that can help chart the way forward. In the first place, they all aim at broad systemic reform, but they focus on the smallscale unit. Organic farming usually works best when the farms are small enough to accommodate the local landscape. Sophisticated renewable energy systems generally create networks of smaller producers rather than one or two enormous powerplants. And sustainable forest use, by definition, is carefully attuned to local ecological realities. There's a second common feature too: each of these goals emphasizes creative use of diversity. A polycultural cropping system, a range of renewable energy technologies, a combination of agroforestry, timber, and tourism—in each case, the idea is to replace a "monoculture approach" with a system that is more diverse and therefore more flexible, more likely to be sustainable over the long term. Small-scale and diverse would appear to be the way to go. You could say that the agenda points towards a high degree of local adaptation.

Humanity has reached a point at which we are dominating—not just particular ecosystems—but the cycles that regulate the basic processes on which all life depends. Our capacity to understand the effects of our interference is growing rapidly. But will we be able to use that understanding productively? Increasingly, it seems, progress on the global level will depend on our ability to reinvent our relationships to the local level—to the particular ecosystems and societies in which we actually live.

Key Sources

GRACE Factory Farm Project website: www.factoryfarm. org.

Vaclav Smil, *Cycles of Life* (New York: Scientific American Library, 1997).

Peter M. Vitousek, et al., *Human Alteration of the Global Nitrogen Cycle, Issues in Ecology*, vol. 1 (Washington, DC: Ecological Society of America, 1997).

Danielle Nierenberg recently completed her M.S. in the agriculture, food, and environment program at Tufts University, in Massachusetts, and is currently working with the WORLD WATCH staff.

From *World Watch*, March/April 2001, pp. 30-38. © 2001 by World Watch (www.worldwatch.org). Reprinted by permission.

Groundwater *Shock*

The Polluting of the World's Major Freshwater Stores

Scientists have shown that the world deep beneath our feet is essential to the life above. Ancient myths depicted the Underworld as a place of damnation and death. Now, the spreading contamination of major aquifers threatens to turn the myth into a tragic reality.

by Payal Sampat

The Mississippi River occupies a mythic place in the American imagination, in part because it is so huge. At any given moment, on average, about 2,100 billion liters of water are flowing across the Big Muddy's broad bottom. If you were to dive about 35 feet down and lie on that bottom, you might feel a sense of awe that the whole river was on top of you. But in one very important sense, you'd be completely wrong. At any point in time, only 1 percent of the water in the Mississippi River system is in the part of the river that flows downstream to the Gulf of Mexico. The other 99 percent lies beneath the bottom, locked in massive strata of rock and sand.

This is a distinction of enormous consequence. The availability of clean water has come to be recognized as perhaps the most critical of all human security issues facing the world in the next quarter-century—and what is happening to water buried under the bottoms of rivers, or under our feet, is vastly different from what happens to the "surface" water of rivers, lakes, and streams. New research finds that contrary to popular belief, it is groundwater that is most dangerously threatened. Moreover, the Mississippi is not unique in its ratio of surface to underground water; worldwide, 97 percent of the planet's liquid freshwater is stored in aquifers.

In the early centuries of civilization, surface water was the only source we needed to know about. Human population was less than a tenth of one percent the size it is now; settlements were on river banks; and the water was relatively clean. We still think of surface water as being the main resource. So it's easy to think that the problem of contamination is mainly one of surface water: it is polluted rivers and streams that threaten health in times of flood, and that have made waterborne diseases a major killer of humankind. But in the past century, as population has almost quadrupled and rivers have become more depleted and polluted, our dependence on pumping groundwater has soared—and as it has, we've made a terrible discovery. Contrary to the popular impression that at least the waters from our springs and wells are pure, we're uncovering a pattern of pervasive pollution there too. And in these sources, unlike rivers, the pollution is generally irreversible.

This is largely the work of another hidden factor: the rate of groundwater renewal is very slow in comparison with that of surface water. It's true that some aquifers recharge fairly quickly, but the average recycling time for groundwater is 1,400 years, as opposed to only 20 days for river water. So when we pump out groundwater, we're effectively removing it from aquifers for generations to come. It may evaporate and return to the atmosphere quickly enough, but the resulting rainfall (most of which falls back into the oceans) may take centuries to recharge the aquifers once they've been depleted. And because water in aquifers moves through the Earth with glacial slowness, its pollutants continue to accumulate. Unlike rivers, which flush themselves into the oceans, aquifers become sinks for pollutants, decade after decade—thus further diminishing the amount of clean water they can yield for human use.

Perhaps the largest misconception being exploded by the spreading water crisis is the assumption that the ground we stand on—and what lies beneath it—is solid, unchanging, and inert. Just as the advent of climate change has awakened us to the fact that the air over our heads is an arena of enormous forces in the midst of titanic shifts, the water crisis has revealed that, slow-moving though it may be, groundwater is part of a system of powerful hydrological interactions—between earth, surface water, sky, and sea— that we ignore at our peril. A few years ago, reflecting on how human activity is beginning to affect climate, Columbia University scientist Wallace Broecker warned, "The climate system is an angry beast and we are poking it with sticks." A similar statement might now be made about the system under our feet. If we continue to drill holes into it—expecting it to swallow our waste and yield freshwater in return—we may be toying with an outcome no one could wish.

Valuing Groundwater

For most of human history, groundwater was tapped mainly in arid regions where surface water was in short supply. From Egypt to Iran, ancient Middle Eastern civilizations used periscope-like conduits to funnel spring water from mountain slopes to nearby towns—a technology that allowed settlement to spread out from the major rivers. Over the centuries, as populations and cropland expanded, innovative well-digging techniques evolved in China, India, and Europe. Water became such a valuable resource that some cultures developed elaborate mythologies imbuing underground water and its seekers with special powers. In medieval Europe, people called water witches or dowsers were believed to be able to detect groundwater using a forked stick and mystical insight.

In the second half of the 20th century, the soaring demand for water turned the dowsers' modern-day counterparts into a major industry. Today, major aquifers are tapped on every continent, and groundwater is the primary source of drinking water for more than 1.5 billion people worldwide (see table, *Groundwater as a Share of Drinking Water Use by Region*). The aquifer that lies beneath the Huang-Huai-Hai plain in eastern China alone supplies drinking water to nearly 160 million people. Asia as a whole relies on its groundwater for nearly one-third of its drinking water supply. Some of the largest cities in the developing world—Jakarta, Dhaka, Lima, and Mexico City, among them—depend on aquifers for almost all their water. And in rural areas, where centralized water supply systems are undeveloped, groundwater is typically the sole source of water. More than 95 percent of the rural U.S. population depends on groundwater for drinking.

A principal reason for the explosive rise in groundwater use since 1950 has been a dramatic expansion in irrigated agriculture. In India, the leading country in total irrigated area and the world's third largest grain producer, the number of shallow tubewells used to draw groundwater surged from 3,000 in 1960 to 6 million in 1990. While India doubled the amount of its land irrigated by surface water between 1950 and 1985; it increased the area watered by aquifers 113-fold. Today, aquifers supply water to more than half of India's irrigated land. The United States, with the third highest irrigated area in the world, uses groundwater for 43 percent of its irrigated farmland. Worldwide, irrigation is by far the biggest drain on freshwater: it accounts for about 70 percent of the water we draw from rivers and wells each year.

Other industries have been expanding their water use even faster than agriculture—and generating much higher profits in the process. On average, a ton of water used in industry generates roughly $14,000 worth of output—about 70 times as much profit as the same amount of water used to grow grain. Thus, as the world has industrialized, substantial amounts of water have been shifted from farms to more lucrative factories. Industry's share of total consumption has reached 19 percent and is likely to continue rising rapidly. The amount of water available for drinking is thus constrained not only by a limited resource base, but by competition with other, more powerful users.

And as rivers and lakes are stretched to their limits—many of them dammed, dried up, or polluted—we're growing more and more dependent on groundwater for all these uses. In Taiwan, for example, the share of water supplied by groundwater almost doubled from 21 percent in 1983 to over 40 percent in 1991. And Bangladesh, which was once almost entirely river- and stream-dependent, dug over a million wells in the 1970s to substitute for its badly polluted surface-water supply. Today, almost 90 percent of its people use only groundwater for drinking.

Even as our dependence on groundwater increases, the availability of the resource is becoming more limited. On almost every continent, many major aquifers are being drained faster than their natural rate of recharge. Groundwater depletion is most severe in parts of India, China, the United States, North Africa, and the Middle East. Under certain geological conditions, groundwater overdraft can cause aquifer sediments to compact, permanently shrinking the aquifer's storage capacity. This loss can be quite considerable, and irreversible. The amount of water storage capacity lost because of aquifer compaction in California's Central Valley, for example, is equal to more than 40 percent of the combined storage capacity of all human-made reservoirs across the state.

As the competition among factories, farms, and households intensifies, it's easy to overlook the extent to which freshwater is also required for essential ecological services. It is not just rainfall, but groundwater welling up from beneath, that replenishes rivers, lakes, and streams. In a study of 54 streams in different parts of the country, the U.S. Geological Survey (USGS) found that groundwater is the source for more than half the flow, on average. The 492 billion gallons (1.86 cubic kilometers) of water aquifers add to U.S. surface water bodies each day is nearly equal to the daily flow of the Mississippi. Groundwater provides the base contribution for the Mississippi, the Niger, the Yangtze, and many more of the world's great rivers—some of which would otherwise not be flowing year-round. Wetlands, important habitat for birds, fish, and other wildlife, are often largely groundwater-fed, created in places where the water table overflows to the surface on a constant basis. And while providing surface bodies with enough water to keep them stable, aquifers also help prevent them from flooding: when it rains heavily, aquifers beneath rivers soak up the excess water, preventing the surface flow from rising too rapidly and overflowing onto neighboring fields and towns. In tropical Asia, where the hot season can last as long as 9 months, and where monsoon rains can be very intense, this dual hydrological service is of critical value.

Groundwater as a Share of Drinking Water Use by Region

Region	Share of Drinking Water from Groundwater	People Served
	(percent)	(millions)
Asia-Pacific	32	1,000 to 1,200
Europe	75	200 to 500
Latin America	29	150
United States	51	135
Australia	15	3
Africa	NA	NA
World		1,500 to 2,000

Sources: UNEP, OECD, FAO, U.S. EPA, Australian EPA.

Numerous studies have tracked the extent to which our increasing demand on water has made it a resource critical to a degree that even gold and oil have never been. It's the most valuable thing on Earth. Yet, ironically, it's the thing most consistently overlooked, and most widely used as a final resting place for our waste. And, of course, as contamination spreads, the supplies of usable water get tighter still.

Tracking the Hidden Crisis

In 1940, during the Second World War, the U.S. Department of the Army acquired 70 square kilometers of land around Weldon Spring and its neighboring towns near St. Louis, Missouri. Where farmhouses and barns had been, the Army established the world's largest TNT-producing facility. In this sprawling warren of plants, toluene (a component of gasoline) was treated with nitric acid to produce more than a million tons of the explosive compound each day when production was at its peak.

Part of the manufacturing process involved purifying the TNT—washing off unwanted "nitroaromatic" compounds left behind by the chemical reaction between the toluene and nitric acid. Over the years, millions of gallons of this red-colored muck were generated. Some of it was treated at wastewater plants, but much of it ran off from the leaky treatment facilities into ditches and ravines, and soaked into the ground. In 1945, when the Army left the site, soldiers burned down the contaminated buildings but left the red-tinged soil and the rest of the site as they were. For decades, the site remained abandoned and unused.

Then, in 1980, the U.S. Environmental Protection Agency (EPA) launched its "Superfund" program, which required the cleaning up of several sites in the country that were contaminated with hazardous waste. Weldon Spring made it to the list of sites that were the highest priority for cleanup. The Army Corps of Engineers was assigned the task, but what the Corps workers found baffled them. They expected the soil and vegetation around the site to be contaminated with the nitroaromatic wastes that had been discarded there. When they tested the groundwater, however, they found that the chemicals were showing up in people's wells, in towns several miles from the site—a possibility that no one had anticipated, because the original pollution had been completely localized. Geologists determined that there was an enormous plume of contamination in the water below the TNT factory—a plume that over the previous 35 years had flowed through fissures in the limestone rock to other parts of the aquifer.

The Weldon Spring story may sound like an exceptional case of clumsy planning combined with a particularly vulnerable geological structure. But in fact there is nothing exceptional about it all. Across the United States, as well as in parts of Europe, Asia, and Latin America, human activities are sending massive quantities of chemicals and pollutants into groundwater. This isn't entirely new, of course; the subterranean world has always been a receptacle for whatever we need to dispose of—whether our sewage, our garbage, or our dead. But the enormous volumes of waste we now send underground, and the deadly mixes of chemicals involved, have created problems never before imagined.

What Weldon Spring shows is that we can't always anticipate where the pollution is going to turn up in our water, or how long it will be from the time it was deposited until it reappears. Because groundwater typically moves very slowly—at a speed of less than a foot a day, in some cases—damage done to aquifers may not show up for decades. In many parts of the world, we are only just beginning to discover contamination caused by practices of 30 or 40 years ago. Some of the most egregious cases of aquifer contamination now being unearthed date back to Cold War era nuclear testing and weapons-making, for example. And once it gets into groundwater, the pollution usually persists: the enormous volume, inaccessibility, and slow rate at which groundwater moves make aquifers virtually impossible to purify.

As this covert crisis unfolds, we are barely beginning to understand its dimensions. Few countries track the health of their aquifers—their enormous size and remoteness make them extremely expensive to monitor. As the new century begins, even hydrogeologists and health officials have only a hazy impression of the likely extent of groundwater damage in different parts of the world. Nonetheless, given the data we now have, it is possible to sketch a rough map of the regions affected, and the principal threats they face (see map, *Groundwater Contamination Hotspots* and table, *Some Major Threats to Groundwater*).

The Filter that Failed: Pesticides in Your Water

Pesticides are designed to kill. The first synthetic pesticides were introduced in the 1940s, but it took several decades of increasingly heavy use before it became apparent that these chemicals were injuring non-target organisms—including humans. One reason for the delay was that some groups of pesticides, such as organochlorines, usually have little effect until they bioaccumulate. Their concentration in living tissue increases as they move up the food chain. So eventually, the top predators—birds of prey, for example—may end up carrying a disproportionately high burden of the toxin. But bioaccumulation takes time, and it may take still more time before the effects are discovered. In cases where reproductive systems are affected, the aftermath of this chemical accumulation may not show up for a generation.

Even when the health concerns of some pesticides were recognized in the 1960s, it was easily assumed that the real dangers lay in the dispersal of these chemicals among animals and plants—not deep underground. It was assumed that very little pesticide would leach below the upper layers of soil, and that if it did, it would be degraded before it could get any deeper. Soil, after all, is known to be a natural filter, which purifies water as it trickles through. It was thought that industrial or agricultural chemicals, like such natural contaminants as rock dust, or leaf mold, would be filtered out as the water percolated through the soil.

But over the past 35 years, this seemingly safe assumption has proved mistaken. Cases of extensive pesticide contamination of groundwater have come to light in farming regions of the United States, Western Europe, Latin America, and South Asia. What we now know is that pesticides not only leach into aquifers, but sometimes remain there long after the chemical is no longer used. DDT, for instance, is still found in U.S. waters even though its use was banned 30 years ago. In the San Joaquin Valley of California, the soil fumigant DBCP (dibromochloropropane), which was used intensively in fruit orchards before it was banned in 1977, still lurks in the region's water supplies. Of 4,507 wells sampled by the USGS between 1971 and 1988, nearly a third had DBCP levels that were at least 10 times higher than allowed by the current drinking water standard.

In places where organochlorines are still widely used, the risks continue to mount. After half a century of spraying in the eastern Indian states of West Bengal and Bihar, for example, the Central Pollution Control Board found DDT in groundwater at levels as high as 4,500 micrograms per liter—several thousand times higher than what is considered a safe dose.

The amount of chemical that reaches groundwater depends on the amount used above ground, the geology of the region, and the characteristics of the pesticide itself. In some parts of the midwestern United States, for example, although pesticides are used intensively, the impermeable soils of the region make it difficult for the chemicals to percolate underground. The fissured aquifers of southern Arizona, Florida, Maine, and southern California, on the other hand, are very vulnerable to pollution—and these too are places where pesticides are applied in large quantities.

Pesticides are often found in combination, because most farms use a range of toxins to destroy different kinds of insects, fungi, and plant diseases. The USGS detected two or more pesticides in groundwater at nearly a quarter of the sites sampled in its National Water Quality Assessment between 1993 and 1995. In the Central Columbia Plateau aquifer, which extends over the states of Washington and Idaho, more than two-thirds of water samples contained multiple pesticides. Scientists aren't entirely sure what happens when these chemicals and their various metabolites come together. We don't even have standards for the many hundred *individual* pesticides in use—the EPA has drinking water standards for just 33 of these compounds—to say nothing of the infinite variety of toxic blends now trickling into the groundwater.

While the most direct impacts may be on the water we drink, there is also concern about what occurs when the pesticide-laden water below farmland is pumped back up for irrigation. One apparent consequence is a reduction in crop yields.

In 1990, the now-defunct U.S. Office of Technology Assessment reported that herbicides in shallow groundwater had the effect of "pruning" crop roots, thereby retarding plant growth.

From Green Revolution to Blue Baby: the Slow Creep of Nitrogen

Since the early 1950s, farmers all over the world have stepped up their use of nitrogen fertilizers. Global fertilizer use has grown ninefold in that time. But the larger doses of nutrients often can't be fully utilized by plants. A study conducted over a 140,000 square kilometer region of Northern China, for example, found that crops used on average only 40 percent of the nitrogen that was applied. An almost identical degree of waste was found in Sri Lanka. Much of the excess fertilizer dissolves in irrigation water, eventually trickling through the soil into underlying aquifers.

Joining the excess chemical fertilizer from farm crops is the organic waste generated by farm animals, and the sewage produced by cities. Livestock waste forms a particularly potent tributary to the stream of excess nutrients flowing into the environment, because of its enormous volume. In the United States, farm animals produce 130 times as much waste as the country's people do—with the result that millions of tons of cow and pig feces are washed into streams and rivers, and some of the nitrogen they carry ends up in groundwater. To this Augean burden can be added the innumerable leaks and overflows from urban sewage systems, the fertilizer runoff from suburban lawns, golf courses, and landscaping, and the nitrates leaking (along with other pollutants) from landfills.

There is very little historical information available about trends in the pollution of aquifers. But several studies show that nitrate concentrations have increased as fertilizer applications and population size have grown. In California's San Joaquin-Tulare Valley, for instance, nitrate levels in groundwater increased 2.5 times between the 1950s and 1980s—a period in which fertilizer inputs grew six-fold. Levels in Danish groundwater have nearly tripled since the 1940s. As with pesticides, the aftermath of this multi-sided assault of excess nutrients has only recently begun to become visible, in part because of the slow speed at which nitrate moves underground.

What happens when nitrates get into drinking water? Consumed in high concentrations—at levels above 10 milligrams (mg) per liter, but usually on the order of 100 mg/liter—they can cause infant methemoglobinemia, or so-called blue-baby syndrome. Because of their low gastric acidity, infant digestive systems convert nitrate to nitrite, which blocks the oxygen-carrying capacity of a baby's blood, causing suffocation and death. Since 1945, about 3,000 cases have been reported worldwide—nearly half of them in Hungary, where private wells have particularly high concentrations of nitrates. Ruminant livestock such as goats, sheep, and cows are vulnerable to methemoglobinemia in much the same way infants are, because their digestive systems also quickly convert nitrate to nitrite. Nitrates are also implicated in digestive tract cancers, although the epidemiological link is still uncertain.

In cropland, nitrate pollution of groundwater can have a paradoxical effect. Too much nitrate can weaken plants' immune systems, making them more vulnerable to pests and disease. So when nitrate-laden groundwater is used to irrigate crops that are also being fertilized, the net effect may be to reduce, rather than to increase production. This kind of over-fertilizing makes wheat more susceptible to wheat rust, for example, and it makes pear trees more vulnerable to fire blight.

In assembling studies of groundwater from around the world, we have found that nitrate pollution is pervasive—but has become particularly severe in the places where human population—and the demand for high food productivity—is most concentrated. In the northern Chinese counties of Beijing, Tianjin, Hebei, and Shandong, nitrate concentrations in groundwater exceeded 50 mg/liter in more than half of the locations studied. (The World Health Organization [WHO] drinking water guideline is 10 mg/liter.) In some places, the concentration had risen as high as 300 mg/liter. Since then, these levels may have increased, as fertilizer applications have escalated since the tests were carried out in 1995 and will likely increase even more as China's population (and demand for food) swells, and as more farmland is lost to urbanization, industrial development, nutrient depletion, and erosion.

Reports from other regions show similar results. The USGS found that about 15 percent of shallow groundwater sampled below agricultural and urban areas in the United States had nitrate concentrations higher than the 10 mg/liter guideline. In Sri Lanka, 79 percent of wells sampled by the British Geological Survey had nitrate levels that exceeded this guideline. Some 56 percent of wells tested in the Yucatan peninsula in Mexico had levels above 45 mg/liter. And the European Topic Centre on Inland Waters found that in Romania and Moldova, more than 35 percent of the sites sampled had nitrate concentrations higher than 50 mg/liter.

From Tank of Gas to Drinking Glass: the Pervasiveness of Petrochemicals

Drive through any part of the United States, and you'll probably pass more gas stations than schools or churches. As you pull into a station to fill up, it may not occur to you that you're parked over one of the most pervasive threats to ground-water: an underground storage tank (UST) for petroleum. Many of these tanks were installed two or three decades ago and, having been left in place long past their expected lifetimes, have rusted through in places—allowing a steady leakage of gasoline into the ground. Because they're underground, they're expensive to dig up and repair, so the leakage in some cases continues for years.

Petroleum and its associated chemicals—benzene, toluene, and gasoline additives such as MTBE—constitute the most common category of groundwater contaminant found in aquifers in the United States. Many of these chemicals are also known or suspected to be cancer-causing. In 1998, the EPA found that over 100,000 commercially owned petroleum USTs were leaking, of which close to 18,000 are known to have contaminated groundwater. In Texas, 223 of 254 counties report leaky USTs, resulting in a silent disaster that, according to the EPA, "has affected, or has the potential to affect, virtually every major and minor aquifer in the state." Household tanks, which store home heating oil, are a problem as well. Although the household tanks aren't subject to the same regulations and inspections as commercial ones, the EPA says they are "undoubtedly leaking." Outside the United States, the world's ubiquitous petroleum storage tanks are even less monitored, but spot tests suggest that the threat of leakage is omnipresent in the industrialized world. In 1993, petroleum giant Shell reported that a third of its 1,100 gas stations in the United Kingdom were known to have contaminated soil and

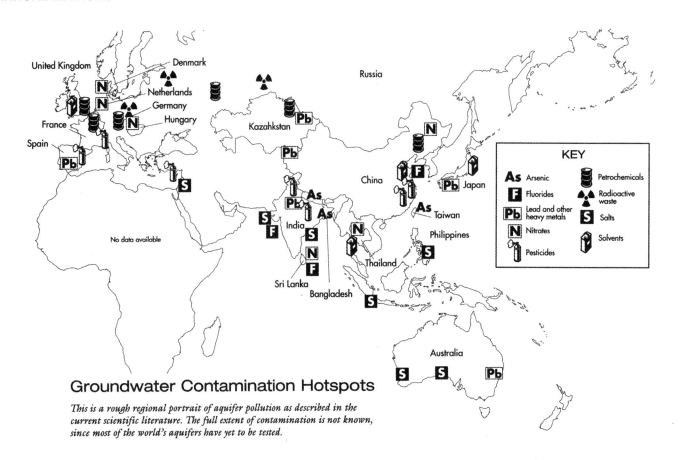

Groundwater Contamination Hotspots

This is a rough regional portrait of aquifer pollution as described in the current scientific literature. The full extent of contamination is not known, since most of the world's aquifers have yet to be tested.

(graphic continues on next page)

groundwater. Another example comes from the eastern Kazakh town of Semipalatinsk, where 6,460 tons of kerosene have collected in an aquifer under a military airport, seriously threatening the region's water supplies.

The widespread presence of petrochemicals in groundwater constitutes a kind of global malignancy, the danger of which has grown unobtrusively because there is such a great distance between cause and effect. An underground tank, for example, may take years to rust; it probably won't begin leaking until long after the people who bought it and installed it have left their jobs. Even after it begins to leak, it may take several more years before appreciable concentrations of chemicals appear in the aquifer—and it will likely be years beyond that before any health effects show up in the local population. By then, the trail may be decades old. So it's quite possible that any cancers occurring today as a result of leaking USTs might originate from tanks that were installed half a century ago. At that time, there were gas tanks sufficient to fuel 53 million cars in the world; today there are enough to fuel almost 10 times that number.

From Sediment to Solute: the Emerging Threat of Natural Contaminants

In the early 1990s, several villagers living near India's West Bengal border with Bangladesh began to complain of skin sores that wouldn't go away. A researcher at Calcutta's

Jadavpur University, Dipanker Chakraborti, recognized the lesions immediately as early symptoms of chronic arsenic poisoning. In later stages, the disease can lead to gangrene, skin cancer, damage to vital organs, and eventually, death. In the months that followed, Chakraborti began to get letters from doctors and hospitals in Bangladesh, who were seeing streams of patients with similar symptoms. By 1995, it was clear that the country faced a crisis of untold proportions, and that the source of the poisoning was water from tubewells, from which 90 percent of the country gets its drinking water.

Experts estimate that today, arsenic in drinking water could threaten the health of 20 to 60 million Bangladeshis-up to half the country's population—and another 6 to 30 million people in West Bengal. As many as 1 million wells in the region may be contaminated with the heavy metal at levels between 5 and 100 times the WHO drinking water guidelines of 0.01 mg/liter.

How did the arsenic get into groundwater? Until the early 1970s, rivers and ponds supplied most of Bangladesh's drinking water. Concerned about the risks of water-borne disease, the WHO and international aid agencies launched a well-drilling program to tap groundwater instead. However, the agencies, not aware that soils of the Ganges aquifers are naturally rich in arsenic, didn't test the sediment before drilling tubewells. Because the effects of chronic arsenic poisoning can take up to 15 years to appear, the epidemic was not addressed until it was well under way.

Groundwater Contamination Hotspots
(continued from previous page)

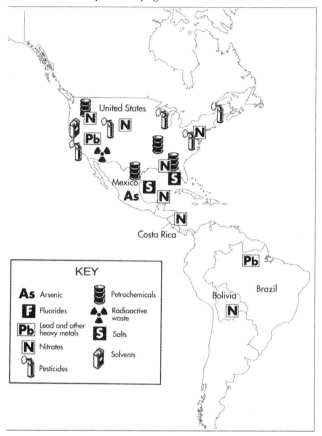

KEY

As Arsenic

F Fluorides

Pb Lead and other heavy metals

N Nitrates

Pesticides

Petrochemicals

Radioactive waste

S Salts

Solvents

Scientists are still debating what chemical reactions released the arsenic from the mineral matrix in which it is naturally bound up. Some theories implicate human activities. One hypothesis is that as water was pumped out of the wells, atmospheric oxygen entered the aquifer, oxidizing the iron pyrite sediments, and causing the arsenic to dissolve. An October 1999 article in the scientific journal *Nature* by geologists from the Indian Institute of Technology suggests that phosphates from fertilizer runoff and decaying organic matter may have played a role. The nutrient might have spurred the growth of soil microorganisms, which helped to loosen arsenic from sediments.

Salt is another naturally occurring groundwater pollutant that is introduced by human activity. Normally, water in coastal aquifers empties into the sea. But when too much water is pumped out of these aquifers, the process is reversed: seawater moves inland and enters the aquifer. Because of its high salt content, just 2 percent of seawater mixed with freshwater makes the water unusable for drinking or irrigation. And once salinized, a freshwater aquifer can remain contaminated for a very long time. Brackish aquifers often have to be abandoned because treatment can be very expensive.

In Manila, where water levels have fallen 50 to 80 meters because of overdraft, seawater has flowed as far as 5 kilometers into the Guadalupe aquifer that lies below the city. Saltwater has traveled several kilometers inland into aquifers beneath Jakarta and Madras, and in parts of the

U.S. state of Florida. Saltwater intrusion is also a serious problem on islands such as the Maldives and Cyprus, which are very dependent on aquifers for water supply.

Fluoride is another natural contaminant that threatens millions in parts of Asia. Aquifers in the drier regions of western India, northern China, and parts of Thailand and Sri Lanka are naturally rich in fluoride deposits. Fluoride is an essential nutrient for bone and dental health, but when consumed in high concentrations, it can lead to crippling damage to the neck and back, and to a range of dental problems. The WHO estimates that 70 million people in northern China, and 30 million in northwestern India are drinking water with high fluoride levels.

A Chemical Soup

With just over a million residents, Ludhiana is the largest city in Punjab, India's breadbasket state. It is also an important industrial town, known for its textile factories, electroplating industries, and metal foundries. Although the city is entirely dependent on groundwater, its wells are now so polluted with industrial and urban wastes that the water is no longer safe to drink. Samples show high levels of cyanide, cadmium, lead, and pesticides. "Ludhiana City's groundwater is just short of poison," laments a senior official at India's Central Ground Water Board.

Like Ludhiana's residents, more than a third of the planet's people live and work in densely settled cities, which occupy just 2 percent of the Earth's land area. With the labor force thus concentrated, factories and other centers of employment also group together around the same urban areas. Aquifers in these areas are beginning to mirror the increasing density and diversity of the human activity above them. Whereas the pollutants emanating from hog farms or copper mines may be quite predictable, the waste streams flowing into the water under cities contain a witch's brew of contaminants.

Ironically, a major factor in such contamination is that in most places people have learned to dispose of waste—to remove it from sight and smell—so effectively that it is easy to forget that the Earth is a closed ecological system in which nothing permanently disappears. The methods normally used to conceal garbage and other waste—landfills, septic tanks, and sewers—become the major conduits of chemical pollution of groundwater. In the United States, businesses drain almost 2 million kilograms of assorted chemicals into septic systems each year, contaminating the drinking water of 1.3 million people. In many parts of the developing world, factories still dump their liquid effluents onto the ground and wait for it to disappear. In the Bolivian city of Santa Cruz, for example, a shallow aquifer that is the city's main water source has had to soak up the brew of sulfates, nitrates, and chlorides dumped over it. And even protected landfills can be a potent source of aquifer pollution: the EPA found that a quarter of the landfills in the U.S. state of Maine, for example, had contaminated groundwater.

In industrial countries, waste that is too hazardous to land-fill is routinely buried in underground tanks. But as these caskets age, like gasoline tanks, they eventually spring leaks. In California's Silicon Valley, where electronics industries store assorted waste solvents in underground tanks, local groundwater authorities found that 85 percent of the tanks they inspected had leaks. Silicon Valley now has more Superfund sites—most of them affecting groundwater—than any other area its size in the country. And 60 percent of the United States' liquid hazardous waste—34 billion liters of solvents, heavy metals, and radioactive materials—is directly injected into the ground. Although the effluents are injected below the deepest source of drinking water, some of these wastes have entered aquifers used for water supplies in parts of Florida, Texas, Ohio, and Oklahoma.

Shenyang, China, and Jaipur, India, are among the scores of cities in the developing world that have had to seek out alternate supplies of water because their groundwater has become unusable. Santa Cruz has also struggled to find clean water. But as it has sunk deeper wells in pursuit of pure supplies, the effluent has traveled deeper into the aquifer to replace the water pumped out of it. In places where alternate supplies aren't easily available, utilities will have to resort to increasingly elaborate filtration set-ups to make the water safe for drinking. In heavily contaminated areas, hundreds of different filters may be necessary. At present, utilities in the U.S. Midwest spend $400 million each year to treat water for just one chemical—atrazine, the most commonly detected pesticide in U.S. groundwater. When chemicals are found in unpredictable mixtures, rather than discretely, providing safe water may become even more expensive.

One Body, Many Wounds

The various incidents of aquifer pollution described may seem isolated. A group of wells in northern China have nitrate problems; another lot in the United Kingdom are laced with benzene. In each place it might seem that the problem is local and can be contained. But put them together, and you begin to see a bigger picture emerging. Perhaps most worrisome is that we've discovered as much damage as we have, despite the very limited monitoring and testing of underground water. And because of the time-lags involved—and given our high levels of chemical use and waste generation in recent decades—what's still to come may bring even more surprises.

Some of the greatest shocks may be felt in places where chemical use and disposal has climbed in the last few decades, and where the most basic measures to shield groundwater have not been taken. In India, for example, the Central Pollution Control Board (CPCB) surveyed 22 major industrial zones and found that groundwater in every one of them was unfit for drinking. When asked about these findings, CPCB chairman D.K. Biswas remarked, "The result is frightening, and it is my belief that we will get more shocks in the future."

Jack Barbash, an environmental chemist at the U.S. Geological Survey, points out that we may not need to wait for expensive tests to alert us to what to expect in our groundwater. "If you want to know what you're likely to find in aquifers near Shanghai or Calcutta, just look at what's used above ground," he says. "If you've been applying DDT to a field for 20 years, for example, that's one of the chemicals you're likely to find in the underlying groundwater." The full consequences of today's chemical-dependent and waste-producing economies may not become apparent for another generation, but Barbash and other scientists are beginning to get a sense of just how serious those consequences are likely to be if present consumption and disposal practices continue.

Changing Course

Farmers in California's San Joaquin Valley began tapping the area's seemingly boundless groundwater store in the late-nineteenth century. By 1912, the aquifer was so depleted that the water table had fallen by as much as 400 feet in some places. But the farmers continued to tap the resource to keep up with demand for their produce. Over time, the dehydration of the aquifer caused its clay soil to shrink, and the ground began to sink—or as geologists put it, to "subside." In some parts of the valley, the ground has subsided as much as 29 feet—cracking foundations, canals, and aqueducts.

When the San Joaquin farmers could no longer pump enough groundwater to meet their irrigation demands, they began to bring in water from the northern part of the state via the California Aqueduct. The imported water seeped into the compacted aquifer, which was not able to hold all of the incoming flow. The water table then rose to an abnormally high level, dissolving salts and minerals in soils that had not been previously submerged. The salty groundwater, welling up from below, began to poison crop roots. In response, the farmers installed drains under irrigated fields—designed to capture the excess water and divert it to rivers and reservoirs in the valley so that it wouldn't evaporate and leave its salts in the soil.

But the farmers didn't realize that the rocks and soils of the region contained substantial amounts of the mineral selenium, which is toxic at high doses. Some of the selenium leached into the drainage water, which was routed to the region's wetlands. It wasn't until the mid-1980s that the aftermath of this solution became apparent: ecologists noticed that thousands of waterfowl in the nearby Kesterson Reservoir were dying of selenium poisoning.

Hydrological systems are not easy to outmaneuver, and the San Joaquin farmers' experience serves as a kind of cautionary tale. Each of their stopgap solutions temporarily took care of an immediate obstacle, but led to a longer-term problem more severe than the original one. "Human understanding has lagged one step behind the inflexible realities governing the aquifer system," observes USGS hydrologist Frank Chapelle.

Some Major Threats to Groundwater

Threat	Sources	Health and Ecosystem Effects at High Concentrations	Principal Regions Affected
Pesticides	Runoff from farms, backyards, golf courses; landfill leaks.	Organochlorines linked to reproductive and endocrine damage in wildlife; organophosphates and carbamates linked to nervous system damage and cancers.	United States, Eastern Europe, China, India.
Nitrates	Fertilizer runoff; manure from livestock operations; septic systems.	Restricts amount of oxygen reaching brain, which can cause death in infants ("blue-baby syndrome"); linked to digestive tract cancers. Causes algal blooms and eutrophication in surface waters.	Midwestern and mid-Atlantic United States, North China Plain, Western Europe, Northern India.
Petro-chemicals	Underground petroleum storage tanks	Benzene and other petrochemicals can be cancer-causing even at low exposure.	United States, United Kingdom, parts of former Soviet Union.
Chlorinated Solvents	Effluents from metals and plastics degreasing; fabric cleaning, electronics and aircraft manufacture.	Linked to reproductive disorders and some cancers.	Western United States, industrial zones in East Asia.
Arsenic	Naturally occurring; possibly exacerbated by over-pumping aquifers and by phosphorus from fertilizers.	Nervous system and liver damage; skin cancers.	Bangladesh, Eastern India, Nepal, Taiwan.
Other Heavy Metals	Mining waste and tailings; landfills; hazardous waste dumps.	Nervous system and kidney damage; metabolic disruption.	United States, Central America and north-eastern South America, Eastern Europe.
Fluoride	Naturally occurring.	Dental problems; crippling spinal and bone damage.	Northern China, Western India; parts of Sri Lanka and Thailand.
Salts	Seawater intrusion; de-icing salt for roads.	Freshwater unusable for drinking or irrigation.	Coastal China and India, Gulf coasts of Mexico and Florida, Australia, Philippines.

Major sources: European Environmental Agency, USGS, British Geological Survey.

Around the world, human responses to aquifer pollution thus far have essentially reenacted the San Joaquin Valley farmers' well-meaning but inadequate approach. In many places, various authorities and industries have fought back the contamination leak by leak, or chemical by chemical—only to find that the individual fixes simply don't add up. As we line landfills to reduce leakage, for instance, tons of pesticide may be running off nearby farms and into aquifers. As we mend holes in underground gas tanks, acid from mines may be seeping into groundwater. Clearly, it's essential to control the damage we've already inflicted, and to protect communities and ecosystems from the poisoned fallout. But given what we already know—that damage done to aquifers is mostly irreversible, that it can take years before groundwater pollution reveals itself, that chemicals react synergistically, and often in unanticipated ways—its now clear that a patchwork response isn't going to be effective. Given how much damage this pollution inflicts on public health, the environment, and the economy once it gets into the water, it's critical that emphasis be shifted from filtering out toxins to not using them in the first place. Andrew Skinner, who heads the International Association of Hydrogeologists, puts it this way: "Prevention is the only credible strategy."

To do this requires looking not just at individual factories, gas stations, cornfields, and dry cleaning plants, but at the whole social, industrial, and agricultural systems of which these businesses are a part. The ecological untenability of these systems is what's really poisoning the world's water. It is the predominant system of high-input agricul-

ture, for example, that not only shrinks biodiversity with its vast monocultures, but also overwhelms the land—and the underlying water—with its massive applications of agricultural chemicals. It's the system of car-dominated, geographically expanding cities that not only generates unsustainable amounts of climate-disrupting greenhouse gases and acid rain-causing air pollutants, but also overwhelms aquifers and soils with petrochemicals, heavy metals, and sewage. An adequate response will require a thorough overhaul of each of these systems.

Begin with industrial agriculture. Farm runoff is a leading cause of groundwater pollution in many parts of Europe, the United States, China, and India. Lessening its impact calls for adopting practices that sharply reduce this runoff—or, better still, that require far smaller inputs to begin with. In most places, current practices are excessively wasteful. In Colombia, for example, growers spray flowers with as much as 6,000 liters of pesticide per hectare. In Brazil, orchards get almost 10,000 liters per hectare. Experts at the U.N. Food and Agricultural Organization say that with modified application techniques, these chemicals could be applied at one-tenth those amounts and still be effective. But while using more efficient pesticide applications would constitute a major improvement, there is also the possibility of reorienting agriculture to use very little synthetic pesticide at all. Recent studies suggest that farms can maintain high yields while using little or no synthetic input. One decade-long investigation by the Rodale Institute in Pennsylvania, for example, compared traditional manure and legume-based cropping systems which used no synthetic

195

fertilizer or pesticides, with a conventional, high-intensity system. All three fields were planted with maize and soybeans. The researchers found that the traditional systems retained more soil organic matter and nitrogen—indicators of soil fertility—and leached 60 percent less nitrate than the conventional system. Although organic fertilizer (like its synthetic counterpart) is typically a potent source of nitrate, the rotations of diverse legumes and grasses helped fix and retain nitrogen in the soil. Yields for the maize and soybean crops differed by less than 1 percent between the three cropping systems over the 10-year period.

In industrial settings, building "closed-loop" production and consumption systems can help slash the quantities of waste that factories and cities send to landfills, sewers, and dumps—thus protecting aquifers from leaking pollutants. In places as far-ranging as Tennessee, Fiji, Namibia, and Denmark, environmentally conscious investors have begun to build "industrial symbiosis" parks in which the unusable wastes from one firm become the input for another. An industrial park in Kalundborg, Denmark diverts more than 1.3 million tons of effluent from landfills and septic systems each year, while preventing some 135,000 tons of carbon and sulfur from leaking into the atmosphere. Households, too, can become a part of this systemic change by reusing and repairing products. In a campaign organized by the Global Action Plan for the Earth, an international nongovernmental organization, thoughtful consumption habits have enabled some 60,000 households in the United States and Europe to reduce their waste by 42 percent and their water use by 25 percent.

As it becomes clearer to decisionmakers that the most serious threats to human security are no longer those of military attack but of pervasive environmental and social decline, experts worry about the difficulty of mustering sufficient political will to bring about the kinds of systemic—and therefore revolutionary—changes in human life necessary to turn the tide in time. In confronting the now heavily documented assaults of climate change and biodiversity loss, leaders seem on one hand paralyzed by how bleak the big picture appears to be—and on the other hand too easily drawn into denial or delay by the seeming lack of immediate consequences of such delay. But protecting aquifers may provide a more immediate incentive for change, if only because it simply may not be possible to live with contaminated groundwater for as long as we could make do with a gradually more irritable climate or polluted air or impoverished wildlife. Although we've damaged portions of some aquifers to the point of no return, scientists believe that a large part of the resource still remains pure—for the moment. That's not likely to remain the case if we continue to depend on simply stepping up the present reactive tactics of cleaning up more of the chemical spills, replacing more of the leaking gasoline tanks, placing more plastic liners under landfills, or issuing more fines to careless hog farms and copper mines. To save the water in time requires the same fundamental restructuring of the global economy as does the stabilizing of the climate and biosphere as a whole—the rapid transition from a resource-depleting, oil- and coal-fueled, high-input industrial and agricultural economy to one that is based on renewable energy, compact cities, and a very light human footprint. We've been slow to come to grips with this, but it may be our thirst that finally makes us act.

"Heaven is Under Our Feet"

Throughout human history, people have feared that the skies would be the source of great destruction. During the Cold War, industrial nations feared nuclear attack from above, and spent vast amounts of their wealth to avert it. Now some of that fear has shifted to the threats of atmospheric climate change: of increasing ultraviolet radiation through the ozone hole, and the rising intensity of global warming-driven hurricanes and typhoons. Yet, all the while, as the worldwide pollution of aquifers now reveals, we've been slowly poisoning ourselves from beneath. What lies under terra firma may, in fact, be of as much concern as what happens in the firmament above.

The ancient Greeks created an elaborate mythology about the Underworld, or Hades, which they described as a dismal, lifeless place completely lacking the abundant fertility of the world above. Science and human experience have taught us differently. Hydrologists now know that healthy aquifers are essential to the life above ground—that they play a vital role not just in providing water to drink, but in replenishing rivers and wetlands and, through their ultimate effects on rainfall and climate, in nurturing the life of the land and air as well. But ironically, our neglectful actions now threaten to make the Greek myth a reality after all. To avert that threat now will require taking to heart what the hydrologists have found. As Henry David Thoreau observed a century-and-a-half ago, "Heaven is under our feet, as well as over our heads."

A Few Key Sources

Francis H. Chapelle, *The Hidden Sea: Ground Water, Springs, and Wells* (Tucson, AZ: Geoscience Press, Inc., 1997).

U.N. Environment Programme, *Groundwater: A Threatened Resource* (Nairobi: 1996).

European Environmental Agency, *Groundwater Quality and Quantity in Europe* (Copenhagen: 1999).

U.S. Geological Survey, *The Quality of Our Nation's Waters—Nutrients and Pesticides* (Reston, VA: 1999).

British Geological Survey et al., *Characterisation and Assessment of Groundwater Quality Concerns in Asia-Pacific Region* (Oxfordshire, UK: 1996).

Payal Sampat is a staff researcher at the Worldwatch Institute.

From *World Watch*, January/February 2000, pp. 10-22. © 2000 by the Worldwatch Institute (www.worldwatch.org). Reprinted by permission.

Water Quality: THE ISSUES

BY MARY H. COOPER

The last three decades have seen dramatic improvements in the quality of America's waterways. Majestic cormorants and herons are again feeding in many rivers, signaling the return of the fish they prey on. Swimmers are taking the plunge into waters that once would have sickened them. Anglers are catching fish that are healthy enough to eat.

"Twenty-eight years ago, the Potomac River was too dirty to swim in, Lake Erie was dying and [Ohio's] Cuyahoga River was so polluted it burst into flames," said Carol M. Browner, administrator of the Environmental Protection Agency (EPA), the agency created in 1970 to implement federal environmental policy.

"Many rivers and beaches were little more than open sewers," Browner continued. "Enactment of the [1972] Clean Water Act dramatically improved the health of rivers, lakes and coastal waters. It stopped billions of pounds of pollution from fouling the water and doubled the number of waterways safe for fishing and swimming. Today, many rivers, lakes, and coasts are thriving centers of healthy communities."[1]

But there is still cause for concern about water pollution in the United States. Even after three decades of federal cleanup efforts, about 40 percent of the nation's streams, rivers, lakes and coastal waters are still too dirty for people to fish or swim in, the EPA reports.[2] That's down just 20 percentage points from the 60 percent recorded in 1972.

The culprits today are less likely to be factories and sewage treatment plants, which were the main focus of the first round of clean-water efforts. Those "point sources" were easy to identify and target because they generally discharged pollutants directly into the water. The Clean Water Act required industries and municipalities to filter out wastes before they reached the water.

Although the most egregious point-source pollution has been reduced, a more insidious form of pollution continues to dirty the nation's waterways—runoff from city streets, suburban construction sites and farms. Contaminated by fertilizers and animal waste, agricultural runoff contains nitrogen, phosphorus and other nutrients, which deprive waterways of the oxygen needed to support aquatic life. Runoff also contains toxins washed into storm sewers from city streets, which kill fish outright and threaten human health.

The extent of impairment to water quality caused by nutrient-polluted runoff is especially evident in estuaries, the bays and tidal rivers that empty into the oceans. Nutrients from as far away as Montana travel down the Missouri and Mississippi rivers, picking up additional pollutants along the way. Emptying into the Gulf of Mexico, they cause a lifeless "dead zone" that spreads over an area that at times equals the state of New Jersey. Similar problems plague the nation's largest estuary, the Chesapeake Bay, where a regional cleanup campaign is trying to stem the impact of nutrient pollution. (*See sidebar*, "Regional Plan Offers Hope for Chesapeake Bay.")

"It's clear that nutrient pollution remains the most daunting challenge in restoring the Chesapeake," says Michael Hirshfield, senior vice president for resource protection at the Chesapeake Bay Foundation, in Annapolis, Md. "We've known for 20 years that the biggest water-quality problem the bay faced was having far too much nitrogen and phosphorus coming in. We've made some modest gains, particularly in dealing with point sources. But when you look at what's happening on the land in the watershed surrounding the bay, it's clear that if we're really going to bring back underwater grasses and have adequate dissolved oxygen and clear water, pretty much every sector of the population and the economy is going to have to do a lot more to control nutrient pollution."

Cleaning up non-point source pollution—which cannot be traced to easily identifiable discharge points—poses the toughest challenge to policy-makers. And proposals for the next generation of programs to achieve the Clean Water Act's goal of making all the nation's waterways "fishable" and "swimmable" are highly controversial.

"Our regulatory systems are set up for fundamentally simple one-cause, one-effect, one-regulation kinds of situations, where you've got a sewage-treatment plant or a factory that's discharging into one little waterbody," Hirshfield explains. "Nutrient pollution, where the effects of lots of non-point sources are both distant from the sources and cumulative, means that allocating responsibility is a real challenge to our sector-by-sector environmental laws."

To speed the process of curbing runoff, President Clinton last year proposed a new set of "non-point source" regulations. The new rules would require states to set pollution limits—known as "total maximum daily loads," or TMDLs—for all the estimated 20,000 waterways that are polluted and establish a timetable for reducing pollution to meet those limits. The states would be responsible for monitoring and enforcing the regulations, and

America's Polluted Waters, 1998

Pollution of the nation's streams, rivers, coastlines, estuaries and lakes is largely caused by non-point pollution, such as agricultural runoff, and the impact of high population density. Waters are considered threatened or impaired if they do not meet one or more state water-quality standards. In some cases, areas not shown to suffer from pollution actually may be polluted but were not identified as impaired.

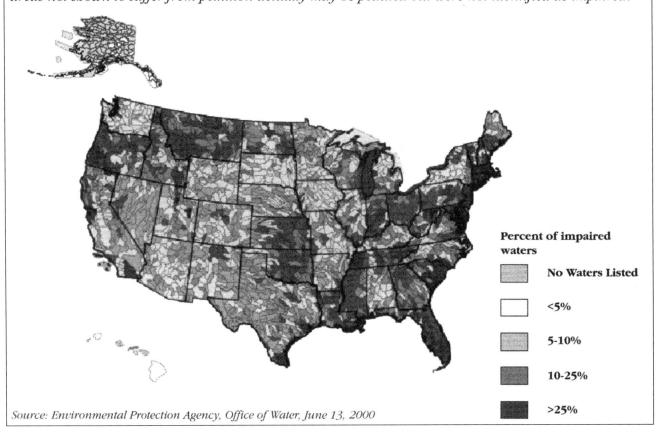

Percent of impaired waters

- No Waters Listed
- <5%
- 5-10%
- 10-25%
- >25%

Source: Environmental Protection Agency, Office of Water, June 13, 2000

EPA would get involved only if states are unable or unwilling to carry out the program.[3]

EPA officials say that the proposed regulations mark a departure from past clean-water efforts. "The TMDL program unquestionably represents a turning point in our nation's Clean Water Act programs," says J. Charles Fox, assistant administrator for EPA and director of the agency's Office of Water. "We've spent the better part of 30 years focusing on discrete point sources of pollution that have been heavily dependent on national and state-level regulations to effect pollution control. Today's challenges are much more diffuse, but the regulatory program is still going to have a very important role in the future as well."

A broad coalition of groups representing farmers, timber companies and builders says the regulations amount to bureaucratic excess. "We believe the agency has well overreached its regulatory authority," says Don Parrish, senior environmental policy specialist for the American Farm Bureau Federation, the country's leading farm lobby. "We don't believe that there's any federal implementation authority for TMDLs; we believe that authority is reserved for the states.

"The way I read the Clean Water Act," Parrish continues, "it gives EPA the authority to permit discharges. It does not give EPA the authority to permit people to operate. We still have a free-enterprise system here."

Because nutrient-laden runoff is the leading culprit in water pollution today, the new regulations would be directed largely toward curbing runoff at the source—such as farms, animal feedlots and timber operations. Farmers might be required to let more land lie fallow, for example, while feedlot operators might have to invest in new technology to store and treat manure to keep it out of waterways. Critics argue that the regulations would drive many operations out of business.

"It's going to be expensive to invest in these new technological fixes," Parrish says. "A lot of small operators are not going to be able to meet those requirements, and they're just going to load their animals on a trailer and take them to the local sale barn and go out of business."

In October, the Senate voted to delay implementation of the new TMDL rules until studies can be done to evaluate their costs and benefits.

As lawmakers take up the next phase of clean-water regulations, these are some of the issues they will consider:

Assessing U.S. Water Quality

About 40 percent of U.S. waters that were assessed in 1998 were not clean enough for fishing and swimming, including more than 290,000 miles of rivers and streams.* Leading pollutants in impaired waters include siltation, bacteria, nutrients and metals. Runoff from agricultural lands and urban areas is the primary source of these pollutants.

Quality of Rivers, Lakes and Estuaries

Waterbody Type	Total Size	Amount Assessed (% of total)	Good (% of assessed)	Good but Threatened (% of assessed)	Polluted (% of assessed)
Rivers (miles)	3,662,255	842,426 (23%)	463,441 (55%)	85,544 (10%)	291,264 (34%)
Lakes (acres)	41,593,748	17,390,370 (42%)	7,927,486 (46%)	1,565,175 (9%)	7,897,110 (45%)
Estuaries (sq. miles)	90,465	28,687 (32%)	13,439 (47%)	2,766 (10%)	12,482 (44%)

*About 32 percent of U.S. waters were assessed.

Note: Percentages may not add to 100% due to rounding.

Source: Environmental Protection Agency, "1998 National Water Quality Inventory Report to Congress," June 2000

Are proposed new rules to curb storm water and agricultural runoff the best way to clean up streams and rivers?

The federal regulatory system constructed by the Clean Water Act and its subsequent amendments marked a clear departure in environmental policy-making. Before the law, water, like land, fell under the jurisdiction of the states. The only exceptions were waterways that pass through more than one state, such as the Missouri River and the Great Lakes.

Congress changed that system because the states had clearly failed to stop the wholesale dumping of industrial pollutants and untreated sewage into the nation's waterways, especially with the economic boom and population increases that followed World War II. The Clean Water Act established a number of programs to clean up the water and shifted the primary responsibility for implementing them from the states to the federal EPA. In most cases EPA has delegated its authority to implement the law to state environmental protection agencies. In the 16 states that are either unable or unwilling to do the job, the federal agency assumes direct responsibility for enforcing its provisions.

Most observers agree that the regulatory approach has worked well to reduce point-source water pollution. It has allowed states that were not eager to offend polluting—but profitable and job-creating—industries to place the blame for the new rules squarely on the shoulders of the federal government. EPA issues the permits defining pollution thresholds to each polluter and is ultimately responsible for enforcing the law and imposing fines and other sanctions to repeat offenders. The states also have received federal funding under the law to help pay for upgrading municipal sewage-treatment plants, the other main point source of water pollution.

But the federal regulatory approach is far more controversial in the context of today's main threat to water quality, non-point source pollution, or runoff. Not only are the polluters harder to identify, they often are less able to bear the cost of mitigating the harm they cause to waterways, often hundreds of miles downstream. Many critics contend that neither EPA nor the states have collected enough scientifically valid data to justify requiring alleged polluters to shoulder the responsibility for fouled waterways.

"After nearly 30 years and $600 billion worth of hit-and-miss technologies, we still don't know what has been achieved or what still needs to be done," writes Richard A. Halpern, director of environmental affairs for the Hudson Institute's Center for Global Food Issues, a conservative think tank in Churchville, Va. "Also, the lack of real-world data leaves activists free to claim that 'our water continues to be poisoned,' and provides them a pretext for demanding an increasingly intrusive role in determining the national lifestyle."[4]

But supporters of the regulatory approach say it's neither necessary nor prudent to await the exact measurement of each and every tainted waterway to take steps to curb pollution. "I will be the first to admit that in many cases more science and more data would be better," says EPA's Fox. "But in the vast majority of cases we have enough information to make a rational decision today." He cites the validity of warnings in the early 1970s that excessive phosphorus levels were causing fish kills in the Great Lakes. "There were scientists at the time who protested that we didn't know exactly how much phosphorus was present. But the trends were unmistakable, and people in the late 1970s had the courage to make a decision to control phosphorus, and it clearly was the right decision."

Critics contend that federal regulations on runoff violate states' jurisdiction over land use. In September, the American Farm Bureau Federation filed a lawsuit in California to block the implementation of federal water-quality regulations that would restrict timber harvests in an effort to reduce runoff into Mendocino County's Garcia River.[5]

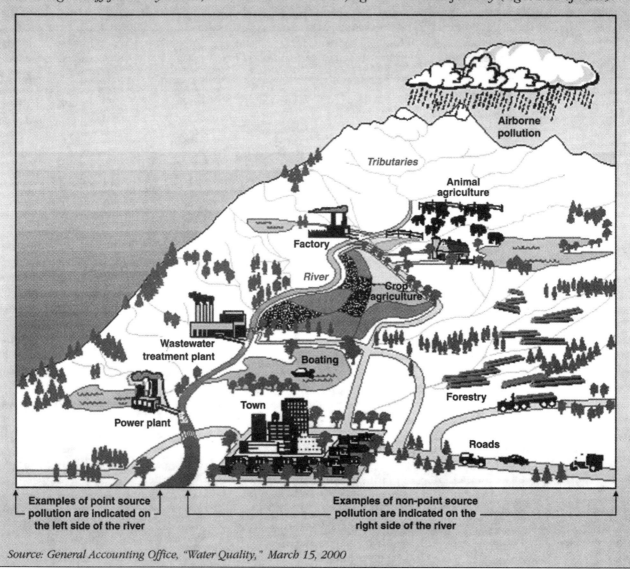

Point and Non-point Sources of Pollution

Factories and sewage-treatment plants (left side of river) were the main focus of U.S. clean-water efforts 30 years ago. But today the biggest remaining causes of water pollution are non-point sources, including runoff from city streets, suburban construction, agriculture and forestry (right side of river).

Airborne pollution

Tributaries

Animal agriculture

Factory

River

Crop agriculture

Wastewater treatment plant

Boating

Forestry

Power plant

Town

Roads

↳ Examples of point source pollution are indicated on the left side of the river ┘

Examples of non-point source pollution are indicated on the right side of the river

Source: General Accounting Office, "Water Quality," March 15, 2000

Removing trees and other vegetation from the land can increase runoff into nearby waterways, but not in an easily measured way, critics say. "As the statute defines total maximum daily loads (TMDLs), they imply a specific daily allocation of pollutant loads," Parrish says. "But you don't have daily loads with agricultural runoff; you only have agricultural runoff in the presence of snowmelt or a rainfall runoff event. And we have a real hard time applying what we believe is very specific, point-source statutory authority to a non-point source that is driven primarily by stormfall events."

Defenders of the regulatory nature of clean-water policies say there is room for accommodation in the proposed new rules. "I don't think anybody has ever said that every farm or every suburban development needs to get a permit," says Hirshfield of the Chesapeake Bay Foundation. He likens the clean-water rules to speed limits on the nation's roadways. "We don't put a machine in each one of our cars that prevents it from going over the speed limit; we post signs, and we have cops. The TMDL approach does not say that everybody needs to have a discharge permit; it says that everybody in a watershed who is contributing to the failure of a waterbody to meet its fishable, swimmable goals has to contribute to the solution."[6]

Can non-regulatory approaches achieve good water quality?

Not surprisingly, the strongest critics of EPA's policies offer a number of solutions to water-quality problems that do not in-

Regional Plan Offers Hope for Chesapeake Bay

Perhaps no body of water better illustrates the nature of today's threats to water quality—and the difficulty of neutralizing them—than the Chesapeake Bay. North America's largest estuary lies at the end of a vast watershed that encompasses six states—Delaware, Maryland, New York, Pennsylvania, Virginia and West Virginia—and the District of Columbia and includes much of the heavily populated East Coast corridor.

When John Smith sailed into the Chesapeake in 1607, he and the English settlers who accompanied him to the New World found a vast body of water that was teeming with oysters, crabs and many species of fish. After almost 400 years of intense development in the bay's watershed, however, industrial pollutants and human waste threatened to kill virtually all the bay's plant and animal species. Since passage of the Clean Air Act in 1970, most factories and sewage-treatment plants have installed technologies that filter out many of these pollutants. But like other waterways around the country, the bay is threatened today primarily by runoff filled with nutrients and chemicals from farms and city streets.

According to the Chesapeake Bay Foundation in Annapolis, Md., efforts to clean up the bay have made only modest improvements. In its annual report on the state of the bay, the group this year gave the bay a score of 28 on a scale of 100, which represents its pristine condition when the Europeans first arrived.[1] That's a 20 percent improvement since 1983, when the bay only scored 23.

Although foundation officials don't expect the estuary to ever return to its earlier condition, they hope to attain a score of 70 over the next two decades. The score is based on 13 factors that help determine the bay's overall health, including water clarity, the presence of underwater grasses and populations of oysters, striped bass and other aquatic species. Of these, only one—the presence of shad—improved last year, but the improvement from a score of 3 to 5 was hardly overwhelming.

The Chesapeake is not alone. "The Chesapeake may be the biggest, most thoroughly studied and best-known estuary in the country," says Michael Hirshfield, the foundation's senior vice president for resource protection. "But if you look at almost any estuary, whether it's Tampa Bay or Long Island Sound, nutrient pollution is a ubiquitous problem. One of the lessons of the last decade or so is that pretty much wherever you look near coastal waters you will see similar issues—point sources from sewage-treatment plants, livestock and agriculture, fallout from cars and power plants, runoff from suburban lawns and even dog poop."

Of all the causes of polluted runoff, the foundation singles out one as the main culprit—sprawl. "Chesapeake Bay restoration is being undermined by sloppy development," said William C. Baker, the foundation's president. "Millions of dollars are being spent to restore wetlands and underwater grasses, yet population growth and its ugly stepchild, sprawl, continue to threaten the bay."[2]

Congress has singled out estuaries for special treatment. On Oct. 25, lawmakers passed a law to establish a national estuary-protection program. It provides $275 million to restore more than a million acres of estuary habitat over the next decade. The new law also provides $40 million a year through fiscal 2005 for continued restoration of the Chesapeake Bay.

Meanwhile, environmental activists are guardedly optimistic about the impact of a new regional agreement to step up efforts to save the bay. By signing the new Chesapeake 2000 agreement in June, Maryland, Virginia, Pennsylvania and the District of Columbia agreed to preserve millions of acres of wetlands, which serve as a vital buffer between runoff and the bay, increase the bay's population of oysters, which filter pollutants from the water, and reduce the rate at which farmland and woodlands are lost to development.

"The Chesapeake Bay agreement and the participation of the three main bay states is pretty remarkable," Hirshfield says. "A drawback of this voluntary and consensus-based approach is that the bay cleanup is taking longer than it should have. We believe that you can get only so far with voluntary, consensus-based goals."

EPA officials say the regional approach shows promise for other waterways threatened by polluted runoff. "We learned in the mid-1980s that the base level of national regulations is not going to result in a clean Chesapeake Bay," says J. Charles Fox, EPA's assistant administrator for the Office of Water. "Defining how much more needed to be done was a job best done in partnership with the federal and state governments, and that's what happened. I think that model is going to be used a lot in the future in our national water programs."

---

[1] Chesapeake Bay Foundation, "The State of the Bay Report 2000."

[2] Baker spoke on Sept. 20, 2000, following the report's release. See Anita Huslin, "Bay Gets Failing Grade, but Progress Is Cited," *The Washington Post*, Sept. 21, 2000.

---

volve federal mandates. The Farm Bureau Federation, for example, says the federal regulatory approach is not appropriate for farmers or other non-point sources of water pollution and calls for a more collaborative effort involving farmers and surrounding communities at the state and local levels.

"What we've tried to do in the agricultural community and what we'd like to see done more specifically is take a closer look at these streams and rivers, and on a more collaborative basis," Parrish says. "We don't mind doing it with the state environmental protection agency or other interest groups within the states, but we would like to see a prioritization of where our water-quality problems are, then apply the resources to those

streams, rivers, lakes and estuaries in such a way that we can systematically address the worst cases first."

Funding for this approach, Parrish says, is available through existing farm bill programs that offer incentives for farmers to invest in clean-water technologies. "We believe that's a more targeted approach," he says. "Our approach would be one in which you actually find your water-quality problems and in a very systematic fashion get the community to buy into the effort and make the hard decisions at the local level to get them addressed."

Some current and former state regulators agree that a more flexible approach is needed. "There's no question about the fact that water-quality problems are site- and situation-specific problems that require site- and situation-specific solutions," says Becky Norton Dunlop, Virginia's secretary of natural resources during the administration of Republican Gov. George Allen and currently vice president for external relations at the conservative Heritage Foundation.

"People from the states understand the national goal to improve water quality should be able to set priorities for each river and move toward achieving clean water in ways that are fully embraced by the communities involved and based on good science," Dunlop continues. "I'm sure there are occasions when good science has been used by EPA, but what we've seen the agency do over the past eight years is use political science to make and implement national policy."

A few industries that are likely to be the focus of EPA's stepped-up effort to curb polluted runoff are taking steps on their own to clean up their operations—and ward off the heavy hand of impending federal regulations. Near the top of the list are so-called concentrated animal-feeding operations. Also known as factory farms, these facilities turn out huge numbers of hogs, cattle and chickens and increasingly are replacing smaller-scale family farms. The concentrations of animals have created vast amounts of manure and the inevitable horror stories. Last year, for example, a waste lagoon burst at a hog farm in Duplin County, N.C., spilling some 2 million gallons of hog manure into a tributary of the Northeast Cape Fear River.[7]

To avert future spills, the National Pork Producers Council has devised the livestock industry's most sweeping clean-water measure to date. The voluntary, industry-funded program employs third-party experts to help hog farmers assess their waste-management practices and invest in technologies to treat and store hog manure.

"Quite frankly, some of our livestock colleagues didn't appreciate what we were doing, but we realized that the states were taking action and setting up regulatory programs that we were concerned may not be affordable and practical for pork producers," says Deborah Atwood, an environmental consultant for the Des Moines, Iowa-based council. "So we decided that all pork producers, of all types and sizes, should have water permits."

Since the council introduced the $1.5-million program in 1997, Atwood says about 1,600 hog farmers have opted to participate. "The program has moved out of the pork industry now into the dairy, turkey, chicken and cattle industries," she says. "They're all realizing that it's very helpful to have a third set of eyes come to their operations and examine them for everything from odor to cleanliness to management. We've discovered that within six months over 67 percent of the problems have been fixed."

Another non-regulatory approach to mitigating water pollution is nutrient trading, a market-based approach that sets a goal for the total amount of nitrogen, phosphorus and other nutrients carried into waterways in runoff. The total amount of allowable pollution in a given waterway is then allocated among the farms, municipalities and other sources that choose to participate in the trading program. Those that can more easily reduce their pollution below their allowable limits earn credits, which they can then sell to polluters that can't afford to eliminate their discharges.

"With trading, whoever has the lowest cost for remediation would have an economic incentive to overcomply with the discharge limits, and in overcomplying they would generate credits," says Paul Faeth, director of the World Resources Institute's economics and population program in Washington, D.C., and the author of a recent study on nutrient trading.[8]

Faeth says the trading approach may greatly reduce the costs associated with further improvements in water quality. "If you go with traditional regulations, it can cost up to $24 for every pound of phosphorus kept out of the water," he says, "but if you go with the trading scheme you can get that down to as little as $2 a pound. We think that one of the key elements in terms of the political acceptance of further movement on water quality is to find ways of keeping those costs down."

Unlike some critics of EPA, however, Faeth says nutrient trading and other non-traditional methods of improving water quality are no substitute for regulations. "We need to move forward with regulation to improve water quality," he says. "But regulations should be coupled with flexibility."

Is tap water safe to drink?

Much of the nation's drinking water comes from lakes and rivers, some of the same waterways that fall under the protection of the Clean Water Act. But about half the nation's drinking water comes from groundwater, or deep underground reservoirs that generally do not mingle with surface waters.

To ensure the safety of the nation's drinking water, Congress passed the Safe Drinking Water Act in 1974. Unlike the Clean Water Act, which attempts to safeguard the cleanliness of all surface waters, the 1974 law generally is limited to regulating the quality of water as it comes out of the tap. In 1996, Congress amended the law to expand its scope to include assessments of water quality in all sources of drinking water.

"The bulk of our attention under the Safe Drinking Water Act is really focused on setting public-health standards for individual contaminants that we believe will protect public health," says EPA's Fox. "We then monitor and enforce the drinking-water treatment facilities and infrastructure to make sure that they can deliver that high-quality water at the tap."

Compared with many parts of the world, where outbreaks of cholera and other waterborne diseases kill about 10 million people each year, the United States has succeeded in safeguarding the purity of its drinking water—so much so that most people take the safety of their tap water for granted.[9]

But the growth of livestock operations and other activities that may pollute runoff into drinking-water supplies poses a health risk that occasionally challenges Americans' complacency. In 1993, for example, 400,000 people in Milwaukee, Wis., got sick, and 100 people died, after drinking municipal tap water that had been contaminated by *Cryptosporidium*, a parasite that is commonly found in livestock waste. Indeed, as many as 7 million Americans get sick from contaminated tap water each year.[10]

In addition to biological toxins, such as bacteria, toxic algae and metals, the water supply is threatened by chemical pollutants, including pesticides, used motor oil and other synthetic contaminants that may be deposited onto city streets and washed into waterways through storm sewers.

Even drugs, including antibiotics used to treat humans and to keep livestock healthy in crowded factory farms, can make their way into tap water.[11] Groundwater also may be contaminated when repeated depositions of toxins seep through soil and rock layers that separate groundwater from the surface.

Over the years, EPA has repeatedly added to the list of contaminants it regulates under the Safe Drinking Water Act. It now contains more than 80 contaminants and is likely to grow even longer as the adverse health effects of chemicals become apparent. The recent discovery in Los Angeles drinking water supplies of chromium 6, a cancer-causing byproduct of an otherwise harmless metal used in aircraft manufacturing and other industries, prompted calls for new standards to address the threat it poses to human health.[12]

Similarly, levels of arsenic, another carcinogen, were found to exceed new standards EPA proposed in May in 95 drinking-water systems in Colorado.[13] Indeed, the National Resources Defense Council (NRDC) estimates that as many as 56 million people in the 25 states that have thus far reported arsenic information to EPA currently are exposed to unsafe levels of arsenic in their tap water.[14]

In fact, scientists have recently determined that some of the chemicals used to purify tap water may be more dangerous than the contaminants themselves. Treating drinking water with chlorine has done much to improve water safety, but the disinfectant has been linked to serious health problems, bladder and colon cancer among them.

To reduce the risk of cancer linked to long-term drinking of chlorinated water, many communities are switching to ammonia. But that change is not without potentially harmful effects: Patients needing kidney dialysis and fish-tank owners will require special treatment of tap water to neutralize the ammonia.

These and other developments in the treatment and quality of drinking water have been brought to the attention of consumers since 1999, when a provision of the 1996 Safe Drinking Water Act Amendments went into effect requiring drinking-water utilities to provide "right-to-know" information to ratepaying consumers. Consumers now receive pamphlets describing the sources, treatment and quality of their tap water.

The law also increased the amount of funding authorized to help the states meet their water-quality responsibilities. But even that may not be enough to enable states to meet the law's stricter standards for drinking water. According to a recent survey by the U.S. General Accounting Office, "over 90 percent of the states predicted that their staffing levels would be less than adequate [to monitor water quality] in the future as a number of new program requirements and complex contaminant regulations take effect.[15]

"Generally speaking, this country has very high-quality drinking water," Fox says. "But we've also learned that we cannot take the quality of that water for granted, even in fairly sophisticated systems like the one Milwaukee had at the time of the *Cryptosporidium* outbreak. If we don't operate these systems very well, and if we aren't very conscious of what pollution can get into our source water, that can have very negative consequences on the quality of our drinking water."

Notes

1. Browner testified on Feb. 23, 2000, before the Senate Agriculture, Nutrition and Forestry Committee.
2. U.S. Environmental Protection Agency, "The Quality of Our Nation's Waters: A Summary of the National Water Quality Inventory: 1998 Report to Congress," June 2000.
3. See Charles Pope, "Clean Water: The Next Wave," *CQ Weekly*, March 18, 2000, pp. 585–586.
4. Richard A. Halpern, "1491 and All That," *American Outlook*, November/December 2000, p. 35.
5. The case, *Pronsolino v. Marcus*, is pending before the 9th U.S. Circuit Court of Appeals.
6. For background, see Mary H. Cooper, "Setting Environmental Priorities," *The CQ Researcher*, May 21, 1999, pp. 425–448.
7. "Spill Caused by Dike Failure," Associated Press Newswires, April 30, 1999.
8. Paul Faeth, "Fertile Ground: Nutrient Trading's Potential to Cost-Effectively Improve Water Quality," World Resources Institute, 2000.
9. Julie McCann, "On tap: The Story of Water: Solid Facts on a Liquid Asset," *National Post*, Oct. 1, 2000.
10. Natural Resources Defense Council, "Drinking Water," www.nrdc.org.
11. See Kathleen Fackelmann, "Drugs Found in Tap Water," *USA Today*, Nov. 8, 2000.
12. See "Chromium 6 Released for Years into L.A. River," The Associated Press, Oct. 1, 2000.
13. See Julia C. Martinez, "Arsenic Levels Too High," *The Denver Post*, Aug. 29, 2000.
14. Natural Resources Defense Council, "Drinking Water," www.nrdc.org.
15. General Accounting Office, "Drinking Water: Spending Constraints Could Affect States' Ability to Meet Increasing Program Requirements," Aug. 31, 2000, p. 4.

It's a breath of fresh air

Thirty years after Earth Day, America is getting its environmental act together

BY DAVID WHITMAN

Shortly after the first Earth Day in 1970, President Richard M. Nixon warned that without far-reaching reforms the United States would, by the year 2000, be "a country in which we can't drink the water, where we can't breathe the air, and in which our children... will not be able to [enjoy] the beautiful open spaces... [of] the American landscape."

Today, as the world prepares to celebrate Earth Day 2000 on April 22, the nation's air is cleaner, its water purer. There is more protected open space in national parks, wildlife refuges, and wilderness areas. Yet surveys indicate that only 14 percent to 36 percent of Americans believe the environment has improved a "great deal" since 1970. And according to a 1999 Roper poll, 56 percent worry that the next 10 years will be "the last decade when humans will have a chance to save the earth from an environmental catastrophe."

Not all the news is good, of course. Urban sprawl, loss of habitat, spoilage of wetlands, and global warming are all serious problems getting worse. And beyond America's borders, the picture is even bleaker. Across the former Soviet Union, environmental contamination is rampant. In Asia, in even some of the most cosmopolitan cities, the air is literally unfit to breathe.

America's record of environmental progress, by contrast, shows that despite the grave problems that persist, there is reason for hope. At the time of the first Earth Day, America was a place where oil-drenched rivers caught fire, loggers lopped down great swaths of national forests, recycling was rare, motorists routinely littered, and fabled icons like the bald eagle were headed toward extinction in the lower

48 states. The Environmental Protection Agency did not exist, and industry and government casually dumped millions of tons of hazardous wastes. In the nation's cities, killer smogs blanketed downtowns, lead emissions addled children's minds, and many municipalities treated urban waterways like open sewers. "In 1970, people were concerned that industrial society was going to choke itself," says George Frampton, chair of the White House Council on Environmental Quality. "The environment is in a lot better condition in the U.S. today."

A look back might start with Thanksgiving Day 1966, when millions of New Yorkers awoke to a surprise far more pungent than the whiff of singed turkey. A smog as thick as a London fog had descended on the city, filled with noxious fumes that left thousands of people with respiratory illnesses literally gasping for air. The smog obliterated chunks of Manhattan's skyline, prompting the shutdown of all 11 of the city's incinerators. Health authorities advised motorists to use their cars only when absolutely necessary; landlords were told to lower thermostats to 60 degrees to curb pollutants.

Several days later, the smog dissipated, but not before an estimated 168 people died as a result—the same number killed in the 1995 Oklahoma City bombing. A mayoral task force appointed to assess the crisis warned that unless polluters were curtailed, New York could become "uninhabitable within a decade."

Three years later, *Life* magazine would report that scientists had amassed "solid experimental... evidence" indicating that by the 1980s urban dwellers "will have to wear gas masks to survive air pollution." Nowhere was the problem more severe

than in Los Angeles, which had been intermittently choked by smog since World War II. Smog attacks dense enough to be dubbed "daylight dim-outs" had prompted local officials to call for public prayer to end the scourge, and the city council had once donned gas masks when fumes infiltrated its chambers.

Under federal standards issued in 1971, the most serious smog emergency requires a "Stage 3" alert, during which health authorities advise everyone in affected areas to remain indoors and minimize physical activity. The last such alert in this country was in 1974, in San Bernardino County, Calif. Then-governor Ronald Reagan was no tree hugger. But he urged residents of the car-crazy Los Angeles basin to avoid driving if possible until the smog abated.

As late as 1979, Los Angeles suffered through 120 days of Stage 1 smog alerts, the level at which the federal government deems the air very unhealthy and advises everyone to avoid rigorous outdoor exercise. Last year, for the first time since record-keeping began in the mid-'50s, Los Angeles did not record one ozone reading high enough to trigger a smog alert. Nationwide, emissions of all but one of the six major air pollutants tracked by the EPA since the 1970 Clean Air Act was enacted have declined.

Fire on the water. One May 1967 night in Kansas City, hundreds of passersby and motorists stopped to gawk at the Kaw River: A giant swath of viscous oil, some 9,000 square feet in all, was aflame, floating down the dirty waterway. The Kaw wasn't the only river to catch fire before the first Earth Day. Even more notorious was Cleveland's Cuyahoga, where drifting oil, picnic tables, and debris gener-

Environmental health indicators

Many ecological trends have improved in the nation—without forestalling economic growth

THE LAND

Protected wilderness
1970: 10 million acres
1997: 104 million acres

National Park System
1970: 30 million acres
1997: 83 million acres

THE AIR

Sulfur dioxide emissions
1970: 31.2 million tons
1997: 20.4 million tons

Lead emissions
1970: 221 million tons
1997: 4 million tons

THE WATER

Oil-polluting incidents reported in and around U.S. waters
1970: 15.2 million gallons
1998: 885,000 gallons

THE FOOD SUPPLY

Conventional pesticide usage
1970: 760 million pounds
1997: 975 million pounds

Percent of U.S. food samples found to contain pesticide residues
1978: 47%
1998: 35%

WASTE DISPOSAL

Municipal solid waste discarded
1970: 121 million tons
1997: 217 million tons

Amount recycled
1970: 8 million tons
1997: 49 million tons

GLOBAL WARMING

Atmospheric carbon dioxide concentrations
1970: 325 parts per million
1998: 367 parts per million

SPENDING ON POLLUTION

Private sector
1972: $32 billion
1994: $72.9 billion

---

ated a five-story-high conflagration in 1969 that inspired a Randy Newman song ("Burn On, Big River").

Thirty years ago, many cities still discharged raw sewage directly into rivers and the oceans. Factories dumped millions of tons of poisonous byproducts into streams, the Great Lakes, and the sea. "People feared that the Great Lakes were becoming vast cesspools of death," says Denis Hayes, national coordinator of the first Earth Day and chair of Earth Day Network 2000. In 1970 alone, the federal government recalled a million cans of tuna for possible mercury contamination. In the nation's capital, fecal coliform counts in the Potomac River rose to 4,000 times the safe level. At the same time, Manhattan's worldly West Side was discharging a mind-numbing 300 million gallons of raw sewage *a day* into the Hudson River—nearly 30 times what the Exxon Valdez released in its infamous, one-time 1989 oil

spill. And long before George Bush made Boston Harbor a campaign issue in 1988, the Standells crooned about the harbor's squalid state in their 1966 rock hit "Dirty Water." ("Well I love that dirty water, Oh, Boston, you're my home....")

Today, the nation's waterways are no longer a waste bucket, mostly thanks to the 1972 Clean Water Act, which prompted a huge drop in uncontrolled "point source" discharges from municipal and industrial polluters. Now two thirds of the nation's waters are safe for fishing and swimming, compared with only a third back then. Where the Kaw once burned, kayakers and swimmers can frolic. And the once foul banks of the Cuyahoga today host a slew of boutiques and bistros.

Deep forest. Last fall, President Clinton announced a ban on logging and road building on more than 40 million acres of national forests—an area larger than Georgia. Even before that, the nation's forests

had grown modestly since the 1970s, and they now contain more timber and fewer small trees. "We have more forested land today in the U.S. than at the turn of the century," says U.S. Forest Service Chief Mike Dombeck.

By contrast, in 1970, the national forests, as Wyoming Sen. Gale McGee put it, had been "depleted to the point that would shame Paul Bunyan." At the Forest Service's behest, contract loggers then relied heavily on clear-cutting—leveling sections of a forest, setting fire to the debris, and "scarifying" the area on occasion by scraping it bare with bulldozers. After an air tour of some clear-cuts, a shaken McGee likened the devastation to a B-52 bombing run, a "shocking desecration that has to be seen to be believed."

All told, loggers in 1970 clear-cut 564,000 acres in national forests. Controversy over clear-cutting peaked in Montana's Bitterroot National Forest, where logging eyesores were headline news in 1970. Between 1967 and 1970, 11,211 acres were clear-cut in the Bitterroot alone, compared with just 8 acres last year. When Gifford Pinchot Jr., the son of Theodore Roosevelt's legendary Forest Service chief, toured the Bitterroot clear-cuts, he declared: "If my father had seen this, he would have cried."

In some cases, the country has taken an ecological step forward only to lurch backward. Nuclear waste sites, for instance, are now more carefully regulated than in earlier decades, when workers at Paducah's Gaseous Diffusion Plant brushed green uranium dust off their lunches. And with the end of the Cold War, there is less new nuclear waste being produced. But because radioactivity is both long lasting and cumulative, the nation's inventory of nuclear waste and the capacity for additional contamination have grown dramatically.

The biodiversity of animals and plants also is in decline, with a few notable exceptions. Several animals that were threatened or near extinction in the lower 48 states at the time of the 1973 Endangered Species Act, including peregrine falcons, bald eagles, gray wolves, and California condors, have had their ranks grow in recent decades. But on the whole, the nation's biodiversity has diminished, as sprawl and alien species have wreaked havoc with native habitat. According to the most recent federal assessment, one third of threatened or endangered species listed by the government are in decline, roughly a quarter are stable, and less than a tenth are improving in status. Dwindling salmon runs in the Pacific Northwest are perhaps the most visible loss since 1970.

In many developing nations, the environment is worsening. Tropical forests have shrunk rapidly, accelerating the irreversible extinctions of rare animals and plants that are heavily concentrated in biological "hot spots" there. Water and air pollution are still deadly plagues: More than 3 million Third World children die each year from preventable waterborne diseases. And the world's oceans are being depleted. From 1975 to 1995, the global marine fish catch nearly doubled.

Eve of destruction? At the time of the first Earth Day, President Nixon, the pope, and Nobel laureates such as Gunnar Myrdal all worried that the Earth was on the verge of environmental calamity. Pope Paul VI in 1970 cautioned that "rivers, lakes, and even the oceans are already polluted to the point where there is reason to fear a real 'biological death' in the near future." Since that time, a series of looming ecodisasters, such as a widening ozone hole and acid-rain die-offs, either have turned out to be exaggerated or been curbed by antipollution regulation. "To some extent the environmental movement has been organized around the self-unfulfilling prophecy, but our critics forget that people and government intervened to change the trend line," says Earth Day Chair Hayes.

The intervention Hayes and fellow Earth Day 2000 participants are pushing now: a shift to cleaner energy sources and reductions in greenhouse gases that contribute to climate change. Numerous scientists have warned that global warming could one day wipe out several island states, displace millions of people in shoreline communities, and create epic droughts and food shortages. But its ultimate impact may depend just as much on nature as on policy changes. Human-induced emissions of carbon dioxide, for example, are dwarfed by the planet's natural carbon cycle, which stores vast amounts of this greenhouse gas in deep oceans and forests. Several decades hence, utilities and industry may be able to bury most man-made carbon dioxide emissions in the deep sea, allowing them to reduce atmospheric concentrations of carbon dioxide.

It is too early to say whether such carbon sequestration strategies are safe and can really arrest global warming. But as the world celebrates Earth Day 2000, it may be worth recalling that doom-and-gloom environmental predictions have proved wrong more often than they have proved right.

With Laura Tangley, Nancy L. Bentrup, and Frank McCoy

Environmental Information Retrieval

ON FINDING OUT MORE

There is probably more printed information on environmental issues, regulations, and concerns than on any other major topic. So much is available from such a wide and diverse group of sources, that the first effort at finding information seems an intimidating and even impossible task. Attempting to ferret out what agencies are responsible for what concerns, what organizations to contact for specific environmental information, and who is in charge of what becomes increasingly more difficult.

To list all of the governmental agencies private and public organizations, and journals devoted primarily to environmental issues is, of course, beyond the scope of this current volume. However, we feel that a short primer on environmental information retrieval should be included in order to serve as a springboard for further involvement; for it is through informed involvement that issues, such as those presented, will eventually be corrected.

I. SELECTED OFFICES WITHIN FEDERAL AGENCIES AND FEDERAL-STATE AGENCIES FOR ENVIRONMENTAL INFORMATION RETRIEVAL

Appalachian Regional Commission
1666 Connecticut Avenue, NW, Washington, DC 20235 (202) 884-7799
http://www.arc.gov

Council on Environmental Quality
Old Executive Office Bldg., Room 360, Washington, DC 20502 (202) 456-6224
http://www.whitehouse.gov/CEQ/

Delaware River Basin Commission
P.O. Box 7360, West Trenton, NJ 08628-0360 (609) 883-9500
http://www.state.nj.us/drbc/drbc.htm

Department of Agriculture
14th and Independence Avenue, SW, Washington, DC 20250 (202) 720-2791
http://www.usda.gov

Department of the Army (Corps of Engineers)
20 Massachusetts Ave., NW, Washington, DC 20314-1000 (202) 761-0660
http://www.usace.army.mil

Department of Commerce
14th and Constitution Ave. NW, Washington, DC 20230 (202) 482-2000
http://www.doc.gov

Department of Defense
Public Affairs, 1400 Defense Pentagon, Room 1E757, Washington, DC 20301-1400 (703) 697-5737
http://www.defenselink.mil/index.html

Department of Health and Human Services
200 Independence Avenue, SW, Washington, DC 20201 (202) 619-0257
http://www.os.dhhs.gov

Department of the Interior
1849 C Street, NW, Washington, DC 20240-0001 (202) 208-3100
http://www.doi.gov
- Bureau of Indian Affairs (202) 208-3711
- Bureau of Land Management (202) 452-5125
- National Park Service (202) 208-6843
- United States Fish and Wildlife Service (202) 208-4131

Department of State, Bureau of Oceans and International Environmental and Scientific Affairs
2201 C Street, NW, Washington, DC 20520 (202) 647-2492
http://www.state.gov/www/global/oes/index.html

Department of the Treasury, U.S. Customs Service
1300 Pennsylvania Avenue, NW, Washington, DC 20229 (202) 927-1000
http://www.customs.ustreas.gov

Environmental Protection Agency (EPA)
401 M Street, SW, Washington, DC 20460 (202) 260-2090
- Region 1, One Congress Street, John F. Kennedy Building, 11th Floor, Boston, MA 02203-0001 (617) 565-3420 (888) 373-7341
  http://www.epa.gov/region01/ (Connecticut, Maine, Massachusetts, New Hampshire, Rhode Island, Vermont)
- Region 2, 290 Broadway, New York, NY 10007-1866 (212) 637-3000
  http://www.epa.gov/region02/ (New Jersey, New York, Puerto Rico, Virgin Islands)
- Region 3, 1650 Arch Street, Philadelphia, PA 19100-2029 (215) 814-5000 (800) 438-2474
  http://www.epa.gov/region03/ (Delaware, District of Columbia, Maryland, Pennsylvania, Virginia, West Virginia)
- Region 4, 4 Forsyth Street, SW, Atlanta, GA 30303-3104 (404) 562-9900
  http://www.epa.gov/region04/ (Alabama, Florida, Georgia, Kentucky, Mississippi, North Carolina, South Carolina, Tennessee)
- Region 5, 77 West Jackson Blvd., Chicago, IL 60604-3507 (312) 353-2000
  http://www.epa.gov/region5/ (Illinois, Indiana, Michigan, Minnesota, Ohio, Wisconsin)
- Region 6, 1445 Ross Avenue, Suite 1200, Dallas, TX 75202 (214) 665-2200 (800) 887-6063
  http://www.epa.gov/region06/ (Arkansas, Louisiana, New Mexico, Oklahoma, Texas)
- Region 7, 726 Minnesota Avenue, Kansas City, KS 66101 (913) 551-7003 (800) 223-0425
  http://www.epa.gov/region07/ (Iowa, Kansas, Missouri, Nebraska)
- Region 8, 999 18th Street, Suite 500, Denver, CO 80202-2466 (303) 312-6312 (800) 227-8917
  http://www.epa.gov/region08/ (Colorado, Montana, North Dakota, South Dakota, Utah, Wyoming)

- Region 9, 75 Hawthorne Street, San Francisco, CA 94105 (415) 744-1500

  http://www.epa.gov/ region09/ (Arizona, California, Hawaii, Nevada, American Samoa, Guam, Trust Territories of Pacific Islands, Wake Island)
- Region 10, 1200 Sixth Avenue, Seattle, WA 98101 (206) 553-1200 (800) 424-4372

  http://www.epa.gov/region10/ (Alaska, Idaho, Oregon, Washington)

Federal Energy Regulatory Commission

825 North Capitol Street, NE, Washington, DC 20426 (202) 208-0000

http://www.ferc.fed.us

Interstate Commission on Potomac River Basin

6110 Executive Boulevard, Suite 300, Rockville, MD 20852-3903 (301) 984-1908

http://www.potomacriver.org

Nuclear Regulatory Commission

One White Flint North, 11555 Rockville Pike, Rockville, MD 20852-2738 (301) 415-7000

http://www.nrc.gov

Susquehanna River Basin Commission

1721 North Front Street, Harrisburg, PA 17102 (717) 238-0422

http://www.srbc.net

Tennessee Valley Authority

400 West Summit Hill Drive, Knoxville, TN 37902 (423) 632-2101

http://www.tva.gov

II. SELECTED STATE, TERRITORIAL, AND CITIZENS' ORGANIZATIONS FOR ENVIRONMENTAL INFORMATION RETRIEVAL

A. Government Agencies

Alabama:

Department of Conservation and Natural Resources

P.O. Box 301450, Montgomery, AL 36130-1450 (334) 242-3486

http://www.dcnr.state.al.us

Alaska:

Department of Environmental Conservation

410 Willoughby Avenue, Suite 105, Juneau, AL 99801-1795 (907) 465-5060

http://www.state.ak.us

Arizona:

Department of Water Resources

500 North 3rd Street, Phoenix, AZ 85004-3226 (602) 417-2400

http://www.adwr.state.az.us

Natural Resources Division

1616 West Adams Street, Phoenix, AZ 85007 (602) 542-4625

Arkansas:

Department of Pollution Control and Ecology

8001 National Drive, Little Rock, AR 72209 (501) 682-0744

http://www.adeq.state.ar.us

Energy Office

1 State Capitol Mall, Little Rock, AR 72201-1012 (501) 682-7325

California:

Conservation Department Resources Agency

801 K Street, MS24-01, Sacramento, CA 95814 (916) 322-1080

http://www.consrv.ca.gov

Environmental Protection Agency

555 Capital Mall, Suite 525, Sacramento, CA 95814 (916) 445-3846

http://www.calepa.ca.gov

Colorado:

Department of Natural Resources

1313 Sherman Street, Room 718, Denver, CO 80203-2239 (303) 866-3311

http://www.dnr.state.co.us

Connecticut:

Department of Environmental Protection

State Office Building, 79 Elm Street, Hartford, CT 06106-5127 (860) 424-3000

http://www.dep.state.ct.us

Delaware:

Natural Resources and Environmental Control Department

89 Kings Highway, Dover, DE 19903 (302) 739-5823

http://www.dnrec.state.de.us

District of Columbia:

Environmental Regulation Administration

2100 Martin Luther King Jr. Avenue, SE, Suite 203, Washington, DC 20020 (202) 404-1167

http://clean.rti.org/state/washdc.htm

Florida:

Department of Environmental Protection

3900 Commonwealth Blvd., M.S.10, Tallahassee, FL 32399-3000 (850) 488-1554

http://www.dep.state.fl.us

Georgia:

Department of Natural Resources

205 Butler Street, SE, Atlanta, GA 30334-4100 (404) 656-3500

http://www.ganet.org/dnr/

Guam:

Environmental Protection Agency

IT&E Harmon Plaza, Complex Unit D-107, 130 Rojas St., Harmon, Guam 96911 (617) 646-9402

Hawaii:

Department of Land and Natural Resources

P.O. Box 621, Honolulu, HI 96809 (808) 587-0400

http://www.hawaii.gov/dlnr/Welcome.html

Idaho:

Department of Lands

P.O. Box 83720, Boise, ID 83720-0050 (208) 334-0200

http://www2.state.id.us/lands/index.htm

Department of Water Resources

1301 North Orchard Street, Boise, ID 83706 (208) 327-7900

http://www.idwr.state.id.us

Illinois:

Department of Natural Resources
> 524 South 2nd Street, Springfield, IL 62701 (217) 785-0075
> http://dnr.state.il.us

Indiana:

Department of Natural Resources
> 402 W. Washington St., Indianapolis, IN 46204-2212 (317) 232-4020
> http://www.state.in.us/dnr/

Iowa:

Department of Natural Resources
> Wallace State Office Bldg., Des Moines, IA 50319 (515) 281-4367
> http://www.state.ia.us/dnr/

Kansas:

Department of Health and Environment
> Landon State Office Bldg., Topeka, KS 66612 (785) 296-1522
> http://www.state.ks.us/public/kdhe/

Kentucky:

Natural Resources and Environmental Protection Cabinet
> Capital Plaza, Frankfort, KY 40601 (502) 564-3350
> http://www.nr.state.ky.us/nrhome.htm

Louisiana:

Department of Environmental Quality
> 7290 Bluebonnet Blvd., Baton Rouge, LA 70810 70817-4401 (225) 765-0741
> http://www.deq.state.la.us

Department of Natural Resources
> 625 North 4th Street, Baton Rouge, LA 70804-9396 (504) 342-2707
> http://www.dnr.state.la.us/index.ssi

Maine:

Department of Environmental Protection
> 17 State House Station, Augusta, ME 04333-0017 (207) 287-7688 (800) 452-1942
> http://www.state.me.us/dep/

Maryland:

Department of Natural Resources
> 580 Taylor Avenue, Tawes State Office Bldg., Annapolis, MD 21401 (410) 260-8400
> http://www.dnr.state.md.us

Massachusetts:

Department of Environmental Management
> 100 Cambridge Street, 19th Floor, Boston, MA 02202 (617) 727-3163
> http://www.state.ma.us/dem/dem.htm

Michigan:

Department of Natural Resources
> Box 30028, Lansing, MI 48909 (517) 373-2329
> http://www.dnr.state.mi.us

Minnesota:

Department of Natural Resources
> 500 Lafayette Road, St. Paul, MN 55155-4046 (612) 296-2549
> http://www.dnr.state.mn.us

Mississippi:

Department of Environmental Quality
> P.O. Box 20305, Jackson, MS 39289 (601) 961-5171
> http://www.deq.state.ms.us/domino/deqweb.nsf

Missouri:

Department of Natural Resources
> P.O. Box 176, Jefferson City, MO 65102 (800) 334-6946
> http://www.dnr.state.mo.us/homednr.htm

Montana:

Department of Natural Resources and Conservation
> 1625 11th Avenue, P.O. Box 201601, Helena, MT 59620-1601 (406) 444-2074
> http://www.dnrc.mt.gov

Nebraska:

Department of Environmental Quality
> P.O. Box 98922, Lincoln, NE 68509 (402) 471-2186
> http://www.deq.state.ne.us

Nevada:

Department of Conservation and Natural Resources
> Capitol Complex, Carson City, NV 89710 (702) 687-5000
> http://www.state.nv.us/cnr/

New Hampshire:

Department of Environmental Services
> 6 Hazen Drive, P.O. Box 95, Concord, NH 03302-0095 (603) 271-3503
> http://www.state.nh.us/des/

Department of Resources and Economic Development
> P.O. Box 1856, Concord, NH 03302-1856 (603) 271-2411
> http://www.dred.state.nh.us

New Jersey:

Department of Environmental Protection
> 401 E. State Street, P.O. Box 402, Trenton, NJ 08625-0402 (609) 292-2885
> http://www.state.nj.us/dep/

New Mexico:

Environmental Department
> 1190 Saint Francis Drive, Santa Fe, NM 87505 (505) 827-2855 (800) 879-3421
> http://www.nmenv.state.nm.us

New York:

Department of Environmental Conservation
> 50 Wolf Road, Albany, NY 12233 (518) 485-8940
> http://www.dec.state.ny.us

North Carolina:

Department of Environment and Natural Resources
> P.O. Box 27687, Raleigh, NC 27611 (919) 733-4984
> http://www.ehnr.state.nc.us/EHNR/

North Dakota:

Game & Fish Department
> 100 North Bismarck Expressway, Bismarck, ND 58501 (701) 328-6300
> http://www.state.nd.us/gnf/

Ohio:

Department of Natural Resources
> 1952 Belcher Drive, Building C-1, Columbus, OH 43224 (614) 265-6565
> http://www.dnr.state.oh.us

Environmental Information Retrieval

Environmental Protection Agency
P.O. Box 1049, Columbus, OH 43216-1049 (614) 644-3020
http://www.epa.state.oh.us

Oklahoma:
Conservation Commission
2800 North Lincoln Boulevard, Suite 160, Oklahoma City, OK 73105-4210 (405) 521-2384
http://www.oklaosf.state.ok.us/~comscom/
Department of Environmental Quality
707 North Robinson, Oklahoma City, OK 73702 (405) 702-6100
http://www.deq.state.ok.us

Oregon:
Department of Environmental Quality
811 S.W. 6th Avenue, Portland, OR 97204-1390 (503) 229-5696
http://www.deq.state.or.us

Pennsylvania:
Department of Environmental Resources
400 Market Street, Harrisburg, PA 17105 (717) 787-2814
http://www.dep.state.pa.us

Puerto Rico:
Department of Natural Resources
P.O. Box 5887, San Juan, PR 00906 (787) 723-3090

Rhode Island:
Department of Environmental Management
235 Promenade Street, Providence, RI 02908 (401) 222-2771
http://www.state.ri.us

South Carolina:
Department of Health and Environmental Control
2600 Bull Street, Columbia, SC 29201 (803) 734-5000
http://www.state.sc.us/dhec/
Department of Natural Resources
1000 Assembly Street, Columbia 29201 (803) 734-3888
http://water.dnr.state.sc.us

South Dakota:
Department of Environment and Natural Resources
Joe Foss Bldg., 523 East Capitol, Pierre, SD 57501 (605) 773-3151
http://www.state.sd.us/state/executive/ denr/denr.html

Tennessee:
Department of Environment and Conservation
401 Church St., 21st Floor, Nashville, TN 37243 (888) 891-8332
http://www.state.tn.us/environment/

Texas:
Natural Resources Conservation Commission
P.O. Box 13087, Austin, TX 78711 (512) 239-1000
http://www.three.state.tx.us

Utah:
Department of Natural Resources
1594 West North Temple, Suite 3710, Salt Lake City, UT 84114 (801) 538-7200
http://www.nr.state.ut.us

Vermont:
Agency of Natural Resources
103 South Main Street, Waterbury, VT 05671-0301 (802) 241-3614
http://www.anr.state.vt.us

Virgin Islands:
Department of Planning & Natural Resources
396-1 Annas Retreat, Foster Bldg., Charlotte Amalie, U.S. Virgin Islands 00802 (340) 774-3320
http://www.gov.vi/pnr/

Virginia:
Secretary of Natural Resources
P.O. Box 1475, Richmond, VA 23212 (804) 786-0044
http://snr.vipnet.org

Washington:
Department of Ecology
P.O. Box 47600, Olympia, WA 98504-7600 (360) 407-6000
http://www.wa.gov/ecology/
Department of Natural Resources
P.O. Box 47000, Olympia, WA 98504-7000 (360) 902-1000
http://www.wa.gov/dnr/

West Virginia:
Division of Natural Resources
1900 Kanawha Blvd. East, Charleston, WV 25305 (304) 558-2771
http://www.dnr.state.wv.us

Wisconsin:
Department of Natural Resources
Box 7921, Madison, WI 53707 (608) 267-7517
http://www.dnr.state.wi.us

Wyoming:
Department of Environmental Quality
122 West 25th Street, Herschler Bldg., Cheyenne, WY 82002 (307) 777-7758
http://deq.state.wy.us

B. Citizens' Organizations

Advancement of Earth & Environmental Sciences
International Association for, Northeastern Illinois University, Geography and Environmental Studies Department 5500 North St. Louis Avenue, Chicago, Illinois 60625 (312) 794-2628
Air Pollution Control Association
1 Gateway Center, 3rd Floor, Pittsburgh, PA 15222 (412) 232-3444
American Association for the Advancement of Science
1200 New York Avenue, NW, Washington, DC 20005 (202) 326-6400
http://www.aaas.org
American Chemical Society
1155 16th Street, NW, Washington, DC 20036 (202) 872-4600
http://acs.org
American Farm Bureau Federation
225 Touhy Avenue, Park Ridge, IL 60068 (847) 685-8600
http://www.fb.com
American Fisheries Society,
5410 Grosvenor Lane, Suite 110, Bethesda, MD 20014-2199 (301) 897-8616
http://www.fisheries.org

American Forest & Paper Association
   1111 19th Street, NW, Suite 800, Washington, DC 20036
   (202) 463-2438
   http://www.afandpa.org
American Forests
   P.O. Box 2000, Washington, DC 20013 (202) 955-4500
   http://www.amfor.org
American Institute of Biological Sciences
   1444 I Street NW, Washington, DC 20005 (202) 628-1500
   http://www.aibs.org
American Museum of Natural History
   Central Park West at 79th Street, New York, NY 10024-
   5192 (212) 769-5100
   http://www.amnh.org
American Petroleum Institute
   1220 L Street, NW, Washington, DC 20005-4070 (202)
   682-8000
   http://www.api.org
American Rivers
   1025 Vermont Avenue NW, Suite 720, Washington, DC
   20005 (202) 347-7550
   http://www.amrivers.org
Association for Conservation Information
   P.O. Box 12559, Charleston, SC 29412 (803) 762-5032
Boone and Crockett Club
   250 Station Dr., Missoula, MT 59801 (406) 542-1888
   http://www.boone-crockett.org
Center for Marine Conservation
   1725 DeSales Street, NW, Suite 600, Washington, DC
   20036 (202) 429-5609
   http://www.cmc-ocean.org
Citizens for a Better Environment
   3255 Hennepin Avenue South, Minneapolis, MN 55331
   (612) 824-8637
   http://www.cbemw.org
Coastal Conservation Association
   4801 Woodway, Suite 220W, Houston, TX 77056 (713)
   626-4222
   http://www.ccatexas.org
Conservation Foundation
   1250 24th Street, NW, Suite 400, Washington, DC 20037
   (202) 293-4800
Conservation Fund
   1800 North Kent Street, Suite 1120, Arlington, VA 22209-
   2156 (703) 525-6300
   http://www.conservationfund.org
Conservation International
   2501 M Street, NW, Suite 200, Washington, DC 20037
   (202) 429-5660 (800) 429-5660
   http://www.conservation.org
Defenders of Wildlife
   1101 14th Street, NW, #1400, Washington, DC, 20005
   (202) 682-9400
   http://www.defenders.org
Ducks Unlimited, Inc.
   One Waterfowl Way, Memphis, TN (901) 785-3825
   http://www.ducks.org
Earthwatch Institute International
   680 Mt. Auburn Street, Watertown, MA 02471 (800) 776-
   0188
   http://www.earthwatch.org

Environmental Action Foundation, Inc.
   6930 Carroll Ave., Suite 600, Takoma Park, MD 20912
   (301) 891-1100
Food and Agriculture Organization of the United Nations (FAO)
   Via delle Terme di Caracalla, 00100, Rome, Italy (396)
   57051
   http://www.fao.org
Friends of the Earth
   1025 Vermont Avenue, NW, Suite 300, Washington, DC
   20005 (202) 783-7400
   http://foe.co.uk
Greenpeace U.S.A.
   1436 U Street, NW, Washington, DC 20009 (202) 462-
   1177
   http://greenpeaceusa.org
International Association of Fish and Wildlife Agencies
   444 North Capitol Street, NW, Suite 544, Washington, DC
   20001 (202) 624-7890
   http://www.teaming.com/iafwa.htm
International Fund for Agricultural Development (IFAD)
   107, Via del Serafico, 00142, Rome, Italy (3906) 54591
   http://www.ifad.org
Keep America Beautiful, Inc.
   1010 Washington Blvd., Stamford, CT 06901 (203) 323-
   8987
   http://www.kab.org
National Association of Conservation Districts
   509 Capitol Court, NE, Washington, DC 20002 (202) 547-
   6223
   http://www.nacdnet.org
National Audubon Society
   700 Broadway, New York, NY 10003 (212) 979-3000
   http://www.audubon.org
National Environmental Health Association
   720 South Colorado Boulevard, 970 South Tower, Denver,
   CO 80246 (303) 756-9090
   http://www.neha.org
National Fisheries Institute
   1901 N. Fort Myer Dr., Suite 700, Arlington, VA 22209 (703)
   524-8880
   http://www.nfi.org/main.html
National Geographic Society
   1145 17th Street, NW, Washington, DC 20036 (202) 857-
   7000
   http://www.nationalgeographic.com
National Parks and Conservation Association
   1717 Massachusetts Avenue, NW, Washington, DC 20036
   (202) 223-6722 (800) 628-7275
   http://www.npca.org
National Wildlife Federation
   8925 Leesburg Pike, Vienna, VA 22184 (703) 790-4000
   http://www.nwf.org
Natural Resources Council of America
   801 Pennsylvania Avenue, SE, No. 410, Washington, DC
   20003 (202) 333-0411
Nature Conservancy
   1815 North Lynn Street, Arlington, VA 22209 (703) 841-
   5300
   http://www.tnc.org

Population Association of America
  721 Ellsworth Drive, Suite 303, Silver Spring, MD 20910
  (301) 565-6710
  http://www.jstor.org

Rainforest Alliance
  650 Bleecker Street, New York, NY 10012 (212) 677-1900
  http://www.rainforest-alliance.org

Save-the-Redwoods League
  114 Sansome Street, Room 605, San Francisco, CA 94104
  (415) 362-2352
  http://www.savetheredwoods.org

Sierra Club
  85 Second Street, 2nd Floor, San Francisco, CA 94105
  (415) 977-5500
  http://www.sierraclub.org

Smithsonian Institution
  1000 Jefferson Drive, S.W., Washington, DC 20560 (202)
  357-2700
  http://smithsonian.org

Society of American Foresters
  5400 Grosvenor Lane, Bethesda, MD 20814-2198 (301)
  897-8720
  http://www.safnet.org

Sport Fishing Institute
  1010 Massachusetts Avenue, NW, Suite 320, Washington,
  DC 20001 (202) 898-0770

*United Nations Educational, Scientific, and Cultural Organiza-
  tion (UNESCO)*
  UNESCO House, 7, Place de Fontenoy, 75352 Paris 07 SP
  France, (331) 45 68 10 00
  http://www.unesco.org

*United Nations Environment Programme/Industry & Environ-
  ment Centre*
  Tour Mirabeau 39-43, quai André Citröen 75739 Paris, ce-
  dex 15, France (331) 44 37 14 50
  http://www. unepie.org

Wilderness Society
  900 17th Street, NW, Washington, DC 20006-2596 (202)
  833-2300
  http://www.wilderness.org

World Wildlife Fund
  1250 24th Street, NW, Washington, DC 20077 (202) 293-
  4800
  http://www.wwf.org

Zero Population Growth, Inc.
  1400 16th Street, NW, Suite 320, Washington, DC 20036
  (202) 332-2200
  http://www.zpg.org

III. CANADIAN AGENCIES AND CITIZENS' ORGANIZATIONS

A: Government Agencies

Alberta:
Alberta Environmental Protection
  23 Legislature Bldg., 10800 97th Avenue, Edmonton, AB
  T5K 2B6 (403) 427-2391
  http://www.gov.ab.ca/env.html

British Columbia:
Ministry of Environment, Lands, and Parks
  337, Parliament Bldgs., Victoria, BC V8V 1X4 (250) 387-
  1187
  http://www.env.gov.bc.ca

Manitoba:
Manitoba Environment
  344 Legislative Bldg., Winnipeg, MB R3C 0V8 (204) 945-
  3522
  http://www.gov.mb.ca/environ/index.html

Manitoba Natural Resources
  Box 22, 200 Saulteaux Crescent, Winnipeg, MB R3J 3W3
  (204) 945-6784
  http://www.gov.mb.ca/natres/index.html

New Brunswick:
Department of Environment
  364 Argyle St., Fredericton, NB E3B 1T9 (506) 453-2558
  http://www.gov.nb.ca/environm/

Department of Natural Resources and Energy
  P.O. Box 6000, Fredericton, NB E3B 5H1 (506) 453-2614
  http://www.gov.nb.ca/dnre/

Newfoundland and Labrador:
Department of Environment and Labour
  Confederation Bldg., PO Box 8700, St. John's, NF A1B 4J6
  (709) 729-2664
  http://www.govt.nf.ca/env/labour/OHS/default.asp

Northwest Territories:
*Department of Resources, Wildlife, and Economic Develop-
  ment, #600 Scotia Centre, Bldg. Box 21, 5102-50 Avenue,
  Yellowknife, NT X1A 3S8 (867) 669-2366 http://
  www.rwed.gov.nt.ca*

Nova Scotia:
*Department of Natural Resources, P.O. Box 698, Halifax, NS
  B3J 2T9 (902) 424-59356*
  http://www.gov.ns.ca/natr/

Department of the Environment
  P.O. Box 2107, Halifax, NS B3J 3B7 (902) 424-5300
  http://www.gov.ns.ca/envi/

Ontario:
Ministry of Natural Resources
  300 Water Street, P.O. Box 7000, Peterborough, ON K9J
  8M5 (705) 755-2000 (416) 314-2000
  http://www.mnr.gov. on.ca/mnr/

Prince Edward Island:
Department of Technology and Environment
  Jones Bldg., 11 Kent Street, 4th Floor, P.O. Box 2000,
  Charlottetown, PEI C1A 7N8 (902) 892-5000
  http://www.gov.pe.ca/te/index.asp

Quebec:
Ministère de l'Environnement et de la Faune
  Edifice Marie-Guyart, 675, boulevard René-Lévesque, Est,
  Québec, PC G1R 5V7 (418) 521-3830 (800) 561-1616
  http://www.mef.gouv.qc.ca

Ministère des Ressources Naturelles
  #B-302, 5700, 4 Avenue Ouest, Charlesbourg, PQ G1H
  6R1 (418) 627-8600
  http://www.mrn.gouv.qc.ca

Saskatchewan:
Saskatchewan Environment and Resource Management
3211 Albert Street, Regina, SK S4S 5W6 (306) 787-2700
http://www.gov.sk.ca/govt/environ/

Yukon Territory:
Council on the Economy & the Environment
A-8E, P.O. Box 2703, Whitehorse, YT Y1A 2C6 (867) 667-5811
http://www.gov.yk.ca
Department of Renewable Resources
Box 2703, Whitehorse, YT Y1A 2C6 (867) 667-5811
http://www.gov.yk.ca

B. Citizens' Groups
Alberta Wilderness Association
Box 6398, Station D, Calgary, AB T2P 2E1 (403) 283-2025
http://www.web.net/~awa/
BC Environmental Network (BCEN)
1672 East 10th Avenue, Vancouver, BC V5N 1X5 (604) 879-2272
http://www.bcen.bc.ca
Canadian EarthCare Society
1476 Water Street, Kelowna, BC V1Y 8P2 (604) 861-4788
http://www.earthcare.org
Ducks Unlimited Canada
Oak Hammock Marsh, Stonewall, P.O. Box 1160, Oak Hammock Marsh, MB R0C 2Z0 (204) 467-3000 (800) 665-3825
http://www.ducks.ca
Federation of Ontario Naturalists
355 Lesmill Road, Don Mills, ON M3B 2W8 (416) 444-8419
http://www.ontarionature.org
L'Association des Entrepreneurs de Service en Environnement du Quebec (AESEQ)
911 Jean-Talon, Est 220, Montreal, PQ H2R 1V5 (514) 270-7110
New Brunswick Environment Industry Association
P.O. Box 637, Stn. A, Fredericton, NB E3B 5B3 (506) 455-0212
http://www.nbeia.nb.ca/index.html
Prince Edward Island Environmental Network (PEIEN)
126 Richmond Street, Charleston, PEI C1A 1H9 (902) 566-4170
http://www.isn.net/~network/index.html
Yukon Conservation Society (YCS)
P.O. Box 4163, Whitehorse, YT Y1A 3T3 (403) 668-6637

IV. SELECTED JOURNALS AND PERIODICALS OF ENVIRONMENTAL INTEREST

American Forests
910 17th Street NW, Suite 600, Washington, DC 20006 (202) 955-4500 (800) 368-5748
http://www.amfor.org
American Scientist
Scientific Research Society, P.O. Box 13975, Research Triangle Park, NC 27709-3975 (919) 549-0097
http://www.amsci.org/amsci/amsci.html
Annual Report of the Council on Environmental Quality
Superintendent of Documents, U.S. Government Printing Office, Washington, DC 20401 (202) 512-1800
http://ceq.ch.doc.gov/nepa/reports/reports.htm

Audubon
National Audubon Society, 700 Broadway, New York, NY 10003 (212) 979-3000
http://magazine.audubon.org
BioScience
American Institute of Biological Sciences, 1444 I St. NW, Suite 200, Washington, DC 20005 (202) 628-1500
http://www.aibs.org
California Environmental Directory
California Institute of Public Affairs, Box 189040, Sacramento, CA 95818 (916) 442-2472
http://www.igc.org/cipa/cipa.html#about
The Canadian Field-Naturalist
Box 35069 Westgate, Ottawa, ON, Canada K1Z 1A2 (613) 722-3050
http://www.achilles.net/ofnc/cfn.htm
Conservation Directory
National Wildlife Federation, 8925 Leesburg Pike, Vienna, VA 22184 (703) 790-4000
http://www.nwf.org/nwf/pubs/considir/index.html
Earth First! Journal
P.O. Box 1415, Eugene, OR 97440-1415 (541) 344-8004
http://host.envirolink.org/ef/
E: The Environmental Magazine
Earth Action Network, P.O. Box 5098, Westport, CT 06881 (203) 854-5559
http://www.emagazine.com
Environment
Heldref Publications, 1319 18th Street, NW, Washington, DC 20036-1802 (202) 296-6267
http://www.heldref.org
Environment Reporter
Bureau of National Affairs, Inc. 1231 25th Street, NW, Washington, DC 2037 (202) 452-4200
http://www.bna.com/prodcatalog/desc/ER.html
Environmental Action Magazine
Environment Action, Inc. 6930 Carroll Ave., Suite 600, Takoma Park, MD 20912-4414 (301) 891-1106
Environmental Science and Technology
American Chemical Society Publications Support Services, 1155 16th Street, NW, Washington, DC 20036 (202) 872-4554 (800) 227-5558
http://pubs.acs.org/journals/esthag/
Focus (bimonthly newsletter)
World Wildlife Fund, 1250 24th Street, NW, Washington, DC 20037 (202) 293-4800
http://www.wwf.org
The Futurist
World Future Society, 7910 Woodmont Avenue, Suite 450, Bethesda, MD 20814 (301) 656-8274 (800) 989-8274
http://www.wfs.org/wfs/
Greenpeace Magazine
Greenpeace USA, 1436 U Street, NW, Washington, DC 20009 (202) 462-1177 (800) 326-0959
http://www.greenpeaceusa.org
Journal of Soil and Water Conservation
Soil and Water Conservation Society, 7515 Northeast Ankeny Road, Ankeny, IA 50021-9764 (515) 289-2331
http://www.swcs.org

Environmental Information Retrieval

Journal of Wildlife Management

The Wildlife Society, 5410 Grosvenor Lane, Suite 200, Bethesda, MD 20814-2197 (301) 897-9770

http://www.wildlife.org/journal.html

Mother Earth News

Sussex Publishers Inc., 49 E. 21st Street, 11th Floor, New York, NY 10010 (212) 260-7210

http://www.MotherEarthNews.Com

National Wildlife

National Wildlife Federation, 8925 Leesburg Pike, Vienna, VA 22184 (703) 790-4510

http://www.nwf.org/nwf/natwild/

Natural Resources Journal

University of New Mexico, School of Law, 1117 Stanford, NE, Albuquerque, NM 87131 (505) 277-4820

http://www.unm.edu/~natresj/NRJ/NRJ.html

Nature

Macmillan Publishers Ltd., Porter South, Grinan Street, London N1 9XW, England (44) 0171 833-4000

http://www.nature.com

Nature Canada

Canadian Nature Federation, One Nicholas Street, Suite 606, Ottawa, ON, Canada K1N 787 (613) 562-3447 (800) 267-40880

http://www. cnf.ca/nc_main.html

Nature Conservancy Magazine

1815 North Lynn Street, Arlington, VA 22209-2003 (703) 841-5300 (800) 267-4088

http://www.tnc.org

Omni

Omni Publications International Ltd., 277 Park Ave., New York, NY 10172 (212) 702-6000

http://www.omnimag.com

Pollution Abstracts

Cambridge Scientific Abstracts, 7200 Wisconsin Avenue, Suite 601, Bethesda, MD 20814-4823 (301) 961-6700 (800) 843-7751

http://www.csa.com

Science

American Association for the Advancement of Science, 1200 New York Avenue NW, Washington, DC 20005 (202) 326-6501

http://www.sciencemag.org

Sierra Magazine

Sierra Club, 85 2nd Street, 2nd Floor, San Francisco, CA 94105-3441 (415) 977-5750

http://www.sierraclub.org/sierra/

Smithsonian

900 Jefferson Drive, Washington, DC 20560 (202) 786-2900

http://smithsonianmag.com

Technology Review

201 Vassar Street, Cambridge, MA 02139 (617) 253-8250

http://www.techreview.com

U.S. News and World Report

2400 N Street, NW, Washington, DC 20037-1196 (202) 955-2000

http://www.usnews.com/usnews/home.htm

The World & I

New World Communications,3600 New York Avenue, NE, Washington, DC 20002 (800) 822-2822 (202) 635-4000

http://www.worldandi.com

SOURCES used to compile this list: Canadian Almanac Directory 1997; Carroll's Federal Directory, April 1997; Carroll's State Directory, February 1997; Congressional Quarterly's Washington Information Directory 1997–1998; Encyclopedia of Associations, 32nd Edition, 1997; Gale Directory of Publications and Broadcast Media, 131st edition; The World Almanac, 1999, Web search engines: Google, Metacrawler.

Glossary

This glossary of environmental terms is included to provide you with a convenient and ready reference as you encounter general terms in your study of environment that are unfamiliar or require a review. It is not intended to be comprehensive, but taken together with the many definitions included in the articles themselves, it should prove to be quite useful.

A

Abiotic Without life; any system characterized by a lack of living organisms.

Absorption Incorporation of a substance into a solid or liquid body.

Acid Any compound capable of reacting with a base to form a salt; a substance containing a high hydrogen ion concentration (low pH).

Acid Rain Precipitation containing a high concentration of acid.

Adaptation Adjustment of an organism to the conditions of its environment, enabling reproduction and survival.

Additive A substance added to another in order to impart or improve desirable properties or suppress undesirable ones.

Adsorption Surface retention of solid, liquid, or gas molecules, atoms, or ions by a solid or liquid.

Aerobic Environmental conditions where oxygen is present; aerobic organisms require oxygen in order to survive.

Aerosols Tiny mineral particles in the atmosphere onto which water droplets, crystals, and other chemical compounds may adhere.

Air Quality Standard A prescribed level of a pollutant in the air that should not be exceeded.

Alcohol Fuels The processing of sugary or starchy products (such as sugar cane, corn, or potatoes) into fuel.

Allergens Substances that activate the immune system and cause an allergic response.

Alpha Particle A positively charged particle given off from the nucleus of some radioactive substances; it is identical to a helium atom that has lost its electrons.

Ammonia A colorless gas comprised of one atom of nitrogen and three atoms of hydrogen; liquefied ammonia is used as a fertilizer.

Anthropocentric Considering humans to be the central or most important part of the universe.

Aquaculture Propagation and/or rearing of any aquatic organism in artificial "wetlands" and/or ponds.

Aquifers Porous, water-saturated layers of sand, gravel, or bedrock that can yield significant amounts of water economically.

Atom The smallest particle of an element, composed of electrons moving around an inner core (nucleus) of protons and neutrons. Atoms of elements combine to form molecules and chemical compounds.

Atomic Reactor A structure fueled by radioactive materials that generates energy usually in the form of electricity; reactors are also utilized for medical and biological research.

Autotrophs Organisms capable of using chemical elements in the synthesis of larger compounds; green plants are autotrophs.

B

Background Radiation The normal radioactivity present; coming principally from outer space and naturally occurring radioactive substances on Earth.

Bacteria One-celled microscopic organisms found in the air, water, and soil. Bacteria cause many diseases of plants and animals; they also are beneficial in agriculture, decay of dead matter, and food and chemical industries.

Benthos Organisms living on the bottom of bodies of water.

Biocentrism Belief that all creatures have rights and values and that humans are not superior to other species.

Biochemical Oxygen Demand (BOD) The oxygen utilized in meeting the metabolic needs of aquatic organisms.

Biodegradable Capable of being reduced to simple compounds through the action of biological processes.

Biodiversity Biological diversity in an environment as indicated by numbers of different species of plants and animals.

Biogeochemical Cycles The cyclical series of transformations of an element through the organisms in a community and their physical environment.

Biological Control The suppression of reproduction of a pest organism utilizing other organisms rather than chemical means.

Biomass The weight of all living tissue in a sample.

Biome A major climax community type covering a specific area on Earth.

Biosphere The overall ecosystem of Earth. It consists of parts of the atmosphere (troposphere), hydrosphere (surface and ground water), and lithosphere (soil, surface rocks, ocean sediments, and other bodies of water).

Biota The flora and fauna in a given region.

Biotic Biological; relating to living elements of an ecosystem.

Biotic Potential Maximum possible growth rate of living systems under ideal conditions.

Birthrate Number of live births in one year per 1,000 midyear population.

Breeder Reactor A nuclear reactor in which the production of fissionable material occurs.

C

Cancer Invasive, out-of-control cell growth that results in malignant tumors.

Carbon Cycle Process by which carbon is incorporated into living systems, released to the atmosphere, and returned to living organisms.

Carbon Monoxide (CO) A gas, poisonous to most living systems, formed when incomplete combustion of fuel occurs.

Carcinogens Substances capable of producing cancer.

Carrying Capacity The population that an area will support without deteriorating.

Glossary

Chlorinated Hydrocarbon Insecticide Synthetic organic poisons containing hydrogen, carbon, and chlorine. Because they are fat-soluble, they tend to be recycled through food chains, eventually affecting nontarget systems. Damage is normally done to the organism's nervous system. Examples include DDT, Aldrin, Deildrin, and Chlordane.

Chlorofluorocarbons (CFCs) Any of several simple gaseous compounds that contain carbon, chlorine, fluorine, and sometimes hydrogen; they are suspected of being a major cause of stratospheric ozone depletion.

Circle of Poisons Importation of food contaminated with pesticides banned for use in this country but made here and sold abroad.

Clear-Cutting The practice of removing all trees in a specific area.

Climate Description of the long-term pattern of weather in any particular area.

Climax Community Terminal state of ecological succession in an area; the redwoods are a climax community.

Coal Gasification Process of converting coal to gas; the resultant gas, if used for fuel, sharply reduces sulfur oxide emissions and particulates that result from coal burning.

Commensalism Symbiotic relationship between two different species in which one benefits while the other is neither harmed nor benefited.

Community Ecology Study of interactions of all organisms existing in a specific region.

Competitive Exclusion Resulting from competition; one species forced out of part of an available habitat by a more efficient species.

Conservation The planned management of a natural resource to prevent overexploitation, destruction, or neglect.

Conventional Pollutants Seven substances (sulfur dioxide, carbon monoxide, particulates, hydrocarbons, nitrogen oxides, photochemical oxidants, and lead) that make up the largest volume of air quality degradation, as identified by the Clean Air Act.

Core Dense, intensely hot molten metal mass, thousands of kilometers in diameter, at Earth's center.

Cornucopian Theory The belief that nature is limitless in its abundance and that perpetual growth is both possible and essential.

Corridor Connecting strip of natural habitat that allows migration of organisms from one place to another.

Crankcase Smog Devices (PCV System) A system, used principally in automobiles, designed to prevent discharge of combustion emissions into the external environment.

Critical Factor The environmental factor closest to a tolerance limit for a species at a specific time.

Cultural Eutrophication Increase in biological productivity and ecosystem succession resulting from human activities.

D

Death Rate Number of deaths in one year per 1,000 midyear population.

Decarbonization To remove carbon dioxide or carbonic acid from a substance.

Decomposer Any organism that causes the decay of organic matter; bacteria and fungi are two examples.

Deforestation The action or process of clearing forests without adequate replanting.

Degradation (of water resource) Deterioration in water quality caused by contamination or pollution that makes water unsuitable for many purposes.

Demography The statistical study of principally human populations.

Desert An arid biome characterized by little rainfall, high daily temperatures, and low diversity of animal and plant life.

Desertification Converting arid or semiarid lands into deserts by inappropriate farming practices or overgrazing.

Detergent A synthetic soap-like material that emulsifies fats and oils and holds dirt in suspension; some detergents have caused pollution problems because of certain chemicals used in their formulation.

Detrivores Organisms that consume organic litter, debris, and dung.

Dioxin Any of a family of compounds known chemically as dibenzo-p-dioxins. Concern about them arises from their potential toxicity as contaminants in commercial products. Tests on laboratory animals indicate that it is one of the more toxic anthropogenic (man-made) compounds.

Diversity Number of species present in a community (species richness), as well as the relative abundance of each species.

DNA (Deoxyribonucleic Acid) One of two principal nucleic acids, the other being RNA (Ribonucleic Acid). DNA contains information used for the control of a living cell. Specific segments of DNA are now recognized as genes, those agents controlling evolutionary and hereditary processes.

Dominant Species Any species of plant or animal that is particularly abundant or controls a major portion of the energy flow in a community.

Drip Irrigation Pipe or perforated tubing used to deliver water a drop at a time directly to soil around each plant. Conserves water and reduces soil waterlogging and salinization.

E

Ecological Density The number of a singular species in a geographical area, including the highest concentration points within the defined boundaries.

Ecological Succession Process in which organisms occupy a site and gradually change environmental conditions so that other species can replace the original inhabitants.

Ecology Study of the interrelationships between organisms and their environments.

Ecosystem The organisms of a specific area, together with their functionally related environments; considered as a definitive unit.

Ecotourism Wildlife tourism that could damage ecosystems and disrupt species if strict guidelines governing tours to sensitive areas are not enforced.

Edge Effects Change in ecological factors at the boundary between two ecosystems. Some organisms flourish here; others are harmed.

Effluent A liquid discharged as waste.

El Niño Climatic change marked by shifting of a large warm water pool from the western Pacific Ocean toward the East.

Electron Small, negatively charged particle; normally found in orbit around the nucleus of an atom.

Eminent Domain Superior dominion exerted by a governmental state over all property within its boundaries that authorizes it to appropriate all or any part thereof to a necessary public use, with reasonable compensation being made.

Endangered Species   Species considered to be in imminent danger of extinction.

Endemic Species   Plants or animals that belong or are native to a particular ecosystem.

Environment   Physical and biological aspects of a specific area.

Environmental Impact Statement (EIS   )A study of the probable environmental impact of a development project before federal funding is provided (required by the National Environmental Policy Act of 1968).

Environmental Protection Agency (EPA)   Federal agency responsible for control of air and water pollution, radiation and pesticide problems, ecological research, and solid waste disposal.

Erosion   Progressive destruction or impairment of a geographical area; wind and water are the principal agents involved.

Estuary   Water passage where an ocean tide meets a river current.

Eutrophic   Well nourished; refers to aquatic areas rich in dissolved nutrients.

Evolution   A change in the gene frequency within a population, sometimes involving a visible change in the population's characteristics.

Exhaustible Resources   Earth's geologic endowment of minerals, nonmineral resources, fossil fuels, and other materials present in fixed amounts.

Extinction   Irrevocable elimination of species due to either normal processes of the natural world or through changing environmental conditions.

F

Fallow   Cropland that is plowed but not replanted and is left idle in order to restore productivity mainly through water accumulation, weed control, and buildup of soil nutrients.

Fauna   The animal life of a specified area.

Feral   Refers to animals or plants that have reverted to a noncultivated or wild state.

Fission   The splitting of an atom into smaller parts.

Floodplain   Level land that may be submerged by floodwaters; a plain built up by stream deposition.

Flora   The plant life of an area.

Flyway   Geographic migration route for birds that includes the breeding and wintering areas that it connects.

Food Additive   Substance added to food usually to improve color, flavor, or shelf life.

Food Chain   The sequence of organisms in a community, each of which uses the lower source as its energy supply. Green plants are the ultimate basis for the entire sequence.

Fossil Fuels   Coal, oil, natural gas, and/or lignite; those fuels derived from former living systems; usually called nonrenewable fuels.

Fuel Cell   Manufactured chemical systems capable of producing electrical energy; they usually derive their capabilities via complex reactions involving the sun as the driving energy source.

Fusion   The formation of a heavier atomic complex brought about by the addition of atomic nuclei; during the process there is an attendant release of energy.

G

Gaia Hypothesis   Theory that Earth's biosphere is a living system whose complex interactions between its living organisms and nonliving processes regulate environmental conditions over millions of years so that life continues.

Gamma Ray   A ray given off by the nucleus of some radioactive elements. A form of energy similar to X rays.

Gene   Unit of heredity; segment of DNA nucleus of the cell containing information for the synthesis of a specific protein.

Gene Banks   Storage of seed varieties for future breeding experiments.

Genetic Diversity   Infinite variation of possible genetic combinations among individuals; what enables a species to adapt to ecological change.

Geothermal Energy   Heat derived from the Earth's interior. It is the thermal energy contained in the rock and fluid (that fills the fractures and pores within the rock) in the Earth's crust.

Germ Plasm   Genetic material that may be preserved for future use (plant seeds, animal eggs, sperm, and embryos).

Global Warming   An increase in the near surface temperature of the Earth. Global warming has occurred in the distant past as the result of natural influences, but the term is most often used to refer to the warming predicted to occur as a result of increased emissions of greenhouse gases. Scientists generally agree that the Earth's surface has warmed by about 1 degree Fahrenheit in the past 140 years.

Green Revolution   The great increase in production of food grains (as in rice and wheat) due to the introduction of high-yielding varieties, to the use of pesticides, and to better management techniques.

Greenhouse Effect   The effect noticed in greenhouses when shortwave solar radiation penetrates glass, is converted to longer wavelengths, and is blocked from escaping by the windows. It results in a temperature increase. Earth's atmosphere acts in a similar manner.

Gross National Product (GNP   ) The total value of the goods and services produced by the residents of a nation during a specified period (such as a year).

Groundwater   Water found in porous rock and soil below the soil moisture zone and, generally, below the root zone of plants. Groundwater that saturates rock is separated from an unsaturated zone by the water table.

H

Habitat   The natural environment of a plant or animal.

Habitat Fragmentation   Process by which a natural habitat/landscape is broken up into small sections of natural ecosystems, isolated from each other by sections of land dominated by human activities.

Hazardous Waste   Waste that poses a risk to human or ecological health and thus requires special disposal techniques.

Herbicide   Any substance used to kill plants.

Heterotroph   Organism that cannot synthesize its own food and must feed on organic compounds produced by other organisms.

Glossary

Hydrocarbons Organic compounds containing hydrogen, oxygen, and carbon. Commonly found in petroleum, natural gas, and coal.

Hydrogen Lightest-known gas; major element found in all living systems.

Hydrogen Sulfide Compound of hydrogen and sulfur; a toxic air contaminant that smells like rotten eggs.

Hydropower Electrical energy produced by flowing or falling water.

I

Infiltration Process of water percolation into soil and pores and hollows of permeable rocks.

Intangible Resources Open space, beauty, serenity, genius, information, diversity, and satisfaction are a few of these abstract commodities.

Integrated Pest Management (IPM) Designed to avoid economic loss from pests, this program's methods of pest control strive to minimize the use of environmentally hazardous, synthetic chemicals.

Invasive Refers to those species that have moved into an area and reproduced so aggressively that they have replaced some of the native species.

Ion An atom or group of atoms, possessing a charge; brought about by the loss or gain of electrons.

Ionizing Radiation Energy in the form of rays or particles that have the capacity to dislodge electrons and/or other atomic particles from matter that is irradiated.

Irradiation Exposure to any form of radiation.

Isotopes Two or more forms of an element having the same number of protons in the nucleus of each atom but different numbers of neutrons.

K

Keystone Species Species that are essential to the functioning of many other organisms in an ecosystem.

Kilowatt Unit of power equal to 1,000 watts.

Leaching Dissolving out of soluble materials by water percolating through soil.

L

Limnologist Individual who studies the physical, chemical, and biological conditions of aquatic systems.

M

Malnutrition Faulty or inadequate nutrition.

Malthusian Theor yThe theory that populations tend to increase by geometric progression (1, 2, 4, 8, 16, etc.) while food supplies increase by arithmetic means (1, 2, 3, 4, 5, etc.).

Metabolism The chemical processes in living tissue through which energy is provided for continuation of the system.

Methane Often called marsh gas (CH^4); an odorless, flammable gas that is the major constituent of natural gas. In nature it develops from decomposing organic matter.

Migration Periodic departure and return of organisms to and from a population area.

Monoculture Cultivation of a single crop, such as wheat or corn, to the exclusion of other land uses.

Mutation Change in genetic material (gene) that determines species characteristics; can be caused by a number of agents, including radiation and chemicals, called mutagens.

N

Natural Selection The agent of evolutionary change by which organisms possessing advantageous adaptations leave more offspring than those lacking such adaptations.

Niche The unique occupation or way of life of a plant or animal species; where it lives and what it does in the community.

Nitrate A salt of nitric acid. Nitrates are the major source of nitrogen for higher plants. Sodium nitrate and potassium nitrate are used as fertilizers.

Nitrite Highly toxic compound; salt of nitrous acid.

Nitrogen Oxides Common air pollutants. Formed by the combination of nitrogen and oxygen; often the products of petroleum combustion in automobiles.

Nonrenewable Resource Any natural resource that cannot be replaced, regenerated, or brought back to its original state once it has been extracted, for example, coal or crude oil.

Nutrient Any nutritive substance that an organism must take in from its environment because it cannot produce it as fast as it needs it or, more likely, at all.

O

Oil Shale Rock impregnated with oil. Regarded as a potential source of future petroleum products.

Oligotrophic Most often refers to those lakes with a low concentration of organic matter. Usually contain considerable oxygen; Lakes Tahoe and Baikal are examples.

Organic Living or once living material; compounds containing carbon formed by living organisms.

Organophosphates A large group of nonpersistent synthetic poisons used in the pesticide industry; include parathion and malathion.

Ozone Molecule of oxygen containing three oxygen atoms; shields much of Earth from ultraviolet radiation.

P

Particulate Existing in the form of small separate particles; various atmospheric pollutants are industrially produced particulates.

Peroxyacyl Nitrate (PAN) Compound making up part of photochemical smog and the major plant toxicant of smog-type injury; levels as low as 0.01 ppm can injure sensitive plants. Also causes eye irritation in people.

Pesticide Any material used to kill rats, mice, bacteria, fungi, or other pests of humans.

Pesticide Treadmill A situation in which the cost of using pesticides increases while the effectiveness decreases (because pest species develop genetic resistance to the pesticides).

Petrochemicals Chemicals derived from petroleum bases.

pH Scale used to designate the degree of acidity or alkalinity; ranges from 1 to 14; a neutral solution has a pH of 7; low pHs are acid in nature, while pHs above 7 are alkaline.

Phosphate A phosphorous compound; used in medicine and as fertilizers.

Photochemical Smog Type of air pollution; results from sunlight acting with hydrocarbons and oxides of nitrogen in the atmosphere.

Photosynthesis Formation of carbohydrates from carbon dioxide and hydrogen in plants exposed to sunlight; involves a release of oxygen through the decomposition of water.

Photovoltaic Cells An energy-conversion device that captures solar energy and directly converts it to electrical current.

Physical Half-Life Time required for half of the atoms of a radioactive substance present at some beginning to become disintegrated and transformed.

Phytoplankton That portion of the plankton community comprised of tiny plants, e.g., algae, diatoms.

Pioneer Species Hardy species that are the first to colonize a site in the beginning stage of ecological succession.

Plankton Microscopic organisms that occupy the upper water layers in both freshwater and marine ecosystems.

Plutonium Highly toxic, heavy, radioactive, manmade, metallic element. Possesses a very long physical half-life.

Pollution The process of contaminating air, water, or soil with materials that reduce the quality of the medium.

Polychlorinated Biphenyls (PCBs) Poisonous compounds similar in chemical structure to DDT. PCBs are found in a wide variety of products ranging from lubricants, waxes, asphalt, and transformers to inks and insecticides. Known to cause liver, spleen, kidney, and heart damage.

Population All members of a particular species occupying a specific area.

Predator Any organism that consumes all or part of another system; usually responsible for death of the prey.

Primary Production The energy accumulated and stored by plants through photosynthesis.

R

Rad (Radiation Absorbed Dose) Measurement unit relative to the amount of radiation absorbed by a particular target, biotic or abiotic.

Radioactive Waste Any radioactive by-product of nuclear reactors or nuclear processes.

Radioactivity The emission of electrons, protons (atomic nuclei), and/or rays from elements capable of emitting radiation.

Rain Fores t Forest with high humidity, small temperature range, and abundant precipitation; can be tropical or temperate.

Recycle To reuse; usually involves manufactured items, such as aluminum cans, being restructured after use and utilized again.

Red Tid e Population explosion or bloom of minute single-celled marine organisms (dinoflagellates), which can accumulate in protected bays and poison other marine life.

Renewable Resources Resources normally replaced or replenished by natural processes; not depleted by moderate use.

Riparian Water Right Legal right of an owner of land bordering a natural lake or stream to remove water from that aquatic system.

S

Salinization An accumulation of salts in the soil that could eventually make the soil too salty for the growth of plants.

Sanitary Landfill Land waste disposal site in which solid waste is spread, compacted, and covered.

Scrubber Antipollution system that uses liquid sprays in removing particulate pollutants from an airstream.

Sediment Soil particles moved from land into aquatic systems as a result of human activities or natural events, such as material deposited by water or wind.

Seepage Movement of water through soil.

Selection The process, either natural or artificial, of selecting or removing the best or less desirable members of a population.

Selective Breeding Process of selecting and breeding organisms containing traits considered most desirable.

Selective Harvesting Process of taking specific individuals from a population; the removal of trees in a specific age class would be an example.

Sewage Any waste material coming from domestic and industrial origins.

Smog A mixture of smoke and air; now applies to any type of air pollution.

Soil Erosion Detachment and movement of soil by the action of wind and moving water.

Solid Wa s t Unwanted solid materials usually resulting from industrial processes.

Species A population of morphologically similar organisms, capable of interbreeding and producing viable offspring.

Species Diversity The number and relative abundance of species present in a community. An ecosystem is said to be more diverse if species present have equal population sizes and less diverse if many species are rare and some are very common.

Strip Mining Mining in which Earth's surface is removed in order to obtain subsurface materials.

Strontium-90 Radioactive isotope of strontium; it results from nuclear explosions and is dangerous, especially for vertebrates, because it is taken up in the construction of bone.

Succession Change in the structure and function of an ecosystem; replacement of one system with another through time.

Sulfur Dioxide (SO^2) Gas produced by burning coal and as a by-product of smelting and other industrial processes. Very toxic to plants.

Sulfur Oxides (SO^x) Oxides of sulfur produced by the burning of oils and coal that contain small amounts of sulfur. Common air pollutants.

Sulfuric Acid ($H2\ SO^4$) Very corrosive acid produced from sulfur dioxide and found as a component of acid rain.

Sustainability Ability of an ecosystem to maintain ecological processes, functions, biodiversity, and productivity over time.

Sustainable Agriculture Agriculture that maintains the integrity of soil and water resources so that it can continue indefinitely.

T

Technology Applied science; the application of knowledge for practical use.

Tetraethyl Lead Major source of lead found in living tissue; it is produced to reduce engine knock in automobiles.

Thermal Inversion A layer of dense, cool air that is trapped under a layer of less dense warm air (prevents upward flowing air currents from developing).

Glossary

Thermal Pollution   Unwanted heat, the result of ejection of heat from various sources into the environment.

Thermocline   The layer of water in a body of water that separates an upper warm layer from a deeper, colder zone.

Threshold Effect   The situation in which no effect is noticed, physiologically or psychologically, until a certain level or concentration is reached.

Tolerance Limit   The point at which resistance to a poison or drug breaks down.

Total Fertility Rate (TFR)   An estimate of the average number of children that would be born alive to a woman during her reproductive years.

Toxic   Poisonous; capable of producing harm to a living system.

Tragedy of the Common   sDegradation or depletion of a resource to which people have free and unmanaged access.

Trophic   Relating to nutrition; often expressed in trophic pyramids in which organisms feeding on other systems are said to be at a higher trophic level; an example would be carnivores feeding on herbivores, which, in turn, feed on vegetation.

Turbidity   Usually refers to the amount of sediment suspended in an aquatic system.

U

Uranium 235   An isotope of uranium that when bombarded with neutrons undergoes fission, resulting in radiation and energy. Used in atomic reactors for electrical generation.

Z

Zero Population Growth   The condition of a population in which birthrates equal death rates; it results in no growth of the population.

Index

221

Index

photovoltaic (PV) cells, 66, 67, 88; PV panels, 71, 89
physics, consumption and, 37–38
phytoplakton, 164, 166
Pimm, Stuart L., 94, 97
Pivot of Civilization (Sanger), 34
Planned Parenthood, 32, 34
plantation, 113
plants: mass extinction of, 117; as providers of human essentials, 120
Project Feederwatch, 129
Poland, decline of farmers in, 147
politics, changes in, impact of, 8–9
pollution, 9; as cause of extinctions, 117, 118; as a threat to Earth's oceans, 163, 165
Population Bomb, The (Ehrlich), 33, 68
population growth, 30–31, 95–96, 148; as cause of environmental degradation, 36–42; control of, 32–35
Population/Consumption (PC) version, 40, 41
poverty: hunger and, 43; malnutrition and, 50–58
Precious Heritage: The Status of Biodiversity in the United States, 100, 101, 110
pregnancy: and nutrition, 52; vitamin A deficiency and, 44
profit-maximizing capitalist, 22
pronatalism, 4
prone-to-obesity myth, 52–56
property rights, the commons and, 137–138
Proposition 9, 80
"public benefits trusts," 72
public health, 56, 170
public lands, 137, 139, 141

Q

quantitative analysis, of consumption, 39

R

radioactive materials, as groundwater pollutant, 194
Raup, David M., 123
Raven, Peter H., 120, 122
Reagan, Ronald, 10, 67
recycled wastewater, 60
red tides, 165, 184
redwood tree, 110
refugees, 45
Reid, W. V., 94
Renewable Portfolio Standards (RPS), 72
Republic of Congo, 15
resource depletion, and consumption, 39
resource efficiency, increase of, 23–25, 27
"reverse logistics," 26
"ripple effects," 16
riverbedfellows, 155–156
Robbins, Mark, 146
Roman Empire, 158
Romania, birth control in, 4
Ryan, John, 39

S

Sacramento Municipal Utility District (SMUD), 89
Safe Drinking Water Act, 202, 203
Safina, Carl, 139
salamanders, 101
Salmonella, 155
salt, as natural ground water pollutant, 193, 195
San Diego Gas & Electric, 79, 81
Sanger, Margaret, 32, 34
satellite imaging, 94
satiation, of consumption, 39, 41

satisfaction, of consumption, 39, 40
Savory, Allan, 26
scarcity myth, 51–52
Schilham, Jan, 24
Schor, Juliet, 40
Schröder, Gerhard, 168
sea trade, 164
seabed mining, 164
second-order effect, 14–15
Section 104(d), of foreign aid legislation of 1978, 33
security anxiety, 6–7
Sempra Energy, 83
sexual health care, of adolescents, in the United Nations, 34
ships, as carrier of species, 112
shrinking, of consumption, 39, 41
shorebirds, 140
shrimp farming, 113
Silent Spring (Carson), 6
Simon, Julian, 93–94, 96, 99
simplicity movement, 40
Singapore, nutritional programs in, 55
Sinking Ark, The (Myers), 93
sinks, 180
smog alerts, 204
snowbirds, 125
sociology, consumption and, 37, 38
soil fertility management, 45
Solar Energy Industries Association (SEIA), 71
solar power, 66–70, 71–73
Solow, Robert, 5
South Asia, hunger in, 43
Southern California Edison (SoCalEd), 79, 81, 82, 83, 85, 86, 87, 89
Soviet Union, 7
Spira/GRACE Project, 157
sprawl, 201
"spreadable acreage," 183
sprinklers, 60
standards, for peasant farmers, 153
state-of-the-shelf technologies, 24
structural adjustment requirements, for peasant farmers, 153
Stuff: The Secret Lives of Everyday Things (Ryan & Durning), 39
sulfur-dioxide emissions, 95
Sumerian farmers, 59
sunlight, 66, 71–73
"Superfund" program, 189
surface fires, 14
Sweden, decline of farmers in, 147
synergism, of environmental change, 13, 16

T

taxes, on food, 56
technology, as cause of environmental degradation, 36
"technological lock-in," 2
textbook optimization, 24
Thatcher, Margaret, 10
Third Revolution, The (Harrison), 95
Tilman, David, 181
Todd, John, 26–27
"total maximum daily loads" (TMDL), 197, 198, 200
totemists, 20
toxic fertility, 180–186
"toxic food environment," 53, 54–55
trade liberalization, 44–45; for peasant farmers, 153
"tragedy of the commons" phenomenon, 136–144
trawling, 165
treadle pump, 61

tree plantation, 113
Triassic extinction, 92, 99, 122
typhoid, 158

U

ultralviolet (UV) radiation, 166
U.N. Population Division, 32
United Nations Framework Convention on Climate Change, 177
United States, 8; coal industry in, 76–77; decline of farmers in, 147; international dominance of, 5; obesity in, 50, 51, 52
unnoticed trend, of environmental change, 13
urban life, 147
U.S. Army Corps of Engineers, 145
U.S. biodiversity, assessing the status of, 100–110
U.S. Department of Agriculture (USDA), 43
utility-scale system, 74–75

V

Values Party, of New Zealand, 9
van Schaik, Carel P., 94
violent conflicts, hunger and, 45
vitamin A deficiency, 58

W

Ward, Peter D., 122
waste: climate concept of, 23, 25; as groundwater pollutant, 193–194
watching, of wildlife, 124–133
water conservation, 158–162
water, food and, 59–61; poverty and, 45
water pollution, in the U.S., 197–203
wealth, environmentalism and, 6
web sites, for information on endangered species, 102
WIC (Women, Infants, and Children), 47
wild and marginal land, poverty and, 45
Wilderness Act of 1964, 121
wildlife-watching tourism, 125
Wilson, E. O., 115, 118, 120, 128
wind power, 74–78
Windforce 10, 75
windmills, 74–75
women: environmentalism and, 6; and hunger, 51–52, 53
World Bank, 33, 161
World Conservation Union (IUCN), 19
World Energy Modernization Plan, 73
World Food Summit, 43; goal of, 47
World Health Organization, 9
world trade, as most dangerous form of environmental decline, 111
World War I, 8
World War II, 8, 32
World Wildlife Fund (WWF), 19, 117, 118, 119, 121
Wuppertal Institute for Climate, Environment, and Energy, 20
Wurmfeld, Charles, 67

Y

yellow fever, 112
Yellowstone National Park, 109
Yosemite National Park, 160

Z

Zazurbin, V., 4
"zero-emissions" mandate, 64

Test Your Knowledge Form

We encourage you to photocopy and use this page as a tool to assess how the articles in *Annual Editions* expand on the information in your textbook. By reflecting on the articles you will gain enhanced text information. You can also access this useful form on a product's book support Web site at *http://www.dushkin.com/online/*.

NAME: _____ DATE: _____

TITLE AND NUMBER OF ARTICLE: _____

BRIEFLY STATE THE MAIN IDEA OF THIS ARTICLE: _____

_____

LIST THREE IMPORTANT FACTS THAT THE AUTHOR USES TO SUPPORT THE MAIN IDEA:

_____

WHAT INFORMATION OR IDEAS DISCUSSED IN THIS ARTICLE ARE ALSO DISCUSSED IN YOUR TEXTBOOK OR OTHER READINGS THAT YOU HAVE DONE? LIST THE TEXTBOOK CHAPTERS AND PAGE NUMBERS:

_____

LIST ANY EXAMPLES OF BIAS OR FAULTY REASONING THAT YOU FOUND IN THE ARTICLE:

_____

LIST ANY NEW TERMS/CONCEPTS THAT WERE DISCUSSED IN THE ARTICLE, AND WRITE A SHORT DEFINITION: